Biological Anthropology and the Study of Ancient Egypt

Biological Anthropology and the Study of Ancient Egypt

EDITED BY
W. VIVIAN DAVIES
AND ROXIE WALKER

HI JAN!
HOPE YOU ENJOY THIS -
LOVE,
Roxie

BRITISH MUSEUM PRESS

Published by British Museum Press
A division of British Museum Publications Ltd
46 Bloomsbury Street, London WC1B 3QQ

British Library Cataloguing-in-Publication Data
A catalogue record for this book is available from the British Library

ISBN 0 7141 09673

Front cover: Predynastic body. From Gebelein.
British Museum, EA 32753.

Back cover: Wooden tomb-model. From Asyut.
Middle Kingdom, about 1900 BC.
British Museum, EA 45197.

Jacket and plate section designed by Andrew Shoolbred

Printed in Great Britain by
Antony Rowe Ltd, Chippenham

CONTENTS

PREFACE

Biological anthropology is the modern collective name for the scientific study of the physical properties of human beings and their associated flora and fauna. With its enormous wealth of well-preserved ancient remains, ranging in date over several millennia, the Nile Valley has long been, and continues to be, a fruitful source of material for such study. As practised today, bioanthropology is far removed from 'mummy studies' in the traditional sense and has a very great deal more to offer, though this is perhaps not as widely appreciated as it should be.

The present volume results from a major international colloquium, entitled 'Biological Anthropology and the Study of Ancient Egypt', which was held in the British Museum in July 1990. Sponsored jointly by The Bioanthropology Foundation and the Department of Egyptian Antiquities, British Museum, the colloquium brought together bioscientists, Egyptologists, and scholars from other related disciplines, to assess the past and present contribution of biological anthropology to knowledge of ancient Egypt and to consider future possibilities. In view of the increasingly specialised nature of the subjects in question, a particular goal of the colloquium was to encourage greater co-operation and communication among the various parties, so that future work might take place within a collaborative framework incorporating all relevant interests.

Almost all the papers delivered at the colloquium are published in this volume, some in modified form. Collectively, they provide a valuable overview of the present state of the subject, with all the major areas of research represented - palaeopathology, dental anthropology, molecular archaeology, archaeobotany and archaeozoology. They include discussions of general methodologies, the presentation of new analytical techniques, and the publication of the results of specific projects (including one devoted to material from a wholly non-Nile Valley source).

On such matters as diet, disease, life expectancy, occupational stress, environmental change, demographic patterning, familial and population relationships, and certain aspects of socio-economic organisation, this volume amply demonstrates the enormous potential of biological anthropology for extracting information that is not otherwise retrievable from the archaeological record but is fundamental to a better understanding of ancient Egypt and indeed of the past in general. It also makes clear that, if this potential is to be fully realised and past errors avoided, the utmost care needs to be exercised in the excavation, documentation and removal of samples; in the subsequent processes of analysis and diagnosis; in the long-term curation of the material; and in ensuring that, at all stages, the possibility of contamination is eliminated. Not least, the material should be published comprehensively and without long delay. Only thus may the biological record (a rich but finite and diminishing resource) be exploited to the fullest advantage of scholarship.

For valuable help rendered to the editors in the preparation of this book, acknowledgments are due to a number of British Museum colleagues: to the late Hannah Lawlor and, especially, to Pat Terry, for typing the text as camera-ready copy; to Christine Barratt and Peter Hayman, for work on the line drawings and photographic plates; to Drs Jeffrey Spencer and Richard Parkinson, for advice and assistance on various matters; and to Joanna Champness and Susanna Friedman of British Museum Press, for seeing the book through to press.

W Vivian Davies
Keeper of Egyptian Antiquities
British Museum

Roxie Walker
President
The Bioanthropology Foundation
Centre de Recherches
Sur les planches
Lignières
Switzerland

AUTHORS' ADDRESSES

Rafael González Antón, Museo Arqueologico y Etnografico de Tenerife, Santa Cruz de Tenerife 38080, Canary Islands, Spain.

George J Armelagos, Department of Anthropology, University of Florida, Gainesville, Florida 32611, USA.

Brenda J Baker, Department of Anthropology, University of Massachusetts, Amherst, Massachusetts 01003, USA.

Piotr Bienkowski, Department of Archaeology and Ethnology, Liverpool Museum, National Museums and Galleries on Merseyside, William Brown Street, Liverpool L3 8EN, UK.

Jane E Buikstra, Department of Anthropology, University of Chicago, 1126 East 59th Street, Chicago, Illinois 60637, USA.

Louis Chaix, Département d'Archéozoologie, Muséum d'Histoire Naturelle, Route de Malagnou, CP 434, CH-1211, Geneva 6, Switzerland.

Della C Cook, Department of Anthropology, Indiana University, Bloomington, Indiana 47405, USA.

A M Deelder, Laboratory of Parasitology, Institute for Tropical Medicine, Medical Faculty, University of Leiden, PO Box 9605, Leiden 2300 RC, The Netherlands.

John Dekker, Human Retrovirus Laboratory, Academic Medical Centre, University of Amsterdam, Meibergdreef 15, 1105 AZ Amsterdam, The Netherlands.

N De Jonge, Laboratory of Parasitology, Institute for Tropical Medicine, Medical Faculty, University of Leiden, PO Box 9605, Leiden 2300 RC, The Netherlands.

Anna Di Rienzo, Division of Molecular Biology and Biochemistry, University of California, Berkeley, California 94720, USA.

John Edmonson, Department of Botany, Liverpool Museum, National Museums and Galleries on Merseyside, William Brown Street, Liverpool L3 8EN, UK.

Mohammad Fawzi Gaballah, Department of Anatomy, Kasr El Ainy Medical Faculty, Cairo University, Cairo, Egypt.

Jan Geelen, Human Retrovirus Laboratory, Academic Medical Centre, University of Amsterdam, Meibergdreef 15, 1105 AZ Amsterdam, The Netherlands.

Barbara Ghaleb, c/o Institute of Archaeology, University College London, 31-34 Gordon Square, London WC1H OPY, UK.

Fernando Estévez González, Museo Arqueologico y Etnografico de Tenerife, Santa Cruz de Tenerife 38080, Canary Islands, Spain.

Jaap Goudsmit, Human Retrovirus Laboratory, Academic Medical Centre, University of Amsterdam, Meibergdreef 15, 1105 AZ Amsterdam, The Netherlands.

M N El Hadidi, The Herbarium, Botany Department, Faculty of Science, Cairo University, Giza 12613, Egypt.

W Benson Harer, Department of Obstetrics and Gynecology, St Bernardine Medical Center, San Bernardino, CA 92404, USA.

R E M Hedges, Oxford University Radiocarbon Accelerator Unit, Research Laboratory for Archaeology and the History of Art, 6 Keble Road, Oxford OX1 3QJ, UK.

Simon W Hillson, Institute of Archaeology, University College London, 31-34 Gordon Square, London WC1H OPY, UK.

Přemysl Klír, Department of Legal Medicine, Postgraduate Medical and Pharmaceutical Institute, Prague, Czech Republic.

F W Krijger, Laboratory of Parasitology, Institute for Tropical Medicine, Medical Faculty, University of Leiden, PO Box 9605, Leiden 2300 RC, The Netherlands.

Carla Kuiken, Laboratory of Parasitology, Institute for Tropical Medicine, Medical Faculty, University of Leiden, PO Box 9605, Leiden, 2300 RC, The Netherlands.

Conrado Rodriguez Martin, Museo Arqueologico y Etnografico de Tenerife, Santa Cruz de Tenerife 38080, Canary Islands, Spain.

R L Miller, 594 Main Street, Northport, NY 11768, USA.

James O Mills, Egyptian Studies Association, University of South Carolina, Columbia, South Carolina, 29208, USA.

Theya I Molleson, Department of Palaeontology, Natural History Museum, Cromwell Road, London SW7 5BD, UK.

Mary Anne Murray, c/o Institute of Archaeology, University College London, 31-34 Gordon Square, London WC1H OPY, UK.

Alena Němečková, Institute of Histology and Embryology, Medical Faculty in Plzeň, Charles University, Czech Republic.

Svante Pääbo, Department of Zoology, University of Munich, Luisenstrasse 14, D-8000 Munich 2, Germany.

Rutger Perizonius, Department of Anthropo-osteology, Rijksuniversiteit Utrecht, Jutfaseweg 7, 3522 HA Utrecht, The Netherlands.

L Stephen Perry, Mullins Library, University of Arkansas, Fayetteville, Arkansas 72701, USA.

Patricia V Podzorski, Lowie Museum of Anthropology, University of California, 103 Kroeber Hall, Berkeley, CA 94720, USA.

Jerome C Rose, Department of Anthropology, University of Arkansas, Fayetteville, Arkansas 72701, USA.

Friedrich G Rösing, Department of Anthropology, Universität Ulm, Liststrasse 3, W-7900 Ulm, Germany.

Delwen Samuel, The McDonald Institute for Archaeological Research, University of Cambridge, 62 Sidney Street, Cambridge CB2 3JW, UK.

Shelly R Saunders, Department of Anthropology, McMaster University, Hamilton, Ontario, Canada.

Lia Smit, Human Retrovirus Laboratory, Academic Medical Centre, University of Amsterdam, Meibergdreef 15, 1105 AZ Amsterdam, The Netherlands.

Eugen Strouhal, Institute for the History of Medicine, Charles University Prague, Katerinská 32, 120 00 Prague 2, Czech Republic.

B A Sykes, Institute for Molecular Medicine, John Radcliffe Hospital, Oxford OX3 9DU, UK.

Willy Woelfli, Institut für Mittelenergiephysik, Eidgenössische Technische Hochschuhe, Zürich, Switzerland.

PALAEOPATHOLOGY AS SCIENCE
THE CONTRIBUTION OF EGYPTOLOGY

George J Armelagos and James O Mills

Egyptology has contributed substantially to the development of palaeopathology. The excellent preservation of the mummified remains from Egypt and the Sudan captured the interest of the public and scientific community alike. The public was fascinated by the discovery of human remains which after thousands of years retained skin and hair, giving these ancient beings an air of immortality. Scientists were impressed with the excellent preservation of soft tissue which allowed them to undertake anatomical and histological analysis in order to understand biological variation and pathological deviation. Much of the interest in ancient diseases and the growth of palaeopathology were stimulated by the impressive discoveries made by scientists working with Egyptian materials. Palaeopathology may now be in a position to repay this intellectual debt by demonstrating how the scientific analysis of disease in mummified remains may in fact fuel renewed interest in Egyptology.

There are four objectives in this paper. The first is to trace the history of palaeopathology and to show the impact that Egyptology has had on its development, the second to discuss the transformation of palaeopathology into a science, the third to demonstrate the importance of a scientific perspective that involves hypothesis testing in the study of ancient disease and the fourth to illustrate how this, the scientific approach to the study of Egyptian and Nubian disease, can enhance Egyptology.

History of Palaeopathology

Palaeopathology has had a long but somewhat disappointing history. Although publications in the field span a 200 year period, the scientific study of palaeopathology did not emerge until recently. Scientists developed remarkable skills of diagnosis and correctly identified many pathological conditions. Unfortunately, the emphasis on diagnosis became an end in itself and palaeopathology remained descriptive.

Egyptian skeletons and mummies have played an important role in the history of palaeopathology. For example, there were sporadic reports on mummies as early as the beginning of the 1700s (Armelagos, Mielke and Winter 1971). Unwrapping of mummies and even anatomical dissection were undertaken during the period. Even before Napoleon's return from Egypt, mummies had long been viewed as curiosities by Europeans. For centuries mummified remains were traded into Europe and used in medical preparations (Smith and Dawson 1924). In fact, the term for these medicinals ('mummiya') gave the human remains their name. However, it was Napoleon's expedition to Egypt at the end of the century (1798-1801) and the resulting publication of *La Description De L'Égypte* (Jomard 1809-1822) which provided the major impetus for heightening interest in Egyptian materials. While the mummies were viewed as curiosities, there were attempts to use mummified remains to answer systematically specific questions. Foremost among these was the scientific inquiry into the process of mummification, especially its historical development and religious significance. Strouhal and Vyhnanek (1979) report that French, British and American researchers were performing autopsies on mummies as early as 1825. In 1852, J N Czermak, a Czech physiologist, used histological techniques to reveal arteriosclerosis in a large artery of a mummy (Strouhal and Vyhnanek 1979).

Apart from the work of classical authors such as Herodotus, Diodorus Siculus, Plutarch, and Porphyry (reviewed by Smith and Dawson 1924), modern treatises on Egyptian mummies began to appear in the nineteenth century.

Included in these works were Granville's 'An essay on Egyptian mummies' (1825) and Thomas Pettigrew's *History of Egyptian Mummies* (1834). The processes involved in mummification continued to capture the interest of Grafton Elliot Smith (e.g. 1906) and played a major role in the development of his theories on diffusion. His research resulted in a monograph on the royal mummies (1912) and an important review on mummies which he co-authored with W R Dawson (Smith and Dawson 1924). Smith's (1915) interests in mummies led to an observation that there were similarities in the process of mummification between Egyptian and Australian mummies. He concluded that these similarities presented indisputable evidence of cultural diffusion from Egypt. He then constructed an elaborate theory of cultural diffusion to explain the influence of Egyptians on the rest of the world (Smith 1923, 1929).

The prevailing diffusionist paradigms of culture change were coupled with the belief that the use of racial types would aid in reconstructing culture history. Researchers became obsessed with identifying the racial affinities in Egyptian populations in order to explain the culture history of the Nile Valley. Observed changes in the archaeological sequences from the Predynastic and Early Dynastic periods were accredited to successive population replacements which culminated in the arrival of a 'Dynastic Race' (Petrie 1917, 1939; Stoessiger 1927). This led to the collection of several large populations, predominantly of Predynastic and Early Dynastic date, and vast amounts of energy were expended in craniometric studies (e.g. Warren 1897; Fawcett 1901; Morant 1925, 1935, 1937; Stoessiger 1927; Davin and Pearson 1924; Woo 1930; Jackson and Cave 1937), which were embraced as an ultimately meaningful tool to determine genetic relatedness. There were early criticisms of the use of population replacement models that relied on racial typology (eg. Batrawi 1945, 1946, 1947; Myers 1905, 1908), and there was convincing evidence of genetic continuity (Berry and Berry 1972; Berry et al. 1968), at least through Middle Kingdom times. Meanwhile, the concern with racial explanations of Egypt's culture history persisted (eg. Vandier 1952; Derry 1956; Emery 1961) and was in part responsible for diverting energy and interest away from biocultural studies within the modern ecological framework. In the 1990s, there has been a rebirth of the concept of race as a means of interpreting world history. An Afrocentric view of world history claims that Africa has been neglected in the writing of world history. Furthermore, the Afrocentric view uses racial affinity of groups and individuals to reinterpret historical developments (Asante 1988; Bernal 1987, 1991; Diop 1991). Thus, the racial identity of the Egyptians and their great individuals as 'Black' (*Newsweek*, 23 September 1991, asks on its cover, 'Was Cleopatra Black?') is used to show African influence on European historical development.

While the principal focus of skeletal research was on discerning racial types and ultimately population movements and their impact on culture history, pioneering palaeopathological inquiry was also being undertaken on Egyptian material. Between 1890 and 1930, researchers were attempting to unravel the history of specific diseases such as syphilis (Lortet 1907), tuberculosis (Smith and Jones 1908) and leprosy (Smith and Derry 1910) and were uncovering evidence of medical practices. Their concern with infectious disease is not surprising, since it represented one of the most significant medical problems of that era. There was little attempt to generalise, even though they were accumulating considerable data. Determining the chronology and geography of specific diseases was the main objective of the palaeopathologist of the period.

Even with the limited focus on geography and chronology, palaeopathology was greatly enhanced through the study of Egyptian material. During this period archaeology and Egyptology were becoming professionally instituted in Europe and the United States. Museums, universities, governments and special societies began sending out regular expeditions to Egypt. Egyptian settlements, located in the ancient floodplain and subsequently buried or destroyed, were all but ignored, while cemeteries and tombs, located in the deserts above flood level, were conspicuous and easily accessible and therefore became the focus of Egyptological inquiry. Apart from the traditional emphasis on monumental architecture and art, a number of large cemeteries were excavated between 1880 and 1930. Emphasis focused on the recovery of artefacts; information from skeletal remains was considered to be of much lesser importance. In most cases skeletal material was not even collected. Crude sexing and ageing of cemetery populations were commonly performed in the field but rarely done quantitatively. (George Mann, 1990, has examined a series of skulls from Qau and Badari and found considerable discrepancies with the original excavators' sexing). Pathological studies were generally limited to extreme examples of pathological abnormalities. Other remains were infrequently described or curated.

Perhaps the single most important event during this period was the completion of the Archaeological Survey of Nubia. In 1907, the Egyptian government finished the great dam at Aswan (now referred to as the Low Dam) (Adams 1977). The reservoir that was created inundated many monuments (including the Temple at Philae) and thousands of burials. Public outrage was so great that a few years later, when the government decided to heighten the dam by seven meters, they realized they could not repeat the fiasco of the destruction of the archaeological sites that had occurred during the earlier construction. The government decided to record all antiquities and examine, describe, photograph and recover all burials that could be found. The recovery of these skeletal populations represents the largest sample of burials that has ever been excavated from an archaeological region (Adams 1977). The examination of the nearly 10,000 burials (Smith and Jones 1908) revealed many interesting diseases and changed the course of palaeopathology.

It was also during this period (1890-1930) that researchers began to apply the new technology of medical science to the study of ancient disease. Soon after Wilhelm Roentgen discovered the X-ray in 1895, mummies were being X-rayed. In 1896, A. Dedekind in Britain and W. Keonig in Germany were using radiographs to interpret mummified remains. Petrie (1898) employed radiography on material from Deshasheh which, according to Sandison (1972), reveals growth arrest/resumption lines in long bones; F J Clendinnen (1898) used it to identify an abnormal bone in the hand of a mummy. But, as Gray (1973) has noted, the pioneering use of radiographic techniques on Egyptian materials waned and was not aggressively re-employed until the 1960s. The sporadic use of advanced medical technology is a theme that was to be replayed throughout the history of palaeopathology.

It was during this same period (1890-1930) that some of the most impressive research in palaeopathology was undertaken using Egyptian material. M A Ruffer, who played a key role in the definition and development of palaeopathology, undertook histological analysis of mummified tissue (1909-1910a). He was able to demonstrate the preservation of normal tissue as well as pathological changes. Ruffer and co-workers showed histological evidence of bilharzia (schistosomiasis) (Ruffer 1910b), Pott's disease (tuberculosis) (Smith and Ruffer 1910), pneumonia (Ruffer 1910a), arteriosclerosis (Ruffer 1910a, 1911a, 1911b) and variola (Ruffer 1914; Ruffer and Ferguson 1911; Smith 1912). A monograph edited by R L Moodie and published in 1921 reprinted many of Ruffer's classic articles. Ruffer was not alone in the study of histological change. Shattock (1909) reported arteriosclerotic changes in the aorta, and Long (1931) reports arteriosclerosis in coronary and renal arteries. Smith and Jones (1908; see also Smith and Dawson 1924) found appendicular adhesions in a Greco-Roman period mummy, and Mitchell (1900) describes unilateral shortening of the femur suggestive of poliomyelitis.

Rare gross pathologies were also first noted during this period. Gallstones (Smith and Dawson 1924), gout (Smith and Dawson 1924), scrotal hernia (Smith 1912) and rectal and vaginal prolapse (Jones 1908) were diagnosed in mummified remains. Achondroplasia (MacIver 1901; Keith 1913, originally misdiagnosed as cretinism by Seligmann 1912; Brothwell 1967:433; Jones 1932; Dawson 1938) and hydrocephaly and associated hemiplegia (Derry 1913) were diagnosed in dry bone. A number of cases of carcinoma were reported during this period (Smith and Dawson 1924; Derry 1909; Smith and Derry 1910). Subsequent diagnoses (Rowling 1961) were made from earlier descriptions (Granville 1825; Ruffer and Willmore 1914). However, these diagnoses have recently been called into question (Brothwell 1967; Rowling 1961; Strouhal 1974; Micozzi 1991).

Other pathologies being recognised and discussed at the time include osseous lesions (osteoarthritis) (Ruffer and Rietti 1912; Ruffer 1918; Smith and Jones 1910; Smith and Dawson 1924), biparietal thinning (Smith 1907), mastoid infection (Derry 1909; Smith and Jones 1910; Smith and Dawson 1924) and patterns and occurrences of dental wear, caries and abscesses (Ruffer 1908, 1920). Although diagnosis of infectious disease is always problematic, researchers at the beginning of the century were able successfully to develop methods of diagnosis. The difficulty in diagnosing a specific infectious disease is that the pathogen does not always leave a distinctive lesion on the skin or bones. Many pathogens cause change on the periosteum (the outer layer of bone), which is a general indicator of inflammation and infection. However, there are diagnostic features (the pattern of skeletal involvement which provides indisputable evidence of specific diseases such as syphilis, leprosy and tuberculosis). During this period, leprosy (Smith and Derry 1910; Smith and Dawson 1924), tuberculosis (Smith and Jones 1908), schistosomiasis (Ruffer 1910b) and smallpox (Ruffer and Ferguson 1911; Smith 1912; Ruffer 1914) were discovered in Egyptian and Nubian material. Although over 10,000 mummies were examined during this period, no evidence of syphilis was ever found (Sandison 1972; Baker and Armelagos 1988).

The several medical papyri and numerous depictions in art provide important insights into ancient Egyptian recognition and conceptions of diseases and their causation, treatment and prevention. Such information is generally not available to palaeopathologists (Sandison 1972) but is vital to a fuller understanding of biocultural processes, especially with regard to the repertoire of existing diseases and the nature of cultural buffering of environmental stresses. The medical papyri list a great number of symptoms and diseases. There is, however, considerable disagreement as to which diseases were being described (cf. Dawson 1953, 1967; Breasted 1930; Ebbell 1937; Leake 1952; and Ghalioungui 1963). Debate also continues as to the degree of understanding Ancient Egyptians had of human physiology, disease etiology and pathogen ecology, and as to the nature and extent of the medical profession and its effectiveness in altering the course of disease (cf. Finlayson 1893; Moodie 1920; Todd 1921; Guiart 1922; Breasted 1930; Dawson 1932, 1953, 1967; Krause 1933; Temkin 1936a, 1936b; Ebbell 1937; Cave 1938; Ranke 1940; Leake 1940, 1952; Hemmeter 1941; Zealand 1951; Riad 1955; Lefebvre 1956; Ghalioungui 1963, 1969; Munch 1965a, 1965b; Hengen 1971; Regoely-Mari 1974). While these are intriguing questions from a biocultural perspective, the texts do not provide information as to the frequency or prevalence of specific diseases (Sandison 1972). Such information can, however, be derived from the study of skeletal populations. A number of studies has made correlations between skeletal evidence of pathologies and the diseases described in medical papyri, but in most cases the goal has been to verify or refute interpretations of the texts, rather than to understand better the effects of disease within the population and its biocultural context.

The primacy of texts and concern with their historical interpretation and veracity in Egyptology have effectively excluded more holistic theoretical considerations which have been developed in the sciences. Dental studies reveal the limited scope of historical questions. A common theme in dental studies has been to determine the antiquity of dentistry in Egypt (eg. Hooton 1917; Anon. 1919; Wienberger 1946; Leek 1967, 1969, 1980; Farrell 1973; Trillou 1976a, 1976b). This question bespeaks an underlying fascination with and delight in demonstrating the precociousness of Egyptian civilisation. In dental studies there has also been an emphasis on royal descent (eg. Harris 1977; Harris and Loutfy 1967; Harris and Weeks 1973), and on the diet and dental health of the elite (Storey 1976; Wall 1976). These biases are in part due to the nature of available data, but are ultimately symptomatic of the historical and elite-centered emphases of traditional Egyptology.

The Rise of Modern Palaeopathology

The 1930s are usually considered to herald the modern era of palaeopathology. The publication of E A Hooton's *The Indians of Pecos Pueblo* in 1930 is often described as the pivotal publication of the era. In this monograph he introduces the palaeoepidemiological approach, which analyses the relationships among the host, the pathogen and the environment. Although Hooton is often cited as initiating the modern era of palaeopathology, in actuality his work had little impact during the 1930s and 1940s. Hooton's major contribution was the use of statistics in presenting his data. It should be pointed out that it is often difficult to reconcile some inconsistencies in Hooton's statistical analyses. Few publications of the era provide information on the observations made or the frequency of a pathology.

There was a number of theoretical developments which set the stage for the modern era of palaeopathology. The first factor was the acceptance of a population perspective. While the individual may be the unit of diagnosis, the population is the unit of analysis. The understanding of the disease process requires consideration of its existence within the context of a population. The second was the realisation that culture is an environmental variable which can affect the disease process. From this perspective a group's technology, social organisation and ideology may play a major role in inhibiting or creating an environment for disease. This consideration of culture has led to a more thorough analysis of human/disease interaction. Thirdly, the concept of pathogenesis was expanded to include other insults which cause disease. Rather than consider pathogens as the sole source of disease, other factors such as trauma, pollutants and psychological and social stresses were seen as potential agents of disease. The full impact of these changes was not realised until the 1950s. The impact of the population perspective on Egyptian palaeopathology experienced further delays.

Why did palaeopathology take so long in accepting the population approach? Undoubtedly, the major factor has been the reliance on the historical perspective. As long as researchers could continue to find earlier occurrences of a pathology or extend the geographic limits of a disease, advances in knowledge continued. When this theme had been played out (when interest in the earliest example of a pathology declines or its geographic boundaries become defined) there is little more said about disease. There were, however, other factors which precluded acceptance of the population concept. For example, the influence of physicians, who comprised the majority of early researchers, may have been another factor slowing its acceptance. Since physicians were striving to improve methods of diagnosis, they often were less concerned with attempting to relate these findings to the broader issues. The reliance on modern technology may have contributed to the limited perspective in another way. In a field that expands its technology as rapidly as medicine, the adoption of new instrumentation gives the illusion of scientific advancement. Abraham Kaplan (1964) in another context refers to this behaviour as the 'law of the instrument' which states that when you give a child a hammer everything in its universe is poundable. In a sense, the palaeopathologist's hammer was medical technology and the historical perspective. Major advances in palaeopathology were often measured by the application of the newest technology rather than by the problems that were being solved. Yet another factor precluding a population perspective has been the loss of adequate population samples due to incomplete collection and loss during 'curation'. All of the above factors combined to focus research on the history of disease.

Three major objectives of modern palaeopathology have emerged from these historical developments. The first objective of palaeopathology remains historical and spatial. Defining the chronology and geography of disease was, and is, the primary objective of most palaeopathological research. The second objective of palaeopathology is to determine the biocultural interactions which occur as a population adapts to its environment. Therefore the analysis of disease can provide important information about the success or failure of its adaptation. The biocultural approach considers not only the effect that culture has on the pattern of disease, but also attempts to understand the impact of disease on the culture. The third objective of palaeopathology is concern for understanding the processes involved in prehistoric disease. Process refers to the production of change. In living organisms, we can often study the biological system as it is actually undergoing transformation in order to unravel the underlying process. In dead populations the understanding of process is much more difficult. The factors which bring about change are inferred from the examination of many individuals in various stages of this transformation and from studying the histological basis underlying this change.

In palaeopathology the study of process can focus on the change that occurs at a number of levels. The analysis can be conducted at the cellular, tissue, organ, organism or population level. For example, biological changes which occur in the skeleton as the result of disease-insult may be studied. In fact, it was the examination of various stages in the development of treponemal infection that helped us understand this disease process in prehistory. Process can also be analysed within the context of the biocultural system. Change in the demographic structure of a population can be studied with respect to this response to infectious disease.

How well have the recent contributions of Egyptian palaeopathology incorporated the biocultural and processual objectives of palaeopathology? The evaluation of the recent contributions of Egyptology is enhanced by the excellent reviews that have been published. Sandison (1968, 1969, 1972, 1980), Rowling (1960, 1961, 1967), Cockburn (1973), Strouhal and Vyhananek (1979) and Brothwell and Powers (1968) have summarised the state of knowledge in the field. Sandison (1970) has also reviewed the histological studies and Gray (1973) has provided a similar analysis of radiographic studies. From these sources it becomes clear that the historical/descriptive themes which guided palaeopathological work in Egypt during the early decades of this century continue as the dominant objectives. The exceptional state of preservation of Egyptian remains continues to allow the application of new technologies and clinical methods to enhance diagnosis and expand and confirm the listing of recognised pathologies. What has been consistently lacking is the systematic application of such methods beyond individual specimens to the level of populations.

The analysis of pathology in prehistoric populations often follows a deductive methodology in which the researcher attempts to reconstruct the factors which caused the skeletal lesions. The palaeopathologist, much like a detective who must reconstruct a crime from the evidence long after the perpetrators have left the scene, has to reconstruct the lifeways of a population long after they lived and died. While the deductive approach has been useful, it has certain inherent limitations. There is often no possible method for falsifying conclusions. There is, however, a methodology which can systematically falsify or exclude alternative hypotheses. J R Platt (1964) suggests that the use of an inductive approach which he calls 'strong inference' can provide a means for hypothesis testing in science. Platt notes that the roots of strong inference derive from the Baconian approach to experimentation. As early as the late 1870s, T Chamberlin, a geologist, argued for the multiple hypothesis testing approach in science. Although inductive inference has been known for many years, it has not penetrated the methodology of palaeopathology.

Strong inference could be very effective if applied to problems in palaeopathology. It would be the means for transforming the discipline into a science. There are four stages in the development of the strong inference approach (Platt 1964:347):

1. Devise alternative hypotheses.
2. Devise experiments to exclude one or more of the hypotheses.
3. Carry out experiments to get 'clean results'.
4. Recycle the procedure.

While strong inference is most effectively applied to sciences with experimental possibilities, it can be useful in non-experimental sciences such as palaeopathology. The application to palaeopathology does require modification; since there is no possibility of carrying out experiments to get 'clean results', the researcher must rely on comparative analysis for 'natural experiments'.

The scientific aspect of the strong inference and the inductive approach can be illustrated by applying it to a model that systematically analyses a number of variables which may affect the disease process. Goodman and Armelagos (1989:226) have devised a model which considers environmental and cultural variables (see Fig. 1). In this model, the environment is not only the source of the essential resources necessary for survival, it also constrains the adaptation of the population if there are limits to necessary resources or if the environment is the source of other stressors. In this context, stressor refers to any factor which inhibits the group's ability to survive or maintain itself. Stressors can be climatological extremes, nutritional deficiencies, predators, parasites, pathogens or other factors which may hinder survival. The cultural system (the technology, social system and ideology) may act as a buffer to the stressors which emanate from the environment. However, the cultural system may be ineffective in buffering the stressors or may actually be the source of additional stressors. The impact of the stressors on the host depends on the resistance of the individual. Those who fail to resist the stressor will suffer from physiological disruption. Since we are dealing with individuals who died long ago, we must rely on the mummified or skeletal remains to assess the extent and significance of that disruption. It is safe to say that if the skeletal structure, which is usually well buffered, is disrupted, then the stressor is quite severe. The skeletal indicators of stress are pathological lesions, changes in growth patterns and ultimately death.

The traditional method of palaeopathological analysis begins with the skeletal indicators of stress and then attempts to deduce the physiological disruption, the host's ability to resist stressors, the effectiveness of the cultural buffering system and the environmental constraints. While this methodology has successfully provided us with a great deal of information on the pattern of disease in prehistory, it does not give a systematic means of testing hypotheses. There are many generalisations, established by the deductive approach, which should be systematically tested. Using the methods

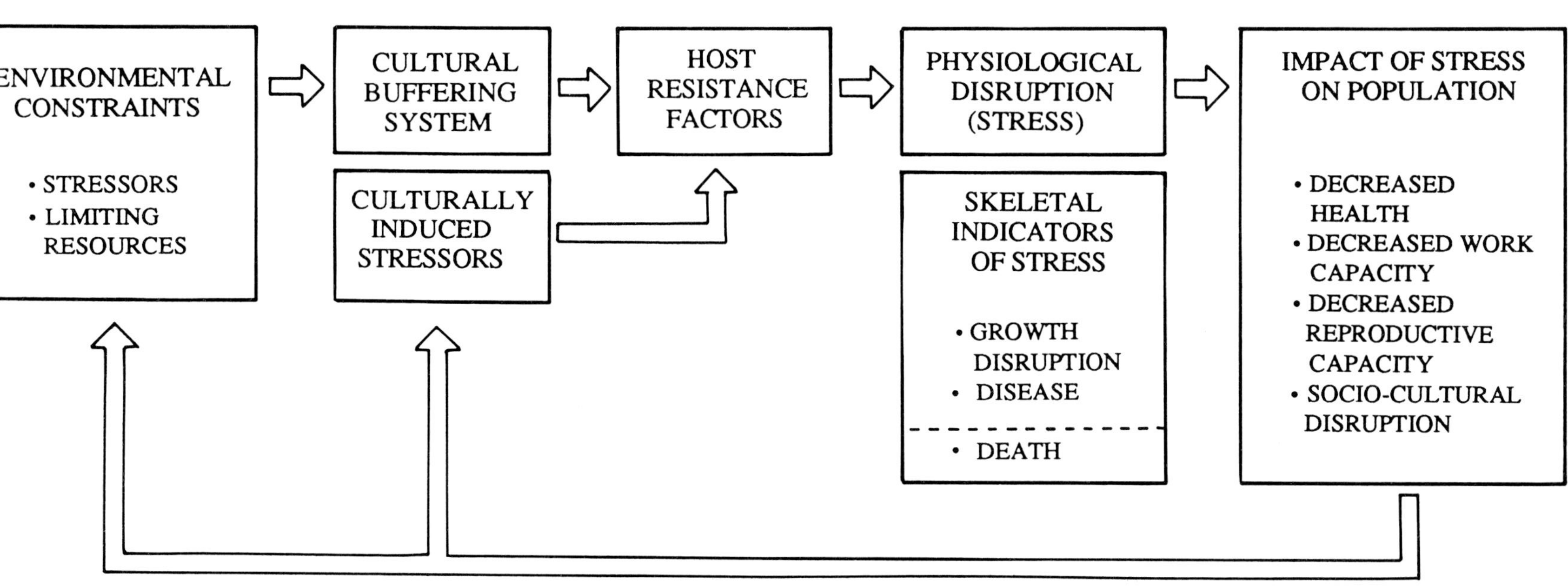

Fig. 1. A general model for the study of stress in skeletal populations (modified from Goodman et al. 1984).

of strong inference, the researcher could test many of the assumptions of palaeopathological research.

Case Study One: Did the Ancient Nubians Anticipate Modern Medicine?

Strong inference can be used to demonstrate how palaeopathologists were able to interpret the serendipitous discovery that the broad spectrum antibiotic tetracycline was used by the Nubians at least 1,400 years ago. Debra Martin, then an anthropology graduate student at the University of Massachusetts, was spending a few days at the calcified tissue laboratory of Henry Ford Hospital in Detroit, Michigan, refining techniques for making thin sections from archaeological bone without first decalcifying it. The technique for preparing a thin section of bone requires hand grinding to a thickness of one hundred microns (a little thicker than the page you are reading). In order to determine the correct and uniform thickness, the thin bone sections are frequently examined under a standard light-transmitting microscope. While preparing to examine a thin section, Martin discovered that the light-transmitting microscope was in use. Another researcher in the laboratory offered her the use of a microscope used to measure fluorescence in bone. To Martin's astonishment, the archaeological bone from the X-Group (Ballana) period (AD 550) showed the unique yellow-green fluorescence characteristic of tetracyclines when exposed to ultraviolet light at the 490 nanometer wave length. Her discovery of tetracycline was as surprising as if someone were to unwrap a mummy and find a pair of Ray-Ban sunglasses strapped to its head.

There are three questions which must be answered if we are to believe that tetracyclines were being used by the ancient Nubians. Is it tetracycline? Is there evidence that the tetracycline was ingested prior to death, or is it the result of post-mortem contamination (Piepenbrink et al. 1983; Piepenbrink 1986)? If it was ingested, what was the source of the antibiotic?

Research clearly suggests that the substance is a form of tetracycline (Bassett et al. 1981). Its yellow-green fluorescence occurs at the expected wave length and the colour is characteristic of tetracyclines. James Booth, who was one of the scientists who originally worked on the commercial application of tetracycline, has been able to extract the antibiotic from archaeological bone from Nubia and has found that it still possesses its ability to kill bacteria.

Was the tetracycline a contaminant of the decay process? It has been argued that most tetracycline labelling is due to post-mortem changes that occur when bone deteriorates and is invaded by bacteria and fungi. When this occurs there is evidence of destruction of the bone and often a diffuse labelling of bone surfaces and especially undermineralised osteons. Margaret Keith and George Armelagos (1988) have shown that in the Nubian archaeological bone the tetracycline labels fully mineralised osteons which are adjacent to unlabelled, undermineralised osteons. In addition, there is no evidence of deterioration of bone or any other indication that tetracycline is the result of post-mortem changes.

Tetracycline is manufactured by a streptomycete, a mould-like bacterium. The organism, which is prevalent in Nubian soil, would have come into contact with the stored millet, barley and sorghum that were major staples of the Nubian diet. The contaminated grain could have been eaten as bread, but we suspect that it was brewed as beer. We know, for instance, that the Egyptians would often bake bread until a crust formed while the centre was still uncooked. The half-baked bread served as a culture medium for airborne yeast, which initiated fermentation (Geller 1989; Kemp 1989). This was then broken up and put into water to continue the fermentation process necessary to make beer. Whatever the source of tetracyclines, the ancient Nubians were receiving therapeutic doses of this antibiotic, which undoubtedly contributed to the low rate of infectious lesions of their bones.

Hummert and Van Gerven (1982) also uncovered evidence of tetracycline labelling in bones recovered from the Christian cemetery at Kulubnarti. In this group, the use was sporadic and the dosage was not very great. Hummert and Van Gerven suggest that differences in pattern of storage related to level of production may account for the low levels of tetracycline ingestion. The environment of Kulubunarti is harsher than that found in Wadi Halfa and this could have decreased the potential for storage. Hummart and Van Gerven also suggest that the grain taxation of the Church would have further reduced storage.

Cook and co-workers (1989) have found similar tetracycline labelled bone from remains recovered from the Dakhla Oasis in Egypt. Similar storage methods are likely to promote the production of tetracyclines. However, these populations did not get the dosage that characterised the Wadi Halfa population and consumption was likely to have been seasonal.

Case Study Two: Did the Sudanese Nubians Experience Nutritional Deficiencies?

The adaptation of the ancient Sudanese Nubians has been studied extensively (Armelagos 1968, 1969; Goodman et al. 1984; Martin et al. 1984). The determination of dietary deficiency is difficult in prehistoric populations. While we have been able to diagnose specific nutritional deficiencies such as scurvy and rickets, there has been little evidence for other dietary problems. Huss-Ashmore et al. (1982), Goodman et al. (1984) and Larsen (1987) have advocated the use of multiple indicators to determine nutritional deficiencies. By analysing pathological conditions, growth, evidence of growth disruption and chemical indicators, nutritional patterns may emerge. Furthermore, it would be possible to apply the strong inference procedures to test one's hypothesis.

Paul Mahler (1986) and Armelagos and co-workers (1972), studying long bone length in a series of Meroitic (350 BC-AD 350), X-Group or Ballana (AD 350-550) and Christian (AD 550-1300) populations from the Wadi Halfa area of Sudanese Nubia, noticed that children who were aged dentally between ages two and six had long bones shorter than expected. At the time it was suggested that this might represent growth retardation associated with weaning. However, the small sample sizes and the fact that the authors were dealing with cross-sectional, as opposed to longitudinal, growth data made the results ambiguous.

If the decrease in long bone length is the result of nutritional deficiency, then there should be other indicators of nutritional inadequacies. For example, porotic hyperostosis has been shown to result from an anaemia that can be nutritionally based. Porotic hyperostosis is defined as an expansion of bone (hyperostosis) in which there is also a very porous appearance to the external covering. The lesion usually occurs on the thin bones of the skull (the roof of the orbits, the frontal bone and the parietals) and is usually a response to anaemia. The middle of these thin bones (the diploe) is an area where red blood cells are manufactured. In response to an anaemia, the red blood cell production increases, causing the diploe to expand. This response results in a thinning of the outer table, exposing trabecular bone, which gives the porous appearance.

Since there are many conditions which cause anaemia (see Mensforth et al. 1978), it is necessary to determine the most likely cause. Hereditary hemolytic anaemia such as thalassaemia and sickle cell anaemia, hereditary nonspirocytic anaemias such as glucose 6-phosphate dehydrogenase deficiency, and even nutritional anaemia such as iron-deficiency anaemia can cause porotic hyperostosis. There are two factors which suggest that iron deficiency anaemia is the most likely cause of porotic hyperostosis in the

Nubian populations. If it were a hereditary anaemia, we would expect to find porotic hyperostosis involving many of the cranial bones and there would be likely to be postcranial changes. In the Nubian population, 21% showed evidence of porotic hyperostosis which was restricted to the upper portion of the orbits. Additionally, the highest frequency is in children under the age of ten and in young adult females.

An analysis of the microstructure of Nubian bone gave us our best evidence that Nubians were suffering from nutritional deficiency. The microscopic structure of Nubian bones has been studied extensively in our analysis of age-related bone loss. Osteoporosis, which occurs frequently in older women in modern populations, was first reported in prehistoric populations from Nubia (Dewey et al. 1969). The analysis of femur cross-sections showed that the Nubians began to lose bone a little earlier than modern populations (bone loss begins at age thirty and continues at a slow and steady rate). In modern populations there is an acceleration of bone loss after menopause, and since women are smaller than men and have less skeletal mass to begin with, they often suffer from pathological fractures. While the rate of loss in Nubian females was similar to that of their modern counterparts, they show no evidence of pathological fracture. Since most Nubians were dead by their sixtieth year, their bone loss was never great enough to create the fractures that occur in modern Americans. (In the United States, 75% of the population live past their sixtieth year, 29% past their eightieth year, and 6% past their ninetieth year).

It was during the study of osteoporosis that Dr Debra Martin (1983; see also Martin and Armelagos 1979) realised that females in their peak reproductive period show premature bone loss. In her analysis of the histological change, Martin was able to show the process involved in premature bone loss. By studying the frequency of resorption spaces and forming osteons in the periosteal, middle and endosteal surfaces, she was able to show the mechanism of bone loss. She demonstrated that females between twenty and twenty-nine years of age had a higher frequency of resorption spaces and forming osteons. Furthermore, the Nubian females showed evidence of bone being resorbed (removed) from the endosteal (inner) surface surrounding the medullary cavity, and while bone cells were being formed on the outer (periosteal) surface, those cells (osteons) were not being mineralised. The calcium that is resorbed in women is passed on to children during lactation.

While the children may have been receiving their mothers' calcium during lactation, they also show evidence of impaired calcium metabolism. Long bone growth seems to be slower when compared with modern standards, and microscopic analyses of femur cross-sections show evidence of bone loss in children as young as two years of age. The resorption of bone may be an attempt by the skeletal system to maintain long bone length at the expense of long bone width. Interestingly, Hummert and Van Gerven (1983) have confirmed this pattern in another Christian Period Nubian population from the island of Kulubanarti. They also show that the bone loss results in bones that are biomechanically stronger than bone showing less loss. However, bones with more loss will lack the reserves necessary to meet other episodes of nutritional deficiency.

The nutritional problems of prehistoric Nubian women and children can be related to a number of factors: reduction in birth spacing, breast-feeding practices, parasites, and diet. Miller and co-workers (in press) have determined, through sophisticated techniques, that schistosomiasis affected a large segment of the population. The reliance on cereal grains (barley, millet and sorghum), which are poor sources of iron and calcium, was undoubtedly a major contributing factor to the dietary deficiency. Christine White (1991), in a study that is destined to be a classic, has measured the isotope ratios of 13C/12C and 15N/14N in bone, mummified skin and hair from the Wadi Halfa populations to assess dietary intake. She concluded that wheat, barley, vegetables and fruit and nuts were the dominant food sources throughout Meroitic, X-Group (Ballana) and Christian Periods. White notes that during the X-Group (Ballana) Period millet and sorghum consumption increases. Furthermore, she found that X-Group (Ballana) individuals show evidence of iron deficiency and higher levels of millet and sorghum consumption.

Multiple hypothesis testing that is the basis of the strong inference approach can provide a model for future work in palaeopathology. The systematic approach of future studies may provide the key to studying the past.

References

Adams, William Y, 1977. *Nubia: Corridor to Africa*, Princeton, Princeton University Press.

Anonymous, 1919. 'Oral Surgery in Egypt During the Old Empire', *Dental Cosmos* 61, 900.

Armelagos, George J, 1968. *Paleopathology of Three Archaeological Populations From Sudanese Nubia*. Ph.D. Thesis, University of Colorado.

Armelagos, George J, 1969. 'Disease in Ancient Nubia', *Science* 163, 255-259.

Armelagos, George J, J H Mielke, K H Owen, D P Van Gerven, J R Dewey and P Mahler, 1972. 'Bone Growth and Development in Prehistoric Population from Sudanese Nubia', *Journal of Human Evolution* 1, 89-119.

Armelagos, George J, James H Mielke and John Winter, 1971. *The Bibliography of Human Paleopathology*, University of Massachusetts, Department of Anthropology, Research Report.

Armelagos, G J, D P Van Gerven, D L Martin, and R Huss-Ashmore, 1984. 'Effects of Nutritional Change in the Skeletal Biology of Northeast African (Sudanese Nubian) Populations', J D Clark and S A Brandt (eds.), in *From Hunters to Farmers*, Berkeley, University of California Press, 132-146.

Asante, Molefe Kete, 1988. *Afrocentricity*, Africa World Press.

Baker, Brenda and George J Armelagos, 1988. 'Origin and Antiquity of Syphilis: A Dilemma in Paleopathology Diagnosis and Interpretation', *Current Anthropology* 29 (5), 703-737.

Basset, E, Margaret Keith, George J Armelagos, Debra L Martin and A Villanueva, 1981. 'Tetracycline-labelled human bone from prehistoric Sudanese Nubia (AD 350)', *Science* 209, 1532-34.

Batrawi, A, 1945. 'The racial history of Egypt and Nubia. Part I. The craniology of Lower Nubia from Predynastic times to the sixth century AD', *Journal of the Royal Anthropological Institute* 75, 81-101.

Batrawi, A, 1946. 'The racial history of Egypt and Nubia. Part 2. The racial relationships of the ancient and modern populations of Egypt and Nubia', *Journal of the Royal Anthropological Institute* 76, 131-156.

Batrawi, A, 1947. 'Anatomical reports', *ASAE*, 47, 97-109.

Bernal, Martin, 1987. *Black Athena*, I, Rutgers University Press.

Bernal, Martin, 1991. *Black Athena*, II, Rutgers University Press.

Berry, A C and R J Berry, 1972. 'Origins and Relationships of the Ancient Egyptians. Based on a Study of Non-metrical Variations in the Skull', *Journal of Human Evolution* 1, 199-208.

Berry, A C and R J Berry, P J Ucko, 1968. 'Genetic Change in Ancient Egypt', *Man* 2, 551-68.

Breasted, J H, 1930. *The Edwin Smith Surgical Papyrus*, Chicago, University of Chicago Press.

Brothwell, D R, 1967. 'Major Congenital Anomalies of the Skeleton: Evidence from Earlier Populations', in D R Brothwell and A T Sandison (eds.), *Diseases in Antiquity*, Springfield, Ill., C C Thomas, 423-43.

Brothwell, D R and R Powers, 1968. 'Congenital malformations of the skeleton in earlier man', in D R Brothwell (ed.), *The Skeletal Biology of Earlier Human Populations*, London, Pergamon Press.

Brothwell, D R and Chiarelli, B (eds.), 1973. *Population Biology of the Ancient Egyptians*, New York, Academic Press.

Cave, A J E, 1938. *The Surgery of Ancient Egypt*, University Leeds Med. Soc. Mag.

Clendinnen, F J, 1898. 'Skiagram of Hand of Egyptian Mummy, Showing Abnormal Number of Sesamoid Bones', *Intercolonial Medical Journal of Australia* 3, 106-9.

Cockburn, T A, 1973. 'Death and Disease in Ancient Egypt', *Science* 181, 470-1.

Cockburn, A, R A Barraco, T A Reyman and W S Peck, 1975. 'Autopsy of an Egyptian Mummy', *Science* 187 (4182), 115-60.

Cook, M, El Molto and C Anderson, 1989. 'Fluorochrome Labelling in Roman Period Skeletons from Dakhleh Oasis, Egypt', *American Journal of Physical Anthropology* 80, 137-43.

Czermack, J, 1852. 'Beschreibung und Mikroskopische Untersuchung zweier ägyptischer Mumien', *Akademie der Wissenschhaften Wien* 9, 427-69.

Davin, A G and K Pearson, 1924. 'On the biometric constants of the human skull', *Biometrika* 16, 328-68.

Dawson, W R, 1932. 'The Earliest Surgical Treatise', *Br. J. Surg.* 20, 34.

Dawson, W R, 1938. 'Pygmies and Dwarfs in Ancient Egypt', *JEA* 24, 185.

Dawson, W R, 1953. 'The Egyptian medical papyri', in *Science, Medicine and History: Essays on the Evolution of Scientific Thought and Medical Practice*, written in honour of Charles Singer, edited and published by E A Underwood, Oxford.

Dawson, W R, 1967. 'The Egyptian Medical Papyri', in D Brothwell and A T Sandison (eds.), *Diseases in Antiquity*, Springfield, Ill., C C Thomas, 98-111.

Derry, D E, 1909. 'Anatomical report', *Archaeological Survey of Nubia*, Bulletin 3.

Dawson, D E, 1913. 'A case of hydrocephalus in an Egyptian of the Roman period', *Journal of Anatomy and Physiology*, London, 47, 436-458.

Dawson, D E, 1956. 'The Dynastic Race in Egypt', *JEA* 42, 80-85.

Dewey, J R, G J Armelagos and M H Bartley, 1969. 'Femoral Cortical Involution in Three Archaeological Populations', *Human Biology* 41, 13-28.

Diop, Cheikh Anta, 1991. *Civilization or Barbarism*, Lawrence Hill Books.

Ebbell, B, 1937. *The Papyrus Ebers: The Greatest Egyptian Medical Document*, Copenhagen, Levin and Munksgaard.

Emery, W B, 1961. *Archaic Egypt*. New York, Penguin Books.

Farrell, E, 1973. 'Dentistry's Past and Future Meet in Egypt', *Journal of the American Dental Association* 86, 553-62.

Fawcett, C D, 1901. 'A second study of the variation and correlation of the human skull with special reference to the Nagada crania', *Biometrika* 1, 408-67.

Finlayson, J, 1893. 'Ancient Egyptian Medicine', *British Medical Journal* 1, 748-62, 1014-16 and 1160-64.

Ghalioungui, P, 1963. *Magic and Medical Science in Ancient Egypt*, London, Hodder and Stoughton.

Ghalioungui, P, 1969. 'Pharaonic Medicine', *J. Med. Liban.* 22, 159.

Goodman, A H and G J Armelagos, 1989. 'Infant and Childhood Morbidity and Mortality Risks in Archaeological Populations', *World Archaeology*, 21 (2), 225-43.

Goodman, A H, D L Martin, and G J Armelagos, 1984. 'Indications of Stress from Bone and Teeth', in M N Cohen and G J Armelagos (eds.), *Paleopathology at the Origins of Agriculture*, Orlando, Academic Press, 13-49.

Granville, A B, 1825. 'An Essay on Egyptian Mummies with Observations on the Art of Embalming among the Ancient Egyptians', *Philosophical Transactions of the Royal Society* 115, 269-319.

Gray, P H K, 1973. 'The Radiography of Mummies of Ancient Egyptians', *Journal of Human Evolution* 2, 51-53.

Guiart, J, 1922. 'La Medecine aux Temps des Pharaons', *Biologie Medicale* 12 (7).
Harris, J E, 1977. 'The Teeth of the Pharaohs Dental Dimensions', *San Fernando Valley Dental Society*, Van Nuys, CA 1, 2-6.
Harris, J and S Loutfy, 1967. 'Pharaohs too had Malocclusion', *Med. News*, P. 4. (Mar.).
Harris, J G and K R Weeks, 1973. *X-Raying the Pharaohs*, New York, Charles Scribner's Sons.
Hemmeter, E, 1941. Médecins et Therapeutique dans L'Egypte Ancienne, *Ciba Revue* 19.
Hengen, O P, 1971. 'Paleomedicine and Ancient Egypt', *Image* 45, 25-33.
Hooton, E A, 1917. 'Oral Surgery in Egypt during the Old Empire', in *Harvard University African Studies* I, 29, 27-32, Cambridge, Harvard University Press.
Hooton, E A, 1930. *The Indians of Pecos Pueblo*, New Haven, Yale University Press.
Hummert, J R and D P Van Gerven, 1982. 'Tetracycline-Labelled Human Bone from a Medieval Population in Nubia's Batn el Hajar (550-1450 AD)', *Human Biology* 54, 355-363.
Hummert, J R and D P Van Gerven, 1983. 'Skeletal Growth in a Medieval Population from Sudanese Nubia', *American Journal of Physical Anthropology* 60, 471-8.
Huss-Ashmore, R, A H Goodman and G J Armelagos, 1982. 'Nutritional Inference from Paleopathology', *Advances in Archaeological Method and Theory* 5, 395-474.
Jackson, J W F and A J E Cave, 1937. 'The Osteology: Report on the Human Remains', in R Mond and O H Myers, *Cemeteries of Armant* I, London, The Egyptian Exploration Society, 144-57.
Jomard, E, 1809-1822. *Description de l'Egypte*, Paris, vols. 1-24.
Jones, E W A H, 1932. 'Studies in Achondroplasia', *Journal of Anatomy and Physiology* 66, 565-77.
Jones, F Wood, 1908. 'The Pathological Report', *Archaeological Survey of Nubia*, Bulletin 2.
Kaplan, A, 1964. *The Conduct of Inquiry: Methodology for the Behavioral Science*, San Francisco, Chandler.
Keith, A, 1913. 'Abnormal Crania - Achondroplastics and Acrocephalic', *Journal of Anatomy and Physiology* 47, 189-206.
Keith, M S and G J Armelagos, 1988. 'An Example of In Vivo Tetracycline Labelling: Reply to Piepenbrink', *Journal of Archaeological Sciences* 15, 595-601.
Krause, A C, 1933. 'Ancient Egyptian Opthalmology', *Bull. Hist. Med.* 1, 258-76.
Larsen, C S, 1987. 'Bioarchaeological Interpretation of Subsistence Economy and Behavior from Human Skeletal Remains', *Advances in Archaeological Method and Theory* 10, 27-56.
Leake, C D, 1940. 'Ancient Egyptian Therapeutics', *Ciba Symposia* 1 (10), 311.
Leake, C D, 1952. *The Old Egyptian Medical Papyri*, Lawrence, University of Kansas Press.
Leek, F F, 1967. 'Reputed Early Egyptian Dental Operation, An Appraisal', in D Brothwell and A T Sandison (eds.), *Diseases in Antiquity*, Springfield, Ill., C C Thomas, 702-5.
Leek, F F, 1969. 'Did A Dental Profession Exist in Ancient Egypt', *Dent. Delin.* 20, 18-21.
Leek, F F, 1980. 'A Third Millennium Dental Profession in Egypt: Fact or Fiction?', *Papers on Paleopathology*, Third European Members Meeting, Paleopathology Association, Caen, France, p. C12.
Lefebvre, G, 1956. *Essai sur la Médecine Egyptienne de l'Epoque Pharaonique*, Paris, Presse Universitaire de France.

Long, A R, 1931. 'Cardiovascular renal disease: report of a case of 3000 years ago', *Archaeological Pathology*, 12, 92, Chicago.
Lortet, L C, 1907. 'Crane Syphilitique et Necropoles Prehistoriques de la Haute-Egypte', *Bulletin de la Societe Anthropologie* (Lyon) 26, 211.
Mahler, P E, 1968. *Growth of Long Bones in a Prehistoric Population from Sudanese Nubia*, MA Thesis, University of Utah, Salt Lake City.
Mann, George E, 1990. 'On the accuracy of sexing of skeletons in archaeological reports', *JEA* 75, 246-49.
Martin, D L, 1983. *Paleophysiological Aspects of Skeletal Remodeling in the Meroitic, X-Group, and Christian Populations from Sudanese Nubia*, Ph D Dissertation, Department of Anthropology, University of Massachusetts, Amherst.
Martin, D L and G J Armelagos, 1979. 'Morphometrics of compact bone: an example from Sudanese Nubia', *American Journal of Physical Anthropology* 51, 571-8.
Martin, D L, G J Armelagos, A H Goodman and D P Van Gerven, 1984. 'The Effects of Socioeconomic Change in Prehistoric Africa: Sudanese Nubia as a Case Study', in Mark N Cohen and George J Armelagos (eds.), *Paleopathology at the Origins of Agriculture*, Orlando, Academic Press, 193-214.
Martin, D L, A H Goodman and G J Armelagos, 1985. 'Skeletal Pathologies as Indicators of Quality and Quantity of Diet', in R Gilbert and James H Mielke (eds.), *The Analysis of Prehistoric Diets*, Orlando, Academic Press.
MacIver, R, 1901. *The Earliest Inhabitants of Abydos*, Clarendon, London.
Mensforth, R P, C O Lovejoy, J W Lallo and G J Armelagos, 1978. 'The Role of Constitutional Factors, Diet, and Infectious Disease in the Etiology of Porotic Hyperostosis and Periosteal Reactions in Prehistoric Infants and Children', *Medical Anthropology* 2, University of Connecticut.
Micozzi, M S, 1991. 'Disease in Antiquity, The Case of Cancer', *Arc. Pathol. Lab. Med.* 115, 838-44.
Miller, R L, G J Armelagos, S Ikram, N De Jonge, F W Krijer, A N Deedler, In Press. 'Paleoepidemiology of *Schistosoma* infection in Sudanese Mummies', *British Medical Journal.*
Mitchell, J K, 1900. 'Study of a Mummy Affected with Anterior Poliomyelitis', *Transactions of the Association of American Physicians* 15, 134-6.
Moodie, Roy L, 1920. 'Primitive Surgery in Ancient Egypt', *Surgical Clinic of Chicago* 4 (2), 349-58.
Moodie, Roy L, 1931. *Studies in Palaeopathology of Egypt*, University of Chicago Press.
Morant, G M, 1925. 'A study of Egyptian craniology from prehistoric to Roman times', *Biometrika* 17, 1-52.
Morant, G M, 1935. 'A study of predynastic Egyptian skulls from Badari based on measurements taken by Miss B N Stoessiger and Professor D E Derry', *Biometrika* 27, 2293-309.
Morant, G M, 1937. 'The predynastic Egyptian skulls from Badari and their racial affinities', in G Brunton (ed.), *Mostagedda and the Tasian Culture*, London, Quaritch, 63-66.
Morse, D F, 1967. 'Tuberculosis', in D R Brothwell and A T Sandison (eds.), *Disease in Antiquity*, Springfield, Ill., C C Thomas, 249-71.
Morse, D F, D R Brothwell and P J Ucko, 1964. 'Tuberculosis in Ancient Egypt', *American Review of Respiratory Diseases* 90, 524.
Munch, I, 1965a. 'Medical Report from Ancient Egypt', *Zahnaertzlichen Mitteilungen* 55, 318.
Munch, I, 1965b. 'Aus der Geschichte der Zahnheilkunde. Die Altägyptischen Medizinischen Papyri', *Zahnaerztlichen Mitteilungen* 55, 216-9 and 318-20.

Myers, C S, 1905. 'Contributions to Egyptian Craniology. II. The Comparative Anthropometry of the Most Ancient and Modern Inhabitants', *Journal of the Royal Anthropological Institute* 35, 80-91.

Myers, C S, 1908. 'Contributions to Egyptian Anthropometry. V. General Conclusions', *Journal of the Royal Anthropological Institute* 38, 99-102.

Petrie, W M F, 1898. *Deshasheh 1897*, Fifteenth Memoir, Egyptian Exploration Fund, London.

Petrie, W M F, 1917. 'Egypt and Mesopotamia', *Ancient Egypt* 26-36.

Petrie, W M F, 1939. *The Making of Egypt*, London, The Sheldon Press.

Pettigrew, T J, 1834. *A History of Egyptian Mummies*, London, Longmans.

Piebenbrink, H, 1986. 'Two Examples of Biogenous Dead Bone Decomposition and Their Consequences for Taphonomic Interpretation', *Journal of Archaeological Sciences* 13, 417-30.

Piepenbrink, H, B Herrmann and P Hoffman, 1983. 'Tetracyclintypische Fluoreszenzen an Bodengelagerten Skeletteilen', *Zeitschrift für Rechtmedizin*, 91, 71-4.

Platt, J P, 1964. 'Strong Inference', *Science* 146 (3612), 347-53.

Ranke, H, 1940. 'Medicine and Surgery in Ancient Egypt', *University of Pennsylvania Bicentennial Conference*, 31-42.

Regoely-Merei, G, 1974. 'Surgery in Ancient Egypt', *Acta. Chir. Acad. Sci. Hung.* 15 (4), 415-25.

Riad, N, 1955. *La Médecine au Temps des Pharaons*, Paris, Librairie Maloine.

Rowling, T J, 1960. *Disease in Ancient Egypt: Evidence from Pathological Lesions Found in Mummies*, MD Thesis, University of Cambridge.

Rowling, T J, 1961. *Proceedings of the Royal Society of Medicine* 54, 409-14.

Rowling, T J, 1967. 'Respiratory disease in Egypt', in D R Brothwell and A T Sandison (eds.), *Disease in Antiquity*, Springfield, Ill., C C Thomas.

Ruffer, M A, 1908. 'Abnormalities and pathology of Ancient Egyptian teeth', *American Journal of Physical Anthropology* 3, 335.

Ruffer, M A, 1909. 'Preliminary note on the histology of Egyptian mummies', *British Medical Journal* 1, 1005.

Ruffer, M A, 1910a. 'Remarks on the histology and pathological anatomy of Egyptian mummies', *Cairo Scientific Journal* 4, 1-5.

Ruffer, M A, 1910b. 'Note on the presence of "bilharzia haematobia" in Egyptian mummies of the Twentieth Dynasty (1250-1000 BC)', *British Medical Journal* 1, 16.

Ruffer, M A, 1911a. 'Histological studies on Egyptian mummies', *Mémoires sur l'Égypte: Institut d'Égypte* 6 (3).

Ruffer, M A, 1911b. 'On arterial lesions found in Egyptian mummies (1580 BC-525 AD)', *Journal of Pathology and Bacteriology* 15, 543-62.

Ruffer, M A, 1914. 'Pathological notes on the royal mummies of the Cairo museum', *Mittheil. z. Gesch. Med. und der Naturwissensch.* 13, 239. (Reprinted in Ruffer, 1921).

Ruffer, M A, 1918. 'Arthritis deformans and spondylitis in Ancient Egypt', *Journal of Pathology and Bacteriology* 22, 152.

Ruffer, M A, 1920. 'Study of abnormalities and pathology of ancient Egyptian teeth', *American Journal of Physical Anthropology* 3, 335-82.

Ruffer, M A, 1921. In R L Moodie (ed.), *Studies in Palaeopathology of Egypt*, University of Chicago Press.

Ruffer, M A and A R Ferguson, 1911. 'An eruption resembling that of variola in the skin of a mummy of the Twentieth Dynasty (1200-1100 BC)', *Journal of Pathology and Bacteriology* 15, 1-3.

Ruffer, M A and A Rietti, 1912. 'On osseous lesions in ancient Egyptians', *Journal of Pathology and Bacteriology* 26, 439.

Ruffer, M A and J G Willmore, 1914. 'Studies in paleopathology: note on a tumor of the pelvis, dating from Roman times (250 AD) and found in Egypt', *Journal of Pathology* 18, 480-4.

Sandison, A T, 1968. 'Pathological changes in the skeletons of earlier populations due to acquired disease and difficulties in their interpretation', in D R Brothwell (ed.), *The Skeletal Biology of Earlier Human Populations*, Pergamon Press, Oxford.

Sandison, A T, 1969. 'Diseases in Ancient Egypt', *Rivista di Antropologia* 56, 225-7.

Sandison, A T, 1970. 'The study of mummified and dried human tissues', in D R Brothwell and E Higgs (eds.), *Science in Archaeology*, 2nd edition, Thames & Hudson, London.

Sandison, A T, 1972. 'Evidence of infective disease', *Journal of Human Evolution* 1, 213-24.

Sandison, A T, 1980. 'Diseases in ancient Egypt', in Aidan and Eve Cockburn (eds.), *Mummies, Disease and Ancient Cultures*, Cambridge University Press.

Seligmann, C G, 1912. 'A cretinous skull of the Eighteenth Dynasty', *Man* 12, 17-18.

Shattock, G S, 1909. 'A report upon the pathological conditions of the aorta of King Merneptah', *Proceedings of the Royal Society of Medicine (Pathological Section)* 2, 122-7.

Smith, G Elliot, 1906. 'A contribution to the study of mummification in Egypt with special reference to the measures adopted during the time of the 21st Dynasty for moulding the form of the body', *Mémoires présentés à l'Institut Égyptien* 5(1), 53.

Smith, G Elliot, 1907. 'The causation of the symmetrical thinning of the parietal bones in ancient Egyptians', *Journal of Anatomy and Physiology* 41, 232.

Smith, G Elliot, 1912. *The Royal Mummies*. Catologue Général des Antiquités Égyptiennes du Musée du Caire, Nos 61051-61100, Service des Antiquités de l'Egypt, Cairo.

Smith, G Elliot, 1915. *Memoires and Proceedings of the Manchester Literary and Philosophical Society* 59, 1-143.

Smith, G Elliot, 1923. *The Ancient Egyptians and the Origin of Civilization*. London/New York.

Smith, G Elliot, 1929. *Human History*, New York, W W Norton.

Smith, G Elliot and W R Dawson, 1924. *Egyptian Mummies*, Dial Press, New York.

Smith, G Elliot and D E Derry, 1910. 'Anatomical Report', *Archaeological Survey of Nubia*, Bulletin 6.

Smith, G Elliot and M A Ruffer, 1910. 'Pott'sche Krankheit an einer Aegyptischen Mumie aus der Zeit der 21 Dynastie (Um 1000 V. Chr.), in K Sudhoff (ed.), *Zur historischen Biologie der Krankheitserreger*, Heft 3, Giessen Leipzig.

Smith, G Elliot and F Wood Jones, 1908. 'Anatomical Report', *Archaeological Survey of Nubia*, Bulletin 1.

Smith, G Elliot and F Wood Jones, 1910. 'Report on the Human Remains', *Archaeological Survey of Nubia*, Vol. 2.

Stoessiger, B N, 1927. 'A study of the Badarian crania recently excavated by the British School of Archaeology in Egypt', *Biometrika* 19, 110-50.

Storey, A, 1976. 'The Diet and Dentition of New Kingdom Pharaohs', in D J Anderson and B Matthews (eds.), *Mastication*, Bristol, John Wright, 5-15.

Strouhal, E, 1974. 'Tumours in the remains of ancient Egyptians', *American Journal of Physical Anthropology* 45, 613-20.

Strouhal, E, 1978. 'Ancient Egyptian Case of Carcinoma', *Bulletin of the New York Academy of Medicine* 54, 290-302.

Strouhal, E and L Vyhnanek, 1979. *Egyptian Mummies in Czechoslovak Collections*, Acta Musei Nationalis Pragae, Vol. XXXV B.

Temkin, Owsei, 1936a. 'Recent Publications on Egyptian and Babylonian Medicine, I', *Bull. Hist. Med.* 4, 247-56.

Temkin, Owsei, 1936b. 'Recent Publications on Egyptian and Babylonian Medicine, II', *Bull. Hist. Med.* 4, 341-7.
Todd, T W, 1921. 'Egyptian Medicine: A Critical Study of Recent Claims', *American Anthropologist* 23, 460-70.
Trillou, J A, 1976a. 'Egyptian Dental Prosthesis, Myth or Reality?' *Chir. Dent. Fr.* 46 (297), 59-60.
Trillou, J A, 1976b. 'Egyptian Dental Prosthesis, Myth or Reality?' *Dent. Cadmos* 44 (10), 52-3.
Vandier, J, 1952. *Manuel d'Archéologie Égyptienne,* Vol. I, Picard, Paris.
Wall, S, 1976. 'Odontology in the Egypt of the Pharaohs', *Nord. Medhist. Arsb.*, 30-41 (Swedish), 195 (English abstracts).
Warren, E, 1897. 'An investigation of the variability of the human skeleton with especial reference to the Nagada Race', *Philosophical Transactions of the Royal Society* 189, 135-227.
Weinberger, B W, 1946. 'Further Evidence that Dentistry was Practised in Ancient Egypt, Phoenicia and Greece', *Bull. Hist. Med.* 20, 188-95.
White, Christine, 1991. 'Isotopic Analysis of Multiple Human Tissues from Three Ancient Nubian Populations', PhD Thesis, University of Toronto.
Woo, T L, 1930. 'A study of seventy-one Ninth Dynasty Egyptian skulls from Sedment', *Biometrika* 22, 65-92.
Zealand, N, 1951. 'Artificial Eyes in Ancient Egypt', *N J Supp.*, 27.
Zimmerman, M R, 1981. 'A Possible Histiocytoma in an Egyptian Mummy', *Arch. Derm.* 117, 364-5.

HEALTH IN PHARAONIC EGYPT

W Benson Harer

The three essentials to sustain life are food, water and shelter. If 'health' is defined as 'the state of the organism when it functions optimally without evidence of disease or abnormality',[1] some amplification of these three essentials is necessary.

The Nile Valley provided these requirements readily. There was a continuous source of pure water. The food supply was both ample and varied with the potential for meeting high standards of nutrition. Barley and emmer wheat were readily cultivated. Diversion of some grain from bread production to beer enhanced the nutritive value by increasing digestible protein and the B complex vitamins.[2] Protein was supplied further by a wide variety of fish, fowl and game along with domesticated swine and cattle. Onions in particular were favoured among the variety of vegetables and fruit which rounded out the diet. Honey was the major sweetener. Mud from the inundation was readily combined with reeds for rudimentary shelter. Even at the dawn of Dynastic Egypt sun-dried mud brick made possible the construction of massive structures such as Abd el Suffian at Hierakonpolis.

The Nile Valley provided such abundance that its dwellers were generally relieved of anxiety regarding acquisition of the essentials. This provided peace of mind and time to devote to religion, politics and massive public works. Thus division of labour into specialised tasks was facilitated, and resources could be diverted to non-essential activities, such as production of elegant jewellery. Arts, crafts, music and dance achieved considerable refinement even in the early dynasties. The ingredients were present for 'Maat', that difficult to define state of the proper order of the universe - the harmonious relationship of man with the rhythms of nature and the Nile.

Certainly the Nile Valley offered the possibility of living to what the Egyptians proclaimed was the ideal life span - 100 years. But if Egypt could be viewed as a Garden of Eden, it had both literal and metaphorical serpents and with them came disease, disability and early death. As the ancient Egyptians themselves observed, only one in a million achieved that ideal longevity.[3]

Many of the factors which reduced their life expectancy can be readily identified. This information comes from three basic sources - pictorial representation, inscriptions, and mummies. These sources often corroborate each other, and further supportive evidence can be derived from modern technical analysis of remains along with data from archaeobotany and archaeozoology.

In the past century there has been a shift in scientific and medical evaluation from gross organ evaluation to examination of cellular activity and currently to focus on the molecular basis of physiologic and pathologic processes. The Human Genome Project now underway is transforming our understanding of disease processes and normal physiology. Already the tip of the long arm of the X-chromosome alone has been shown to be the site of thirty disease-related genes.[4] Thus, genetic make up is a major determinant of health as a co-factor in cancer, cardiovascular disease, obesity, etc. Accordingly, a favourable genetic status should be added to the criteria for a long and healthful life.

It is probable that the most common spontaneous, non-lethal chromosomal dominant genetic mutation is that inducing achondroplasia. That appears to have been the case in ancient Egypt, too. Currently this spontaneous mutation occurs in fewer than one in 10,000 births. In such cases, both parents have normal chromosomes and normal stature. As this mutation is an autosomal dominant, half of the offspring of an achondroplastic will manifest the condition.[5] The similar but extremely rare diastrophic dwarfism is caused by an autosomal recessive gene. That means that both normal-appearing parents must provide a recessive gene in order for diastrophism to manifest itself. The differences between these two conditions can be readily discerned by physical examination. However, the distinctions (such as 'hitchhikers thumb' in diastrophism) are sufficiently subtle that they would not be evident in depictions of dwarfs from pharaonic times.[6] Nevertheless, a knowledgeable anthropologist confronted with the skeletons could make this distinction. Dr Fawzia Hussien has studied the skeleton of an Old Kingdom dwarf and confirmed the diagnosis of achondroplasia.[7]

Accurate maternal and infant mortality statistics have been developed only since the mid-nineteenth century. Numerous studies since then have shown that in the absence of modern medical care there is a world-wide pattern of maternal-related mortality of about 1% and a neonatal loss on the order of 20% during the first month of life.[8,9] It seems reasonable to assume a similar situation in ancient times. The mixed health benefit of the XX-chromosome is apparent from modern analysis of Third World countries, which generally show infant mortality of males exceeding females. The toll of maternal mortality is so great, however, that the overall life expectancy of women historically has been less than that for men.[10] The reduction of maternal mortality in developed countries to one in 10,000 (0.0001%) has been a major contribution to life expectancy of women in modern times. It now exceeds that of men by at least six years. In the days of the pharaohs life expectancy was at best in the order of thirty-five to thirty-nine years for males and three to five years less for women. Females who survive the child-bearing years appear then to have a life expectancy as long as or even longer than men.[11,12] Childbirth carries other non-lethal disadvantages. Egypt has provided the earliest known examples of vaginal-vesical fistula[13] and uterine prolapse.[14]

Statuary demonstrating hydrocephalus is known from the Old Kingdom and representative mummies have been reported. Weaknesses such as umbilical and inguinal hernia are depicted in Old Kingdom tombs, such as that of Ankhmahor at Saqqara.

Evidence of soft tissue tumours, both benign and malignant, is scarce and often unconvincing. Granville described a left ovarian mass in 1825, but none has been recorded since.[15] Skin is the best preserved soft tissue in mummies, but only a few small benign growths have been reported.[16] This seems surprising in view of ancient Egyptians' exposure to the sun, which induces skin cancer. Certain bone lesions may reflect metastatic carcinoma, but primary bone tumours are best attested.[17,18]

Toxins from the poisonous cobra and viper were well known and feared. The scorpion sting was dreaded for its pain, but would rarely prove fatal.[19] Lead has been sporadically identified at high levels in mummies.[20] While Egypt was not known to have its own lead deposits, occasional finds of lead objects are known from predynastic times.[21] A lead drinking straw from Tel el-Amarna, now in the British Museum, could certainly provide its owner with significant toxicity.[22]

Trauma is depicted in both military and occupational settings. Evidence of healed injuries as well as fatal ones is provided by mummies and skeletons. At Abu Simbel a soldier of Ramses II is shown in the camp having his leg treated. The tomb of Ipuy at Dier el Medina is famed for its occupational scenes including those of trauma and of a victim of Pott's Disease working a shadouf.

Infections were a major source of both morbidity and mortality. Representations of withered limbs and occasional similarly affected mummies have been postulated as evidence of poliomyelitis. The paucity of this material and lack of inscriptional support make this hypothesis very unlikely as they fail to conform to the epidemiologic character of the disease. The same observations hold regarding skin changes attributed to smallpox. There actually is no convincing evidence of any specific viral disease.

Tuberculosis is corroborated by both pictorial evidence of Pott's Disease and classic alteration of vertebral architecture in numerous mummies. Leprosy cannot be shown before the Graeco-Roman period.[23] *Chlamydia trachomatis*, a major modern cause of blindness and fatal neonatal pneumonia, may well have been the venereal disease of ancient Egypt, but there is no evidence to connect it specifically with portrayals of blindness.

Parasites are the infectious agents best demonstrated in mummy tissue. Stigmata of schistosomiasis such as gynecomastia and abdominal distention from ascites are shown in Old Kingdom tomb decorations such as those in the Tomb of Ti at Saqqara. Recent use of ELISA testing has further confirmed antibodies to schistosoma in a predynastic adolescent.[24] The related liver fluke *Fasciola hepatica* was also found in Manchester Mummies 21470 and 21471. *Strongyloides* was demonstrated in Manchester Mummy 1777. *Dracuncula* or guinea worm was detected in Mummy 1770 at Manchester. The ancient verb for 'spin' has recently been linked to the possible ancient treatment of this worm.[25]

Several parasites are shared by both man and domesticated animals.[26] Swine carry both *Trichinella* and *Taenea solium* (tape worm). Both have been found in mummies. The pet dog could share *Trichuris* (whip worm) and *Ascaris* (round worm) as well as *Echinococcus*, the dog tape worm. These, too, have been identified in mummy tissues. *Ascaris* can also be carried by pet cats.[27, 28, 29]

Those who survived into advanced age suffered the same maladies of time which afflict modern men. Atherosclerosis has been demonstrated as well as arthritic changes. The hieroglyphic sign of 'old man' shows a hunched man leaning on a staff - an accurate reflection of osteoporosis as confirmed by mummy bones. The most striking problem, however, is dental attrition. The ancient bread was stone ground, and while nourishing, it was also abrasive. This produced steady erosion of the enamel and underlying dentine ultimately leading to infection.[30]

The manifestations of disease, disability and death were apparent to the ancient Egyptians. They coped with them by resourceful mechanisms still current today - medicine, magic and religion. Egyptian physicians were the most renowned of the ancient world. In a population of believers, magic is a potent force. The religious establishments and trappings of the pharaohs remain awesome to this day. Modern theologists may regard ancient pantheism as unsophisticated, but it was undeniably utilitarian. When all failed and the inevitable death arrived, their religion provided for everlasting life in the next world. Preservation of the body through mummification and careful burial served that purpose. Those burials with their accoutrements now provide us a with window to view health in ancient Egypt.

---

1. *Steadman's Medical Dictionary*, 22nd ed., Williams and Wilkins Co., Baltimore, 1972.

2. Katz, S and Voigt, M M, 'Bread and Beer', *Expedition* 28:2 (1986), 23-4.

3. Lichtheim, M, *Ancient Egyptian Literature*, Vol III, *The Late Period*, Univ. Calif. Press, Berkeley, 1980, 199.

4. Watson, J D and Cook-Deegan, R M, 'Genome Project and International Health', *Journal of the American Medical Association* 263:24 (1990), 3322-4.

5. Bergsma, D (ed.), *Birth Defects Compendium*, The National Foundation March of Dimes, NY, 1979, 34.

6. Nyhan, W L, 'Structural Abnormalities', *Clinical Symposia, Ciba-Geigy* 42:2 (1990), 5-7.

7. Hussien, F J, Communication at 'Biological Anthropology and the Study of Ancient Egypt', Colloquium, British Museum 4-6 July, 1990.

8. Kerr, J M M, Johnstone, R W and Phillips, M H, *Historical Review of British Obstetrics and Gynaecology*, E&S Livingstone Ltd., London, 1954, 257-85.

9. Committee on Population and Demography, *Reports* 1-5, National Academy Press, Washington, DC, 1981.

10. Angel, J L, 'Ecology and Population in the Eastern Mediterranean', *World Archaeology* 4 (1972), 88-105.

11. Masali, M and Chiarelli, B, 'Demographic Data on the Remains of Ancient Egyptians', *Journal of Human Evolution* 1 (1972), 161-9.

12. *Vital Statistics of the US 1930-1983*, Vol 2. Washington DC, US Department of Health and Human Services.

13. Smith, G E and Jones F W, *Archaeological Survey of Nubia - Report 1907-1908*, Vol. II, *Human Remains*, Cairo, 1910, 267.

14. Derry, D E, 'Note on Five Pelves of Women of the Eleventh Dynasty in Egypt', *Journal of Obstetrics and Gynaecology of the British Empire* 42:3 (1935), 490-8.

15. Granville, A B, 'An Essay on Egyptian Mummies', *Phil. Trans. Royal Soc.*, London, 1825, 269-316.

16. Zimmerman, M R, 'A Possible Histiocytoma in an Egyptian Mummy', *Archives of Dermatology* 117 (1981), 364-5.

17. Strouhal, E, 'Tumours in the Remains of Ancient Egyptians', *American Journal of Physical Anthropology* 45 (1976), 613-620.

18. Retsas, S (ed.), *Paleo-Oncology: the Antiquity of Cancer*, Farrand Press, London, 1986.

19. Frazier, C A, 'Diagnosis and Treatment of Insect Bites', *Clinical Symposia, Ciba* 20 (1968), 3.

20. Nielsen O V, Grandjean P and Shapiro I M, 'Lead Retention in Ancient Nubian Bones and Teeth and Mummified Brains', in A R David (ed.), *Science in Egyptology*, Manchester University Press, 1986, 25-33.

21. BM 32138 (small figurine of a woman).

22. BM 55148, 55149.

23. Manchester, K, 'Tuberculosis and Leprosy in Antiquity: An Interpretation', *MASCA Journal* 4:1, 22-30.

24. Deelder, A M, Miller, R L, de Jonge, N, and Krijger, R W, 'Detection of schistosome antigen in mummies', *Lancet* 335 (1990), 724-5.

25. Miller, R L, '*Dqr*, Spinning and treatment of Guinea worm in P Ebers 875', *JEA* 75 (1989), 249-54.

26. Beaue, P C, Jung, R C, and Cupp, E W, *Clinical Parasitology*, Lea and Febiger, Philadelphia, 1984, 12-13.

27. David, A R (ed.), *The Manchester Museum Mummy Project*, Manchester, 1979, 95-102.

28. David, A R and Tapp, E, *Evidence Embalmed*, Manchester University Press, 1984, 70, 81-83.

29. Thompson, P, Lynch, P G and Tapp, E, 'Neuropathological Studies on the Manchester Mummies', in A R David (ed.), *Science in Egyptology*, Manchester University Press, 1979, 95-102.

30. Fleming, S, Fishman, B, O'Connor, D and Silverman, D, *The Egyptian Mummy: Secrets and Science*, University Museum, Philadelphia, 1980.

WHAT DISEASES PLAGUED ANCIENT EGYPTIANS? A CENTURY OF CONTROVERSY CONSIDERED

Jane E Buikstra, Brenda J Baker and Della C Cook

'Il ne semblera peut-être pas sans intérêt d'avoir pu retrouver, dans ces nécropoles, des lésions pathologiques permettant de reconnaître, à ces époques lointaines, des traces de deux maladies qui dominent encore aujourd'hui toute la pathologie par les ravages qu'elles exercent: la tuberculose et la syphilis. Je n'ai pu que signaler ici ces faits à l'attention du monde savant, j'espére pouvoir y revenir plus tard pour en faire une étude complète.'

Fouquet 1897:379

'Voilà, pour le moment, les résultats de nos recherches sur des squelettes dont la date remonte à la plus haute antiquité égyptienne. Nous pensons que, par le examen de ces photographies, la conviction de l'Académie sera faite sur deux points importants, qui concernent la pathologie des temps préhistoriques, savoir, l'existence de la *syphilose* et de la *tuberculose*.'

Zambaco-Pacha 1900:64

'Mais cependant, pour nous, comme pour un certain nombre de spécialistes qui ont étudié cette pièce c'est la syphilis que l'on doit incriminer.'

Lortet et Gaillard 1909:43

'I do not think there can be any doubt whatsoever that the injuries described by Professor Lortet and Dr. Fouquet as the results of syphilis were really produced long after death and burial and that the damage was done by small beetles.'

Elliot Smith 1908a:524

'I have continued the work upon Cemetery No. 5 on Biga . . . One young woman (Cemetery No. 5, Grave 91,a) presented a perfect picture of tuberculous disease of bone.'

Wood Jones 1908:38

'The discovery of a case of tuberculosis in the Biga cemetery is exceptionally interesting because there are no records whatever of the existence of this disease in Ancient Egypt. In my examination of the human remains found by Dr. Reisner in the Ancient Empire cemetery at the Giza Pyramids . . , I found the skeleton of a small child with the typical lesion of advanced hip disease which may have been tubercular.'

Elliot Smith 1908b:35

'Dr. A. R. Ferguson, Professor of Pathology in the Cairo School of Medicine, has made a careful examination of these and other pathological specimens obtained by us, and has given his deliberate opinion that none of them are the lesions of tuberculosis.'
'Under these circumstances we are bound to admit that no case of definite tubercular disease has yet been found in an ancient Egyptian of any date.'

Elliot Smith and Wood Jones 1908:41

'Es ist nichts Neues, die Entdeckung eines Falles von Pottscher Krankheit in Überresten aus Altägypten anzukündigen. Wir glauben jedoch, daß unser Fall das erste echte Beispiel jener Krankheit ist, das an ägyptischen Mumien gefunden wurde . . .'

Elliot Smith and Ruffer 1910:9

'Tuberculosis has been described as a disease of ancient Egypt, but the distinction has certainly not been made between the very common manifestations of spondylitis deformans, and true tubercular caries of the spine.'

Wood Jones 1910:264

Introduction

The infectious disease load of ancient Egyptians was a subject of intense debate during the first decade of the twentieth century. Syphilitic changes, described by Fouquet (1897), Zambaco-Pacha (1897; 1900) and Lortet (1907; Lortet and Gaillard 1909), were identified by Elliot Smith (1908a) as a much less portentous phenomenon, post-mortem alteration due to burrowing beetles, an opinion echoed by Gangolphe (1912). Fouquet's early attributions of tuberculosis (Fouquet 1897) were also questioned, with Wood Jones (1910) arguing convincingly that Fouquet's cases instead resembled spondylitis deformans, a form of degenerative arthritis.

Elliot Smith and Wood Jones in turn faced serious medical challenges to their initial identifications of tuberculosis, leading them to publish triumphant pronouncements, followed by a retraction within the same year (Elliot Smith 1908b; Elliot Smith and Wood Jones 1908; Wood Jones 1908, see above). By the close of the decade, however, dissection and histological study of Nesperehân's mummy by Elliot Smith and Ruffer (1910) established conclusively the presence of tuberculosis in ancient Egypt. This diagnosis, based upon angular curvature of the spine in association with a psoas abscess, was accepted by both medical and anthropological authorities (Klebs 1917; Ruffer 1921; Moodie 1923; Williams 1929; Pales 1930; Derry 1938; Cave 1939; Garcia Frias 1940; Rowling 1961; Morse, Brothwell and Ucko 1964; Wells 1964; Morse 1967; Sandison 1968).

Although the Nesperehân example documented the presence of tuberculosis in ancient Egypt, the debate over syphilis lingered. By 1930, however, most scholars concluded, as did Pales, 'En sorte que, pas plus l'examen du squelette que celui de la peau et des organes des anciens Egyptiens, ne permet de démontrer ou plus simplement de mettre en doute l'existence de la syphilis en Egypte' (Pales 1930: 202).

Scholarly interest in the history of syphilis and tuberculosis has continued throughout the twentieth century, with the rich archaeological record of ancient Egypt figuring significantly in ongoing discussions. Both historians and palaeopathologists have emphasised these two maladies, in part because they leave relatively clear evidence in bone. Moreover, each is one of a group of closely related infectious diseases that have long afflicted humankind. Tuberculosis presents human, bovine and avian forms and is linked taxonomically to leprosy and the 'atypical mycobacterial' infections. Similarly, venereal (and congenital) syphilis - whose New vs. Old World origin is still hotly debated - is one of four disorders collectively known as the treponematoses. The other three treponemal infections are pinta, yaws and endemic syphilis. As indicated in the following discussion, theories concerning the development and diversification of these important disease clusters require careful consideration of historical, cultural and ecological factors.

Goals for Contemporary Studies of Health and Disease in Ancient Egypt

Today, in the closing decade of the twentieth century, a number of issues faced by earlier scholars continues to affect our ability to characterise health status and disease patterning in ancient Egypt. Certainly, the presence of soft tissues and written records makes the Nile valley an exceptionally attractive context for palaeopathological investigations. Even so, current studies are constrained by several factors: 1) the way in which scholars approach the archaeological record, 2) the difficulty in distinguishing the effects of ante-mortem disease processes from post-mortem changes, and 3) a lack of diagnostic rigor.

Selective excavation, collection and curation of human remains, as well as incomplete contextual documentation, are factors that severely limit the skeletal biologist's ability to develop rigorous differential diagnoses and to generate population-based estimates of health status. In the early twentieth century, researchers expressed concern over the 'wholesale destruction of the pathological specimens of the past, which has been going on for over one thousand years' and the 'hundreds of mummies and dried human corpses . . . removed from Egyptian tombs' for which 'there exist only a few very imperfect records' (Ruffer and Rietti 1912: 439-440). In response, Ruffer and Rietti (1912: 439) called for a halt to exportation of skeletons and mummies until examined by experts. Then, in discussion of a number of remains transferred to them by several prominent archaeologists, they reported that 'the best pathological specimens will be deposited in the Museum of the Medical School at Cairo, but a large number of diseased bones will remain over, which will be sent to any recognized pathological Institute, for study and comparison with examples of the same disease occurring at the present time' (Ruffer and Rietti 1912: 440). Such selective retention and dispersal of collections were typical of the period.

Although the comparative approach continues to be important in the development of differential diagnoses, researchers today would underscore the importance of maintaining the integrity of collections: curating grossly pathological and nonpathological materials together in a single location and retaining materials thought to be unexceptional by contemporary standards. The intervening years have heightened scholarly concern for the excavation of representative samples, conservation of complete collections and documentation of contexts. However well-informed by current scientific methods, researchers simply cannot anticipate future technical advances and changing problem orientations. Even the most up-to-date scientist making selections of specimens based upon contemporary wisdom may be imposing severe restrictions upon scholars of the future. Bourke makes this position clear: 'Many specimens have been described in the past showing gross and typical pathological changes. In most the subsequent re-examination has cast doubts on, or even changed the original diagnosis.' (Bourke 1971a: 372).

As underscored by Bourke (1971a) and illustrated later in this paper, differential diagnoses in ancient materials frequently require population-based, epidemiological perspectives. Diagnostic specificity is too often compromised by missing provenience data, biased collection strategies and lost materials. The possibility of returning to old collections with new technology and observational procedures, e.g., computerised axial tomography and methods for DNA extraction, could settle many of the debates that have raged for nearly a century. The extensive Egyptian archaeological record simply cannot be considered an inexhaustible resource. Responsible excavation requires provision for the study and long-term curation of remains.

A second matter for concern is diagenetic mimicry, as in the mistaken identification of beetle burrows as syphilitic disease by Lortet (1907; Lortet and Gaillard 1909). Although distinguishing the signs of ante-mortem disease processes from post-mortem destruction due to the action of such forces as water, ground pressure and insect infestation is a matter of concern to any palaeopathologist, such determinations appear to be particularly problematic in Egyptian materials. This difficulty may result from a predilection of local fauna for human tissues, at times aggravated by a tendency for those who study health and disease to be unfamiliar with the signs of diagenesis. As discussed later in this paper, distinctions between true bony responses to infectious disease and post-mortem artefacts remain problematic in contemporary studies.

Differences between ante-mortem and post-mortem changes may also be obscured in mummified soft tissues. The recent controversy over diagnoses of alkaptonuric ochronosis, a condition characterised by calcification of intervertebral discs and articular cartilages, serves to illustrate this point. Radiographically-identified vertebral disc radiolucencies, once thought to represent ochronotic calcifications (Zorab 1961; Simon and Zorab 1961; Wells and Maxwell 1962), are now considered artefacts introduced during the embalming process (Gray 1967; Bourke 1971a; Gardner and Griffin 1971; Vyhnanek and Strouhal 1976; Wallgren, Caple and Aufderheide 1986; Braunstein et al. 1988). Using spectographic procedures, Stenn and co-workers (Stenn, Milgram, and Lee et al. 1977; Lee and Stenn 1978) have attempted to reopen discussion through the identification of ochronotic pigment in the hip joint of a 1500 BC Egyptian mummy. This diagnosis has, however, been questioned by Wallgren, Caple and Aufderheide (1986), who utilized the powerful nuclear magnetic resonance technique (NMR), a recently elaborated, sophisticated procedure for chemical analysis of organic compounds. Both Wallgren et al. (1986), based upon chemical analysis, and Gray (1967), using an epidemiologic approach, conclude that the radiopacities are most likely due to the invasion of intervertebral spaces by natron.

Other examples of pseudopathology in mummified remains include mold mimicry of erythrococytes (Sandison 1968) and post-mortem arterial rifts that resemble aneurysms (Wells 1967). Kilgore (1989) has recently suggested that May's (1897) diagnosis of rheumatoid arthritis in a Fifth Dynasty mummy may be a misinterpretation of joint distortion produced as an artefact of soft tissue dessication. Care must also be taken to differentiate micro-organisms introduced after death from those that represent ante-mortem disease processes (Zimmerman 1976). Thus, both hard and soft tissues must be carefully scrutinized for the presence of diagenetic changes.

A third issue of concern in characterising ancient health is rigour in the development of objective differential diagnoses. Comparisons across time and space, especially attempts to link ancient diseases with modern syndromes, require the development of strategies for standard description of true pathological changes, as emphasised by Ortner and Putschar (1981). A rigorous and objective comparison of observed examples with expected patterns should follow such a crucial initial step. This contemporary call for scientific rigour echoes Elliot Smith's (1908a: 522) 'surprise' in finding that, following a thorough description of post-mortem changes, Dr Fouquet then selected, without apparent justification, three specimens with 'so-called syphilitic lesions [that] seem to fall into the category of post-mortem damage . . .' Elliot Smith (1908a: 522) argues that the 'unbiased reader' will have great difficulty replicating Fouquet's distinctions. More recently, Bourke (1971a: 372) has called into question Elliot Smith and Derry's (1910) objectivity in the course of a diagnosis of a rectal tumour without consideration of possible alternatives. Objectivity and replicability remain key goals for any investigation of ancient disease.

Test Cases: Tuberculosis and Syphilis in Ancient Egypt

Today's discussions of both syphilis and tuberculosis could benefit from population-based perspectives and renewed diagnostic rigour. While the fact that tuberculosis-like pathology did exist in ancient Egypt is well established (Elliot Smith and Ruffer 1910; Morse, Brothwell and Ucko 1964; Satinoff 1972), important issues remain unresolved.

Because additional identifications of tuberculosis have been made since Morse, Brothwell and Ucko's 1964 review, a new synthesis - including critical and rigorous evaluation of both previously-reported and new cases - is desirable. Whenever possible, complete descriptions of all available hard and soft tissues should be presented, as well as appropriate demographic information.

The discussion of isolated cases, sometimes even single vertebrae, without reference to demographic, environmental or cultural contexts makes it exceedingly difficult, if not impossible, to consider fully the natural history of tuberculosis in Egypt. Although a complete synthesis is beyond the scope of this paper, in later sections we provide a critical review of recently described cases and call attention to newly discovered examples of tuberculosis-like pathology.

Following a thorough re-evaluation, there are several important questions that should be addressed, beginning with the antiquity of tuberculosis in Egypt. While Early Dynastic attributions seem relatively secure (Elliot Smith and Derry 1910; Derry 1938), the question of Predynastic cases remains unresolved. Shore (1936), for instance, reports a Predynastic case from Hierakonpolis. Morse, Brothwell and Ucko (1964) assign a Predynastic date to an ankylosed mass of thoracic vertebrae from Petrie's Nagada excavations. These rare examples in isolated vertebrae are supplemented by Predynastic artistic representations such as the wooden figure from the Musées Royaux du Cinquantenaire located in Brussels and kyphotic clay figures recovered from within Predynastic vessels (Haneveld 1980; Morse, Brothwell and Ucko 1964). Most such early finds are ambiguous and from disputed contexts, providing slender evidence indeed for Predynastic tuberculosis (Haneveld 1980; Hare 1967).

None of the Egyptian materials, however, approach in age the earliest Old World specimens purported to represent tuberculosis. The most ancient example is the widely cited case published by Bartels in 1907. This Neolithic skeleton, dating to 5000 BC, was excavated from a grave near Heidelberg and is considered tubercular by many workers (Sudhoff 1910; Garcia Frias 1940; Morse, Brothwell and Ucko 1964; Morse 1967; Steinbock 1976; Manchester 1983). The fourth and fifth thoracic vertebrae are collapsed and ankylosed with T6, creating - as Ortner and Putschar (1981) state - 'an angulation often, but not exclusively, seen in spinal tuberculosis'. Williams (1929: 870) more explicitly suggests that this case represents a compression fracture, a position echoed by Haneveld (1980). A review of detailed published accounts suggests that such scepticism is warranted. Unfortunately it appears that re-study may not be productive in this instance, since Klebs (1917: 263), who viewed the remains early in this century, reported that 'the vertebrae in the skeleton here examined were in such a state of disintegration' that he found the argument for tuberculosis not 'altogether convincing'.

More credible, however, is a Neolithic example from the Arene Candide cave near Liguria, Italy. With dates ranging between 4000 and 3500 BC, the skeleton of a male approximately fifteen years old closely resembles an expected tubercular pattern (Formicola, Milanesi and Scarsini 1987). Equally impressive are a young adult male and a child from the site of Bab edh-Dhra in Jordan (Ortner 1979). Dating to the Early Bronze Age component (3150-3000 BC) and contemporary with the first undisputed cases from Early Dynastic Egypt, the Bab edh-Dhra remains bear witness to the widespread nature of the disease by the beginning of the third millennium BC.

While Egyptian examples may not be the earliest cases of tuberculosis, the Egyptian materials *are* important as the presence of mummified soft tissue provides a unique opportunity to examine the natural history of the disease. It has been hypothesised that human tuberculosis developed from the bovine form, which in humans has a predilection for bone. This emphasis upon bony involvement may be due to the fact that the bovine form tends to enter humans via the digestive tract rather than by inhalation. Primary respiratory foci are more commonly associated with human-to-human transmission. Thus, within a series of remains, the ratio of skeletal to pulmonary lesions emerges as an important issue. Egyptian mummified materials can contribute uniquely to this avenue of inquiry if canopic jars and visceral packets are investigated along with other tissues and if frequency data on bone and pulmonary tissues are generated from the same populations. Examinations of pulmonary tissues have revealed the presence of anthracosis, pneumoconiosis and silicosis, as well as evidence of emphysema, pneumonia and tuberculosis (Ruffer 1910; Long 1931; Shaw 1938; Cockburn et al. 1975; Tapp et al. 1975; Zimmerman 1976; Cockburn and Cockburn 1980; Walker et al. 1987).

It may also be possible to investigate the natural history of tuberculosis through the isolation of ancient mycobacteria themselves and the use of techniques drawn from molecular biology (Shoemaker 1986; Shoemaker et al. 1986; Young et al. 1985a, 1985b). To date, Zimmerman (1979) provides the only example of acid-fast bacilli recovered from ancient Egyptian material that has been interpreted as firm evidence of tuberculosis. Zimmerman's example derives from the remains of a five year old child, an intrusive burial recovered from the tomb of Nebwenenef and probably dating to the early Christian era (c. 1000 BC to AD 400) (Zimmerman 1979: 604).

The course taken by tuberculosis in Egypt may therefore serve as a model for the development of this malady in the Old World. Such information is also relevant to current debates concerning the presence of a tuberculosis-like pathology in the Americas. Although the presence of pre-contact tuberculosis was disputed until Allison and co-workers (1973) identified acid-fast bacilli in an ancient Andean mummy (c. AD 700), it is now clear that Indian peoples in the Americas suffered from a tuberculosis-like pathology for over a millennium prior to the Columbian entrada. While both hard and soft tissue remains document the presence of the disease in South America, numerous examples of Pott's deformity from North America confirm the widespread nature of the condition (Buikstra 1981; Buikstra and Williams 1991). Interpretations of the origin and the natural history of American tuberculosis-like pathology can benefit from observations made upon Egyptian materials. A common protocol that includes explicit definitions of descriptive terminology is requisite to such an effort.

Although ancient Egyptian examples of syphilis remain as unconvincing as they did in Pales' (1930) day, the identification of any treponematosis in ancient Egyptian remains is of great interest in the continuing controversy about the history of this notorious disease cluster. Hudson (1964) has argued for an African origin of the treponematoses, developing a model whereby slavery and other population movements in antiquity carried yaws from Sub-Saharan Africa to the north, where it became endemic treponematosis. From northern Africa the disease was borne to urban centres, where factors of hygiene, housing, and clothing, in Hudson's unitarian view, permitted the transformation of endemic treponematosis to venereal syphilis. The absence of any uncontested evidence for endemic syphilis, yaws, or venereal/congenital syphilis in the region prior to the report of lesions in eighteenth century Bedouin remains (Goldstein, Arensburg and Nathan 1976) is an impediment to this scenario. Similarly, there is little evidence from the literature on Egyptian remains for the evolutionary history of the treponematoses proposed by Guerra (1978), who posits that all four diseases were present in the Americas, while only yaws and endemic syphilis existed in the ancient Old World. In contrast, arguments favouring New World origins of the treponematoses (for example Baker and Armelagos 1988; Crosby 1969; 1972; Grmek 1989; Suzuki 1984) find support in the lack of skeletal evidence from ancient Egypt.

Keeping key issues - curation, diagenesis, differential diagnosis and soft-tissue correlates - in mind, we would now like to illustrate two diagnostic models for investigations of infectious disease in Egyptian material. In so doing, we are developing themes that largely arise out of our prior work on the differential diagnosis of tuberculosis and similar mycotic infections (Buikstra 1976, 1981; Baker, Kealhofer and Richards 1989) and the treponemal diseases (Cook 1976). In closing, we shall re-evaluate the diagnosis of tuberculosis for several previously reported cases and also comment briefly upon some new examples from Abydos and Saqqara.

Differential Diagnosis in Skeletal Materials

An examination of the course taken by infectious disease in ancient remains must be firmly grounded in careful description of abnormal remodelling processes. Given the relative abundance of bone in comparison to soft tissue, even among Egyptian interments, the investigation of skeletal pathology is of fundamental importance.

At the bare bones level of analysis, an initial task is simply to recognise the degree to which bone has been added or subtracted. Disease may stimulate bone resorption or proliferation, or a combination of both bone loss and gain. Shape differences may also result from abnormal growth patterns when juveniles are affected (Ortner and Putschar 1981: 36-37).

Consideration should then be given to the pattern taken by abnormal processes in the individual, followed by analysis of the demographic profile of affected individuals compared to an appropriate reference sample. Ideally, this should be a representative sample of contemporaneous remains. An awareness of cultural and environmental factors that may favour one disease over another is also useful. For example, occupational information is clearly relevant when distinguishing tuberculosis from fungal infections caused by environmental pathogens, such as blastomycosis or histoplasmosis. Even though all three forms of pathology may present similar skeletal lesions, tuberculosis can be distinguished on epidemiologic grounds since patterns of risk vary by age, geographic location and lifestyle. Also important, especially in the Old World, are ancient use patterns for animal products that might have served as hosts for mycobacteria. Contaminated milk, meat and faeces are variously important as vectors in bovine and, in the latter two instances, avian tuberculosis.

A necessary first step in the discussion of any disease in skeletal remains is distinguishing ante-mortem disease processes from post-mortem artifacts. Well worth heeding is the advice offered by Elliot Smith (1908a) and Ruffer (1914): to focus upon changes that produce clear evidence of ante-mortem bony reaction. Even such simple technology as a hand lens may reveal post-mortem processes such as the 'little grooves produced by the scraping of the beetles' that Elliot Smith (1908a: 524) used to discount Fouquet and Lortet's identifications of syphilis. While it is recognized that some diseases that may affect the skeleton typically progress so rapidly that death can occur prior to clear evidence of bony response, e.g. metastatic processes, these instances are relatively rare. Thus, the argument that cranial lesions without bony reaction represent cases of rapidly progressive syphilis (Lortet and Gaillard 1909: 43) are as unconvincing now as they were in Elliot Smith's day (Elliot Smith 1908a: 522). Unequivocal evidence of remodelling should be considered a prerequisite to most identifications of ante-mortem lesions.

Following the identification of true abnormal remodelling, it is useful to keep in mind a number of distinctions that may assist in establishing differential diagnoses. Table 1 summarises several such distinctions, with emphasis upon features that are useful in discussions of tuberculosis and the treponematoses, a constellation of diseases that includes both venereal and non-venereally transmitted forms of syphilis, as well as yaws. Unless otherwise noted, the following discussion is drawn largely from previously reported work by Buikstra (1976), Cook (1976) and Ortner and Putschar (1981).

TABLE 1

FORMS OF BONY RESPONSE

Proliferative vs. Resorptive vs. Mixed
Focal vs. Diffuse
Generalized vs. Specific
Symmetrical vs. Asymmetrical
Mature vs. Juvenile
Intramembranous vs. Endochondral Formation
Secondary to Soft Tissue Influence
Articular vs. Diaphyseal
Axial vs. Peripher
Shape
Size

Of key importance in the identification of chronic diseases is the determination of primary reactions as opposed to secondary or healing responses. In tuberculosis, where the spine and peripheral joint surfaces are common sites for lesions, primary reactions are likely to be resorptive, followed by ankylosis during a secondary, healing phase in individuals who live sufficiently long for the disease to reach this stage. In studies of ancient afflictions, it is important to remember that modern infections are often halted through effective chemotherapy prior to the full bony involvement observed in earlier times. Thus, in the case of tuberculosis, more extensive involvement and higher frequencies of ankylosis are to be expected in ancient human groups. The frequently cited model by Morse (1961; 1967), which is based primarily on recent clinical experience should, therefore, be considered a minimalist perspective (Buikstra 1976; Ortner and Putschar 1981).

In contrast to the primary resorptive lesions expected in tuberculosis, the micro-organisms responsible for treponematosis more commonly stimulate proliferative responses, usually on the periosteal or non-articular surfaces of limb long bones and on the cranial vault. Characteristic granulomatous reactions of the skull lead to pathognomonic 'caries sicca' (Virchow 1858; Cook 1976; Hackett 1976; Ortner and Putschar 1981). Resorption of the facial bones also may occur in yaws and in tertiary syphilis, usually affecting the vomer, the nasal bones, the hard palate, the inferior nasal concha and the lateral walls of the nasal cavity. Leishmaniasis and leprosy may present similar patterns (Ortner and Putschar 1981).

Bony responses may be localized or generalized, depending upon the nature of the insult and the extent of effective immunological response. Tuberculosis, for example, may affect appendicular skeletal elements, but usually in asymmetrical fashion. Treponematosis tends to produce symmetrical, systemic bony responses, a pattern that differs from the localized periostitis expected in instances of local trauma.

Immature bone presents different patterns of expression for infectious disease than mature specimens. Thus, the spinal lesions of tuberculosis tend to be centrally located in juveniles, while those of adults will present either under the anterior longitudinal ligament or in association with the disk. This pattern reflects the distinctive vascular emphases in immature and mature osseous tissue. Similarly, congenital expressions of venereal syphilis differ markedly from those in adults. Endochondral growth is severely affected in infants with congenital syphilis, leading to the accumulation of calcified cartilage near the growth plate. These symmetrical long bone lesions (osteochondritis) are not pathognomonic, given that similar growth disturbances appear in a number of conditions. Given the fragile nature of metaphyseal bone at the growth plate, such changes have little chance of discovery in ancient remains. Facial features, as well as dental stigmata, result from the predilection of the organisms for the mucosa of the mouth, nose and pharynx, a condition labelled 'snuffles' or syphilitic rhinitis in infected infants. The presence or absence of such cranio-dental complexes are frequently used to bolster arguments concerning treponematosis in antiquity.

Syphilitic dactylitis and tuberculous spina ventosa are additional examples of age-related changes that can be explained by developmental processes. These conditions, characterised by thinned cortices and enlargement of tubular bones of the hands and feet, are more common in infants and children due to the presence of hemopoietic marrow throughout the shaft in juveniles (Ortner and Putschar 1981).

Bones formed from cartilage may respond differently to specific stressors from bone developed from membrane. This distinction is prominent in diagnoses of dysplasias, such as achondroplasia, and endocrine disorders, such as pituitary dwarfism.

The influence of soft tissue on the expression of skeletal pathology becomes especially important in distinguishing between tubercular and syphilitic cranial lesions. Tuberculosis, with its predilection for hematogenous tissue, tends to localize within the diploë and produces lesions which expand at the expense of the inner table of the skull. By contrast, treponemal infection commonly presents as a periosteal reaction, seldom affecting the inner table (Ortner and Putschar 1981). An affinity for cooler tissues (Knox et al. 1976) may explain the concentration of these periosteal reactions in areas such as the vault and the anterior crest of the tibia in treponematosis. Alternatively, trauma as a predisposing factor may explain the observed treponemal vault and tibia predilection (Musher and Knox 1983).

The association of lesions with articular surfaces, as opposed to long bone shafts or diaphyses, is another important diagnostic distinction. In tuberculosis, resorptive areas tend to localise within the spinal column and asymmetrically on the articular surfaces of limb long bones. Periostosis can occur, but is relatively rare. By contrast, treponematosis is typically characterised by periosteal reaction on the limb long bones. Although Charcot's joints - articular destruction secondary to neural degeneration - do occur in tertiary venereal syphilis, these lesions are relatively infrequent and largely confined to weight-bearing surfaces.

The distinctive forms of bony response illustrated in Table 1 are fundamentally important in the development of differential diagnoses, including those appropriate for the treponematoses and for tuberculosis. Two model diagnostic procedures are presented here. The first is a formal analysis, presented as a key diagram (Fig. 1) that isolates tuberculosis from other conditions that typically present spinal lesions. As illustrated in Fig. 1, the parameters useful in distinguishing osseous tuberculosis from similar disease expressions are derived from the attributes illustrated in Table 1 and discussed above. A formal diagnostic procedure such as this is most easily developed for diseases in which bone lesions are relatively rare, thus contrasting with the second example, which differentiates the treponematoses from other skeletally similar conditions. The second, more explicitly epidemiological approach is ideal in circumstances where skeletal lesions typically occur in high frequency.

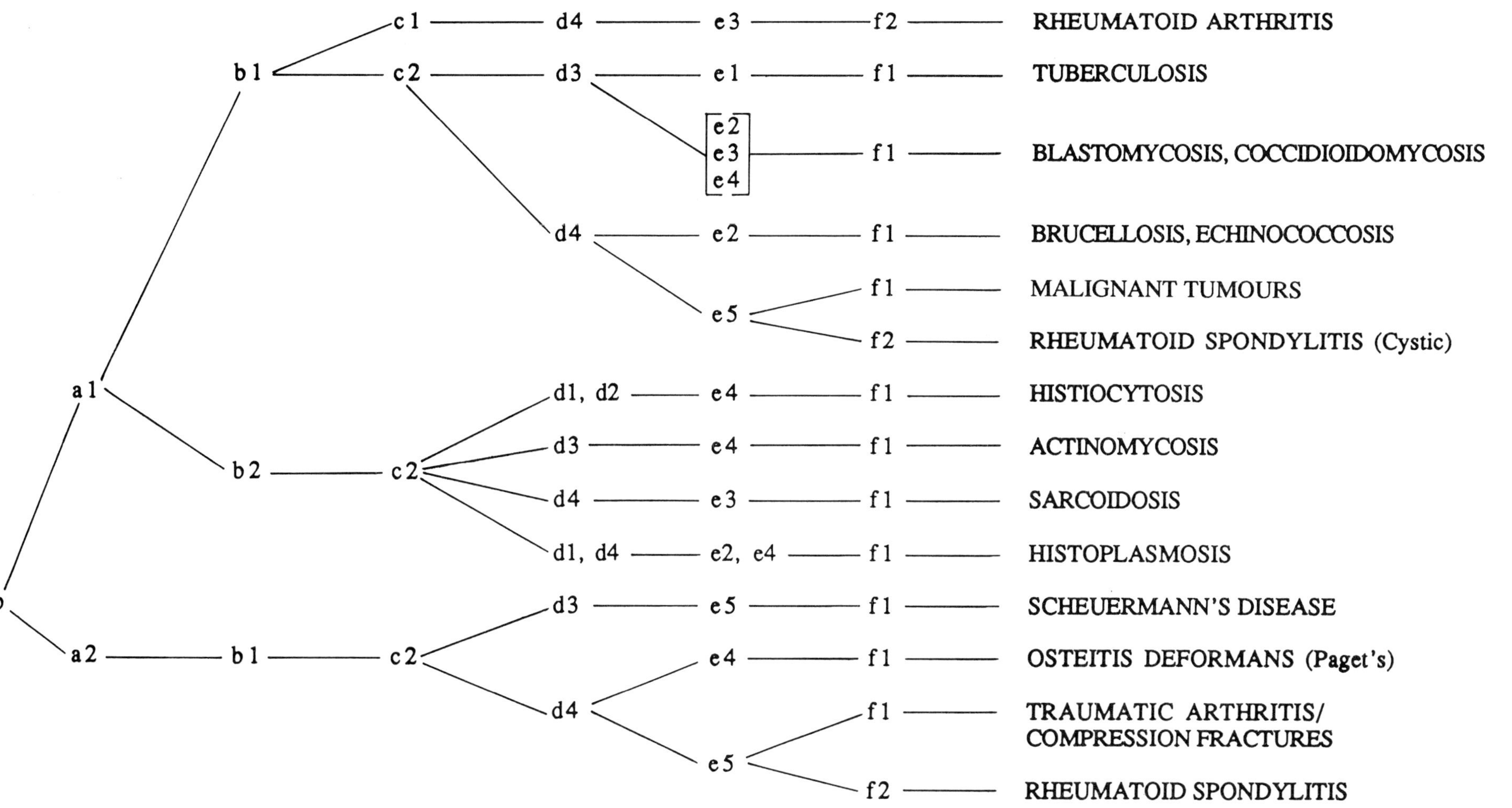

Fig.1. Key diagram useful for differential diagnosis of vertebral diseases. Variable definitions: (P) Common or 'root' feature: types of pathology presenting vertebral symptoms. (A) Specific skeletal response: a1, resorptive lesion present; a2, resorptive lesion absent. (B) Primary locus for skeletal lesions: b1, vertebral involvement characteristic or common; b2, vertebral involvement rare. (C) Primary focus within spinal column: c1, cervical; c2, thoracic/lumbar or nonspecific. (D) Age of maximum visibility in death sample: d1, birth-2.9 years; d2, 3-11.9 years; d3, 12-34.9 years; d4, 35+ years or nonspecific. (E) Site of most frequent extra-vertebral involvement: e1, articular surfaces of limb long bones; e2, diaphyses of limb long bones; e3, hands and feet; e4, skull; e5, nondiagnostic. (F) Ankylosis: f1, uncommon except as healing response; f2, common in primary phases of pathological change.

TABLE 2

DISEASES THAT MAY CAUSE LESIONS IN THE SPINAL COLUMN

Rheumatoid Arthritis	Actinomycosis
Tuberculosis	Sarcoidosis
Blastomycosis	Histoplasmosis
Coccidioidomycosis	Scheuermann's Disease
Brucellosis	Paget's Disease
Echinococcosis	Traumatic Arthritis
Malignant Tumours	Histiocytosis

Rheumatoid Spondylitis

Although the list of possible modern diseases that may produce angular changes in the spinal column is extensive (Table 2), consideration of a few basic dimensions of variability permits an effective differential diagnosis (after Buikstra 1976). These dimensions include: 1) primary skeletal response, 2) the most common locations of skeletal lesions, 3) the primary focus of lesions within the spinal column, 4) the sites of most common extra-vertebral involvement, 5) the age(s) of expected maximum mortality, and 6) ankylosis.

Fig. 1 illustrates a key diagram useful in differential diagnosis of tuberculosis-like pathology. Although this formal analysis somewhat simplifies the highly variable processes of complex diseases, it is a useful heuristic device for evaluating ancient skeletal samples. While an absence of vertebral lesions does not preclude the presence of tuberculosis in an individual, it is unlikely that tuberculosis would fail to present this pattern in adequate samples from affected populations.

Since the anticipated model for tuberculosis includes a primary resorptive response - a focal destructive lesion - with ankylosis only in a healing phase, rheumatoid spondylitis (ankylosing spondylitis), compression fractures, Paget's disease and Scheuermann's disease can be eliminated from further consideration. In fact, it was this fundamental distinction between the primary, focal destructive changes expected in tuberculosis and the 'characteristic bony outgrowths' so frequently found in ancient Egyptian and Nubian materials that led Wood Jones (1910: 277) to discount Fouquet's (1897: 371) early identifications of tuberculosis.

A high frequency of vertebral lesions distinguishes tuberculosis from histiocytosis, actinomycosis, sarcoidosis and histoplasmosis. The thoraco-lumbar emphasis expected for tuberculosis effectively distinguishes it from juvenile and adult forms of rheumatoid arthritis. Tuberculosis's typically elevated young adult mortality eliminates from further consideration such conditions as brucellosis, echinococcosis, malignant tumours, and rheumatoid spondylitis (cystic form). Since extra-vertebral involvement in tuberculosis emphasises asymmetrical expression on the joint surfaces of limb long bones, tuberculosis is further distinguished from blastomycosis and coccidioidomycosis.

This exercise illustrates a diagnostic strategy for distinguishing the bony expressions of tuberculosis from similar conditions. Such an approach is most appropriately applied in population contexts, where variable disease expressions can be assessed. As in this example, an ability to consider age-related mortality patterning (variable 'd' in Fig. 1) is crucial to the overall success of the method. Diagnoses based upon cases isolated from reference populations obviously are weakened by the inability to develop epidemiological perspectives. As forcefully stated by Bourke (1971a: 368), 'The techniques of population biology must be increasingly applied to the skeletal remains from Ancient Egypt and Nubia.'

Differential Diagnosis: Periosteal Reaction and Treponematosis
A diagnostic process appropriate for frequently occurring disease expressions is the approach developed by Cook (1976). This model was generated to identify the probable cause of periosteal reaction in large samples of archaeologically-recovered Woodland (2000-900 BP) Indian remains from west-central Illinois. Periosteal reaction has an overall prevalence of slightly more than 50% in these samples. Such high frequencies in association with pathognomonic changes in the cranio-facial skeleton have led many researchers, including Cook, to posit the presence of treponematosis in North America prior to 1492 (see Baker and Armelagos 1988 for a review of this evidence).

A consideration of clinically documented, modern conditions that may produce periosteal reaction (or periostosis, *sensu* Ortner and Putschar 1981: 39) identifies four broad classes of pathology: 1) developmental disturbances, including dysplasias and nutritional disorders, 2) inflammatory diseases, 3) trauma, and/or 4) tumours. Given the high prevalence of periostosis in Woodland samples, a further selection is made of those diseases that are expected to present periosteal reactions in more than 1% of a death sample. This procedure narrows the list of expected forms of pathology to osteomyelitis, treponemal infections, tuberculosis and other mycobacterial infections, fungal infections, scurvy, rickets (in the healing stage) and trauma. To further refine the pattern fit, two epidemiological features are considered: 1) expected prevalence within a death sample, and 2) anticipated age-associated patterning. As indicated in Table 3, venereal syphilis is uniquely defined by an overall low prevalence and bimodal age-association. Based upon an overall high prevalence, the age-regressive nature of active bony response and the age-accumulative nature of healed lesions, endemic syphilis and yaws are identified as the most likely models for the west-central Illinois example (Cook 1976, 1984).

TABLE 3

EXPECTATIONS FOR PERIOSTEAL REACTIONS OF COMMON PREVALENCE IN AN ARCHAEOLOGICALLY RECOVERED SKELETAL SAMPLE

Disease	*Prevalence*	*Age Patterning in Mortality Sample*
Scurvy	Common	Juvenile Emphasis
Rickets	Common	Juvenile "
Osteomyelitis	Relatively Low	Juvenile "
Tuberculosis	Relatively Low	Young Adult "
Fungal Infections	Variable	Age Accumulative
Trauma	Common	Age Accumulative
Treponemal Infections		
Yaws	Relatively High	Age Accumulative
Endemic Syphilis	Relatively High	Age Accumulative
Venereal Syphilis	Relatively Low	Bimodal

Cook's more recent study of circular caries in North American prehistoric remains raises the intriguing possibility that this dental condition may have resulted from congenital treponematosis (Cook 1990). A few isolated examples of pre-Columbian New World juveniles presenting skeletal evidence for congenital treponematosis have also been reported, e.g. Ortner and Putschar (1981: 207-209), Frankenberg and Albertson (1988: 286-287). Powell (1988: 162), following Grin (1953, 1956), emphasises the fact that probability of congenital transmission is closely linked to the level of treponemal pathogens in maternal blood during pregnancy. Grin, basing his conclusions on clinical observations, argues that while risk of transplacental transmission is significantly lower under conditions of endemic syphilis, congenital transmission is still theoretically possible, especially for females in primary or secondary stages of the disease. It follows that both venereal and congenital expressions could occur, albeit in low frequency, in situations where treponematosis is endemic. Individuals who had acquired the disease relatively recently would be the most likely sources of infection. Further, Grin links relatively high rates of neonatal mortality with 'diaplacental infection' in mothers afflicted with endemic syphilis (Grin 1953: 41-42). This model for treponemal transmission could explain both the pattern of circular caries reported by Cook and the rare cases of congenital syphilis observed in the archaeological record. Further, it requires little imagination to see such a flexible pathogen producing severe epidemics, when transmitted to 'virgin soil' populations of (consenting?) adults under the conditions prevalent in Europe during the late fifteenth century.

Enthusiasm for theories based on the few possible examples of congenital syphilis in New World materials must, however, be tempered by the absence of Hutchinson's incisors and Moon molars in prehistoric American remains. These distinctive lesions of the permanent dentition are expected in populations experiencing the venereal and congenital forms of treponematosis. While ample evidence for skeletal treponematosis exists in prehistoric American samples, Hutchinson's incisors and Moon molars only appear in post-contact contexts, e.g. Jacobi et al. (nd). Systematic searches of prehistoric skeletal series fail to identify any examples of these pathognomonic dental features (Cook 1990). Similarly, it must be remembered that Grin did not conclusively demonstrate congenital transmission in endemic treponematosis, thus rendering his model both provocative and problematic.

Although both the tuberculosis and the treponematosis diagnostic models illustrated here were developed during studies of ancient New World populations, they are also appropriate for Old World contexts where representative samples of remains are available. Since both strategies are objective and based upon standardised observational procedures, they fully satisfy Elliot Smith's (1908a) call for replicable approaches that are not prone to subjective interpretation.

Syphilis: Review and Re-evaluation

In marked contrast to the New World, evidence for treponematosis in ancient Egyptian remains is exceedingly sparse. While negative evidence is by definition difficult to evaluate (Ortner and Putschar 1981: 206), a scarcity of convincing examples bolsters the argument that treponematosis was absent in ancient Egypt. A careful review of all putative syphilis cases indicates that each can readily be assigned to either diagenetic change or to one of the multitude of diseases that could mimic treponematosis in isolated examples. This statement echoes the comments of Charles Fournier, the pre-eminent syphilis specialist of his day, on Zambaco-Pacha's claims for congenital syphilis in a skeleton that presented generalised diaphyseal swelling of the long bones: 'que nombre de maladies peuvent déterminer sur les os des lésions très voisines comme aspect macroscopique de celles qui dérivent de la syphilis' (Fournier, cited by Zambaco-Pacha 1900: 65).

The degree of confidence expressed for diagnoses of syphilis at the turn of the century varied from Fouquet's (1897) detailed descriptions and circumspect statements to the enthusiastic endorsements of Zambaco-Pacha (1897; 1900) and Lortet (1907; Lortet and Gaillard 1909). Retrospective evaluations are rendered difficult by a number of factors. While it is clear that Fouquet and Zambaco-Pacha considered some of the same materials, including both Négadah sud #10 and El-Amra #32, as reported by Fouquet (1897: 359-363), the failure of Zambaco-Pacha to provide site names, specimen numbers and illustrations makes it exceedingly difficult to reconcile the work of these two observers. One nearly complete set of remains that appears to have been evaluated by both Fouquet and Zambaco-Pacha is reported as Négadah sud #10 by Fouquet (1897: 352-359) and simply as being from Abydos by Zambaco-Pacha (1897: 379-380). No provenance is specified in Zambaco-Pacha's 1900 article, although it appears that the same specimen is also described there (Zambaco-Pacha 1900: 60-61). Based on hypertrophic metaphyseal lesions and united right tibia and fibula, Zambaco-Pacha (1897: 379) first identified the condition as tertiary syphilis, revising his opinion to emphasise congenital syphilis in 1900 (p. 62). Fouquet (1897: 359), with studied reticence, indicated that he could not provide evidence in conflict with Zambaco-Pacha's diagnosis of syphilis.

Fouquet's (1897) discussion explicitly attempts to distinguish between ante-mortem and post-mortem change. Several cases from Kawamil (26F, 31F, 5) are said to reflect insect activity, while syphilis is identified with four specimens: Négadah sud #10 and #32; El-Amra #4 and Kawamil #40. Fouquet (1897: 379-380) reports that Gangolphe concurs with a diagnosis of syphilis for El-Amra #4, a twenty-year-old male with destructive lesions localised within the external table of the frontal bone. No such concurrence is evident, however, in Gangolphe's (1912) later publication.

Of the specimens reported by Fouquet (1897) and Zambaco-Pacha (1897; 1900), few present clear evidence of ante-mortem processes. One of the ancient Egyptian skulls viewed in the Musée de Boulaq by Zambaco-Pacha (1897: 379) is said to have a round, destructive lesion on the left parietal that penetrated to the internal table and is surrounded by repair tissue. The second suggestive example, Négadah sud #10 (Fouquet 1897, as described above and apparently also discussed in both Zambaco-Pacha publications), presents nodular, dense, proliferative lesions localised in the metaphyseal regions of the limb long bones. The absence of diaphyseal periosteal reaction and the osteophytic fusion of the right tibia to the fibula appears more consistent with hereditary multiple cartilaginous exostosis (diaphyseal aclasia, Ortner and Putschar 1981: 373) than with syphilis. Unconvincing are the two examples (#1, #4) from Rôda cited by Lortet (Lortet and Gaillard 1909), given the explicit absence of repair tissue.

While it appears that scepticism over identifications of syphilis in ancient Egyptian remains, as expressed by Elliot Smith (1908a) and Gangolphe (1912), was justified, it would be extremely interesting to re-examine those few cases reported as showing ante-mortem changes, if they can be located. Unfortunately, the ambiguity surrounding provenance and place of curation renders this desirable goal improbable.

The limitations imposed by incomplete descriptions are well-illustrated by Hussein's (1949: 12) report of a single tibia from the Early Dynastic site of Helwan that shows lamellar periostitis. This case, presented without reference to the remainder of this skeleton or to other remains recovered from the same cemetery, is unconvincing given that many other conditions may, in such isolated examples, resemble the bony expressions of treponematosis.

Other tantalizing examples of possible treponematosis derive from Rösing's (1990) report of remains from Old Kingdom contexts at Qubbet el Hawa. All six long bones from the lower limbs of an elderly male are said to present periosteal reaction consistent with a diagnosis of yaws. Rösing notes that this Nubian site is outside the normal range for yaws and does not consider other possible treponemal diseases such as endemic or venereal syphilis. A second isolated tibia from a young adult female with similar symptoms is considered to be an advanced case of osteitis, due to the localised nature of the response (Rösing 1990: 88-89).

The negative evidence for treponematosis in Egypt does speak rather loudly. For example, in his careful study of 109 primarily Predynastic crania from the University of California collections, Leigh (1934: 33) comments that in these remains there were 'no dental stigmata of congenital syphilis . . . nor any lesion of the calvarium or facial bones'. Although he fails to provide any details, Møller-Christensen (1969) reports that his examination of more than 4000 Egyptian mummies revealed no evidence for syphilis. Similarly, Derry (1907) investigated platycnemia in a series of more than 400 individuals from Predynastic Egypt. To those familiar with New World collections, it is quite surprising to find no discussion of anterior bowing, swelling, or pronounced periostosis presenting difficulties in observing platycnemia, and more remarkable to encounter Derry's observation that:

> 'In a normal European bone as much as 60 per cent. of the whole outer surface of the bone, measured between the anterior and posterior borders, may be in front of the interosseous line; while in many cases amongst the Egyptian bones fully 60 per cent., and in some cases as much as 63 per cent., is behind the ridge.'
>
> Derry (1907:125)

Clearly, the rare lesions that show similarities to treponematosis in ancient Egyptian remains are not sufficient to produce the pattern of highly prevalent, chronic symmetrical periostosis seen in New World samples or even in the Bedouin materials reported by Goldstein et al. (1976). A convincing case for the presence of treponematosis in either ancient Egypt or Nubia has yet to be made.

Tuberculosis: Review, Re-evaluation and New Cases

Review and Re-evaluation

The following discussion, which concentrates upon the question of tuberculosis in ancient Egypt, amply illustrates the importance of re-study and re-evaluation of earlier finds. The cases initially presented here are Nubian examples that are described by Morse, Brothwell and Ucko (1964: 531-534), and are also reviewed in Morse's later synthesis (Morse 1967). Emphasis is placed upon evaluating earlier diagnoses, correcting misidentified vertebrae and questioning certain archaeological associations. Diagnostic observations by Dr John Price, a radiologist who serves as a consultant to the Department of Palaeontology at the British Museum of Natural Hist ry, offer important insights.

These materials were originally part of the holdings of the Museum of the Royal College of Surgeons. Following serious damage to the collections during World War II, the surviving specimens were removed to the British Museum of Natural History (Morse, Brothwell and Ucko 1964; Brothwell 1969, and see now also Molleson in this volume). The remains reviewed here include the cases identified as 20A, 182B and 182C, as well as three specimens labelled 182E. Only two of the three 182E examples are available for study. The following summarises observations by the senior author (JEB) and, when so indicated, those of Dr John Price (JP), whose notes are on file at the British Museum of Natural History. For reference, Figures 8 and 9 from Morse, Brothwell and Ucko (1964) are reproduced here as Plates 1,1 and 1,2. For clarity, the original figure designations only will be referred to in the following discussion.

20A: This case is reported to include the first six thoracic vertebrae, with 'destruction of much of the bodies of thoracic vertebrae 1, 2, and 3 . . .' (Morse, Brothwell and Ucko 1964: 533). The full specimen is available for study and appears to include only five vertebrae, C7-T4, rather than T1-T6. The final thoracic unit presents a divided spinous process, which may have led Morse, Brothwell and Ucko to infer the presence of an additional complete unit. As emphasised by Morse, Brothwell and Ucko, the first three vertebral bodies (C7-T2 JEB) were heavily eroded by the disease process, which was followed by kyphosis and ankylosis. Although tuberculosis is a possible diagnosis, Price argues that congenital fusion must be considered:

> 'There is wedging and fusion of two vertebral bodies (T1-T2-JEB) and a partial fusion with that above it (C7-JEB). The bony fusion appears to extend to involve the pedicles. The posterior margins of all the vertebral bodies are sharply delineated and have not been disrupted. I suspect that the appearances are due to *congenital fusion* and maldevelopment of the vertebral bodies. Tuberculosis is an alternative diagnosis but the internal anterior wall of the spinal canal and the partial fusion of the adjacent vertebrae with a fairly visible vertebral plate tend to favour the former diagnosis.'
>
> (John Price, notes on file at the British Museum of Natural History)

As noted by Morse, Brothwell and Ucko (1964), the degree of fusion for the ring epiphyses suggests that this individual died in late adolescence. The bony mass appears essentially as figured in Morse, Brothwell and Ucko (1964: Figure 8, upper right).

182B and 182C: These specimens are correctly identified and described by Morse, Brothwell and Ucko (1964: 532, Figure 8). Specimen 182B, shown on the upper left, includes the last thoracic through the fourth lumbar vertebrae. Extensive focal destructive change followed by bony regeneration is apparent. Fused ring epiphyses indicate that this individual died as an adult, as did the person represented by Specimen 182C, illustrated in the lower portion of Figure 8 (Morse, Brothwell and Ucko 1964: 532). Perhaps younger than 182B, based upon clear delineation of ring epiphyses in nonpathological vertebrae, Specimen 182C includes the last thoracic vertebra, all lumbar vertebrae, and the sacrum. Focal resorption, kyphosis and fusion centring in the lumbar region are readily observable. Although diagnostic rigor is somewhat limited by the incomplete nature of the remains, evidence for tuberculosis-like pathology is clearly present.

182E: Morse, Brothwell and Ucko (1964: 533, Figure 9) illustrate three different vertebral specimens representing at least two and more likely three individuals, all listed under the same catalogue number. Only two specimens are available for study: the bony mass illustrated on the upper left (UL) and the two thoracic units shown on the lower right (LR). The remaining specimen, an isolated lumbar vertebra illustrated on the upper right (UR), would appear to be the least convincing prospective case of tuberculosis. As stated by Morse, Brothwell and Ucko (1964: 533), '. . . this cavity is more suggestive of a bone cyst or destructive neoplasm than of tuberculosis . . .' Bourke (1971a), Rowling (1960, as cited in Bourke 1971a) and Strouhal (1991) also concur with this diagnosis.

Morse, Brothwell and Ucko (1964: 533) conclude that the missing specimen (UR) is the one described by Derry in 1909. Excavated from the Middle Nubian component of the Gennari cemetery south of Aswan, these remains are attributed by Derry to a twenty-one-year-old woman who showed degeneration of the lumbar vertebrae. There was no evidence of tuberculosis-like pathology elsewhere in her body.

'The first three vertebrae of the series were involved in the disease, which consisted in an ulceration of the centra of the vertebrae. The inflammatory process had completely destroyed the body of the second lumbar vertebra, which was firmly ankylosed to the first vertebra. The latter, owing to the loss of support from below, had fallen forward, so that its upper surface faced anteriorly, and it rested upon the surface of the third. The centrum of the latter was much eaten away by the disease, and formed with the superimposed centrum of the first lumbar, and the pedicles and laminae of the second, a large abscess cavity. When these three diseased vertebrae were placed in position with the other lumbar and lower dorsal vertebrae, a most striking picture of the acute curvature incident to Pott's disease was revealed.'

(Derry 1909: 31-32)

Derry also reports this case, using identical language, in his later overview, summarising it as a twenty-one-year-old female, lumbar vertebrae 1-3 affected, '2nd entirely destroyed, 3rd centrum much eaten away' (Derry 1938: 198). Cave (1939: 150, Figure 6) and Elliot Smith and Dawson (1924: 157, Figure 63) cite this example as evidence for tuberculosis in ancient Egypt, illustrating the same isolated lumbar vertebra figured by Morse, Brothwell and Ucko (1964: 533, Figure 9, UR). The identification of the Gennari example described by Derry with the isolated lumbar vertebra illustrated by Cave (1939), Elliot Smith and Dawson (1924) and Morse, Brothwell and Ucko (1964), is problematic and requires critical review. First, the isolated lumbar vertebra attributed to the Derry report by Cave (1939), Elliot Smith and Dawson (1924), and Morse, Brothwell and Ucko (1964) appears somewhat more heavily remodelled than one might expect in a very young adult. Neither does Derry's description of morphological changes fit the Morse, Brothwell and Ucko (1964) case.

In fact, the ankylosed mass figured by Morse, Brothwell and Ucko (1964, Figure 9) on the upper left is a more likely candidate for Derry's tuberculosis case. Comprising portions of the eighth thoracic through first lumbar units, incorrectly identified by Morse and co-workers (1964: 534) as 'upper thoracic,' this specimen illustrates the presence of degenerative changes that long predated death. The marked kyphosis occurred prior to the fusion of vertebral centra to the neural arches, as demonstrated by the fact that the unfused centrum of T8 has rotated anteriorly and fused to the bodies of T12 and L1 (see Plate 2, 1-3). The marked kyphosis generally conforms to Derry's description and reflects a disease process that began in childhood, prior to approximately seven years of age when neural arches typically fuse to centra (Bass 1987). The fact that the remaining neural arches have fused developmentally to appropriate centra affirms the chronic nature of the malady. While the disease process may have affected chronological expectations for the appearance of ossification centres, the absence of ring epiphyses or other evidence of secondary centres - as well as overall size - suggests that death occurred in late childhood or early adolescence. This may be an underestimate, however, given the somewhat eroded nature of the remains and the likelihood that the disease may have slowed skeletal maturation.

As indicated in Plate 2, the centra of T9, T10 and T11 are heavily remodelled and eroded by an infectious process. Focal lesions and sclerotic emargination are present on the centra of T12 and L1. Even though ankylosis has occurred, bone proliferation is relatively minor. Slight scoliosis to the right is illustrated in Plate 3, 1, while a disproportionately expanded neural canal for L1 is shown in Plate 3, 2. The age at death for the specimen reported by Derry (1909) is twenty-one years, somewhat older than the estimate reported here for 182E (UR). Similarly, Derry (1909) reports that the major resorptive focus appears at the level of L1-L3, rather than T9-T11. Even so, the pathological changes in these remains conform more closely to Derry's original description than does that for the isolated lumbar vertebra identified by Cave (1939), Elliot Smith and Dawson (1924), and Morse, Brothwell and Ucko (1964).

While this case (Morse, Brothwell and Ucko 1964, Figure 9) clearly represents a pattern common to Pott's disease, Price cautions that staphylococcal infections or brucellosis are possible alternatives. Price is concerned that the new bone formation evident anteriorly and laterally is not characteristic of tuberculosis (John Price, notes on file at the British Museum of Natural History). On the other hand, proliferation in healing phases is not unexpected in ancient examples.

The remaining 182E case, correctly identified by Morse, Brothwell and Ucko (1964: 534, Figure 9, LR) as 'lower thoracic,' includes the eleventh and twelfth thoracic units. These vertebrae present primary destruction with little proliferative change, the details of the resorptive process being essentially as previously reported.

Although a link between one of the 182E specimens and Derry's Gennari context may never be unequivocally established, this exercise amply illustrates the need to maintain collections for re-study. Further, the problems raised by the presence of isolated specimens, incomplete and often without appropriate documentation, serve well to illustrate the limitations that selective collection strategies impose upon the diagnosis of ancient health patterns.

Recent Studies

Subsequent to Morse's synthesis, Strouhal (1976-77) reported two possible cases of tuberculosis from the Old Kingdom component of Naga-ed-Dêr. The first instance involves 'osteolytic lesions' of the cranium in a twelve to thirteen-year-old adolescent; the second, cranial and vertebral lesions (T4-T5) identified in a twenty-five-year-old male. Strouhal (1976-77: 46-48) suggests that the two most probable diagnoses are tuberculosis and Hand-Schüller-Christian disease (histiocytosis). He also notes that these cranial changes present a patterning similar to those reported as syphilis by Lortet (1907) for the Rôda skull.

Strouhal's comparison with the Rôda skull is apt, since the problem of diagenetic mimicry is apparent in both Naga-ed-Dêr examples. Few, if any, of the cranial lesions illustrated in Strouhal's numerous photographs (Figs. 1-9; 14-29) present convincing evidence of bony reaction. The vertebral changes from the young adult male *are* clearly ante-mortem, but are most likely the result of a compression fracture. Neither the photographs (Figs. 30 and 31) nor the radiograph (Fig. 32) illustrate the 'cavity with pathological contents' that Strouhal believes initiated the sequence of events leading to angular kyphosis (Strouhal 1976-77: 41-44). A traumatic origin appears most probable.

Diagenetic mimicry also affects Strouhal's more recent study of tumors in four Late Period crania from the sites of Saqqara and Abusir (Strouhal and Vyhnánek 1981, 1987). While the first two cases, adult females from Saqqara - C32 and C36 - are possible examples of ante-mortem disease processes, the third and fourth, adult males from Abusir - 204 a/A/78 and 204 h/A/78 - should be reconsidered. The absence of bony reaction in the Abusir examples suggests that at least some of the lesions are post-mortem artefacts rather than the result of ante-mortem disease processes. Indeed, Strouhal and Vyhnánek's evidence for tumors in these two cases strongly resembles that put forward for syphilis by Lortet (1907; Lortet and Gaillard 1909) and contemporaries at the turn of the century.

Two examples of tuberculosis-like pathology recently described by Strouhal (1987, 1989, 1991) are much less problematic. One of these derives from a Middle Kingdom tomb at Abusir (900/1/84) and the other is from an early Christian burial at the Nubian site of Sayala (K 87). The twenty-two to twenty-four-year-old male from Sayala presents extensive focal resorption, kyphosis and fusion, centring in lower thoracic vertebrae, with evidence for the pathology extending throughout the thoracic and lumbar regions (Strouhal 1987, 1989, 1991). The Abusir remains are those of a middle-aged male whose lower thoracic vertebrae present a kyphotic angulation of approximately 130 degrees (Strouhal 1987, 1991). In these studies, Strouhal's rich descriptions, including information about archaeological contexts and the number of contemporaneous remains that lack focal, resorptive lesions, provide convincing evidence of tuberculosis.

Strouhal (1991) also provides a useful, contextually sensitive review of Morse's earlier studies. Morse, for example, had confused the case reported by Abadir (1953) with that described by Watermann (1960: 170-171). The former originated at Giza, while the latter is said to be on display at the museum of Helwan. Strouhal reports the context for the Helwan specimen as a prehistoric grave, although Watermann's attribution is early historic ('frühgeschichtlichen Grab', Watermann 1960: 170). Both Morse (1967: 264) and Strouhal (1991: 20) emphasise atypical neural arch involvement in the Helwan example, which Watermann (1960: 170) tentatively identifies with 'an old tuberculous process' ('einen alten tuberkulösen Prozeß'). This case, not fully prepared at the time of Watermann's observations, could be usefully re-examined using a rigorous differential diagnosis.

Strouhal's enthusiasm for a diagnosis of tuberculosis in the middle-aged male excavated from a Fifth Dynasty context at Giza (Abadir 1953: 85; Hamada and Rida 1972: 256) appears, however, unwarranted. The lower thoracic kyphosis apparent in this specimen is best attributed to compression fractures of T10 and T12, as previously diagnosed by Abadir (1953) and Hamada and Rida (1972).

Other possible examples of tubercular lesions have been reported by Armelagos (1968: 189, 195-197), who limits his discussion to a general description of lesion form and location. Excavated from the Nubian X-Group cemetery 24I3, an adult male (Burial 34) presents infectious lesions at the L5-S1 joint. A fifty-year-old female (Burial 40) from the same cemetery shows extensive destructive changes followed by osteophyte development within the lumbar and thoracic regions. Re-study of these specimens would be required to establish a differential diagnosis.

New Cases

Examples of tuberculosis-like pathology have been reported for Saqqara (Walker 1991) and Abydos (Baker 1990; Baker, Kealhofer and Richards 1989). Since Abydos and Saqqara were two of the largest and most important cemeteries in ancient Egypt, the perspective developed by their study is especially important. Significant, too, is the fact that these are rare examples of projects undertaken to investigate non-elite tombs. Careful excavation and complete recovery at both sites ensure that skeletal series can be evaluated from a population-based perspective. Since extensive reports for total samples will appear elsewhere, this discussion will provide only a brief overview.

Saqqara, the Tomb of Iurudef: The tomb of Iurudef is located south of the Unas causeway in the New Kingdom section of the Saqqara necropolis. Saqqara served as the burial ground for the major city of Men-nefer (Memphis) from the Archaic Period through the early Christian (Coptic) period, a span of approximately four thousand years. This huge cemetery extends along the western desert of the Nile Valley from just south of modern Cairo to the site of Dahshur, nearly twenty miles farther south. Preservation of human soft tissue is often poor, but skeletal preservation is usually quite good, except when burial chambers are located near the water table (Raven 1991; Walker 1991).

Iurudef was a major functionary of the household of Princess Tia, sister of Pharaoh Rameses II of the Nineteenth Dynasty (New Kingdom) and her husband, also named Tia. Iurudef was granted the distinction of being permitted to construct his tomb within the precinct of the tomb of the two Tias. It appears that Iurudef and some of his family were indeed buried here, although the original burials were robbed in antiquity, possibly within a few years of the final interment. The tomb was then re-used for multiple burials of at least sixty lower-status individuals during the tenth or ninth centuries BC.

The intrusive remains constitute a discrete social stratum of individuals who performed work requiring marked physical labour but who were able to afford some burial goods, which serve as diagnostic temporal markers. Rarely have such non-noble, non-royal people been studied in detail.

As reported by Walker (1991), the remains of a nineteen to twenty-year-old female (IUR-85-51) present extensive spinal degeneration (Plate 3, 3-4). The bodies of the vertebrae between the eighth thoracic and first lumbar present focal lytic areas, kyphotic collapse and ankylosis. Resorption on the right side of thoracic vertebrae, as illustrated in Plate 3, 4 suggests the presence of a paravertebral abscess. Kyphosis is marked, having reduced significantly a stature that would have been 'strikingly tall' for the population, given the length of her long bones. In addition to the spinal lesions, Walker notes an absence of robusticity linked to limited mobility and unusually mild dental wear as an indication of an invalid's life. She had been buried without a coffin, her legs extended, with a child (IUR-85-52) on her lap.

Walker (1991) also describes a child of four to five years (IUR 85-25), who, she believes, may have suffered from hydrocephaly. Important to the present discussion are lesions of the cranial vault that appear tubercular. Although tuberculosis of the skull is relatively rare in modern clinical experience, juveniles are most commonly affected. As in this example, lesions are characteristically small, multiple and affect both tables. Plate 4, 1-2, illustrates one of the cranial lesions, located on the right parietal. There is obvious evidence of bony reaction, thus ensuring that this is the result of true ante-mortem disease processes. The small size of the lesion and its conformation clearly resemble the pattern expected for cranial tuberculosis, as described by Ortner and Putschar (1981).

Abydos: Abydos is a large, multicomponent site, 160 km (about 100 miles) north of Luxor (ancient Thebes) and 10 km (6 miles) west of the Nile River. The site covers an area of approximately 13.5 square km (5.2 square miles; O'Connor 1979: 46) and includes a complex of towns and temples on the alluvial plain and a series of cemeteries in the low desert. Use of the cemeteries at Abydos began in the Predynastic period. During the First and Second Dynasties (c. 3100-2686 BC), Abydos served as the royal necropolis. One of the royal tombs was later identified with that of Osiris, God of the Underworld, thus enhancing the ritual importance of the site. Today, Abydos is best known for the well-preserved New Kingdom temples of Seti I (c. 1318-1304 BC) and Ramses II (c. 1304-1237 BC).

The Pennsylvania-Yale Expedition to Abydos, under the co-direction of Drs David O'Connor and William Kelly Simpson, has conducted research at the site since 1967 (O'Connor 1967). Controlled burial excavation was a focus of the archaeological investigation of the Northern Cemetery, directed by Janet Richards (1988, 1989) in 1988. The sixty burials discovered in 1988 date from the Middle Kingdom to the Roman period (c. 2040 BC to AD 395). During this span of more than two thousand years unrestricted burial of non-royal individuals was permitted in the Northern Cemetery (Richards 1989). Burial contexts ranged from interments in tomb shafts and/or their chambers to pits and coffins placed in the sand.

Because a physical anthropologist (BJB) was a member of the field staff, osteological data could be recorded in situ. In addition, thirty-four relatively intact individuals (sixteen males, ten females, one adult of indeterminate sex, and seven subadults) were examined at the field house during a ten-day period of study at the end of the 1988 season (Baker, Kealhofer and Richards

1989; Baker 1990). All but one of these individuals exhibit evidence of disease, including developmental, degenerative, traumatic and infectious lesions.

The skeleton of a young adult male (E840N780 B9), twenty-two to twenty-five years old, provides possible evidence of tuberculosis. He was interred in a wooden, mummiform coffin placed in the sand near the coffins of another adult male, an adult female and an infant. These burials date to the late New Kingdom or Third Intermediate Period (c. 1186-646 BC; J. Richards, personal communication 1990). The right hip bone presents a focus of infection around the acetabulum and anterior iliac spine (Plate 5, 1). Both the lateral and medial aspects are remodelled extensively (Plate 5, 1-2). Even though the proximal right femur is not affected, the lesions on the hip bone could have resulted from a tubercular abscess of the psoas muscle. Although Schmorl's nodes, lesions of degenerative or traumatic origins, can be identified on the bodies of T10-L2, there is no evidence of tuberculosis-like infection in the vertebral column. Most of the long bone diaphyses display mild to moderate periosteal reaction; the distal half of the left fibula presents the most severe manifestation. In addition, the young man's skull shows cribra orbitalia as well as dental pathology. The buccal half of the maxillary left second molar was destroyed by dental caries and an abscess in this area caused extensive resorption and ante-mortem loss of the third molar. It is obvious that several distinct disease processes contributed to the ill health of this individual.

The seven complete subadults analysed at Abydos present a complex diagnostic puzzle (Baker, Kealhofer and Richards 1989; Baker 1990). These children, less than four years of age at death, were interred in the sand, in pits or in simple wooden coffins, with amulets and/or beads dating them to the late New Kingdom or Third Intermediate Period (c. 1186-646 BC; J. Richards, personal communication 1990). Cranial and postcranial bone destruction is extensive in these remains (Plate 5, 3-4). Because the massive areas of destruction present in the subadults are not always accompanied by bony reaction, it is difficult to distinguish post-mortem damage from actual pathology. The pattern and extent of involvement accompanying the destruction, and especially the healing observed in some individuals, suggest the presence of a disease active at the time of death (Baker 1990).

Cranial destruction occurs in conjunction with moderate to severe cribra orbitalia in all seven subadults. The best evidence for bony reaction associated with cranial destruction is found in a child three to four years of age (E840N780 B8). Extensive osteolysis on the occipital has been nearly obliterated by blastic activity (Plate 5, 3). Postcranial manifestations of pathology in this child and the other Abydos subadults include pitting of the vertebral centra and periosteal reaction on the ribs, clavicles, long bones and pelves. In Plate 5, 4, periosteal reaction is obvious adjacent to the area of destruction of the right ilium of an infant approximately eleven to fourteen months of age (E840N780 B5). Flaring of the long bone metaphyses and bowing of the weight-bearing elements occur in most of the Abydos subadults. Metacarpals and metatarsals consist only of thin cortical shells of normal diameter.

In a differential diagnosis, the possibility that many areas of destruction unaccompanied by bony reaction are post-mortem in origin must be considered, particularly in fragile subadult materials. Rapidly progressive destruction in acute diseases, such as primary tuberculosis, histoplasmosis, leukemia, infectious brucellosis or histiocytosis, however, does not always elicit a proliferative response. Access to state-of-the-art equipment and facilities in Egypt would expedite analysis of human remains and aid in clarifying issues of ante-mortem versus post-mortem changes.

Despite such ambiguity, it appears that tuberculosis can be confidently eliminated from consideration in these Abydos subadults (Baker 1990). The extent and pattern of cranial involvement are atypical, while key features anticipated in tuberculosis, such as vertebral collapse, cortical expansion of long bone metaphyses, and spina ventosa, are absent. More likely diagnoses, given this patterning, would be cryptococcosis, histoplasmosis and histiocytosis (Baker 1990). While the pattern of lesions, the age group affected and the apparent duration of the disease correspond most closely with a diagnosis of histiocytosis (Baker, Kealhofer and Richards 1989; Baker 1990), further study is planned to refine the distinction between the results of ante-mortem disease processes and post-mortem events.

The flaring metaphyses and bowing noted in most Abydos subadults suggest the possibility of rickets. Early evidence for rickets in Egypt has interesting cultural implications. Severe rickets occurs in Moslem women and in children of both sexes in modern groups for whom the custom of purdah is expressed most vigorously (Resnick and Niwayama 1988: 2101). The seclusion of women is shared among many ethnic groups in the Mediterranean and Asia Minor, and it may well predate the Moslem religion. Evidence of rickets might be expected in women and children from subgroups among ancient Egyptians who first began to adhere to customs concerning clothing, architecture, division of labour, and constriction of domestic activities that reflect the practice of purdah. Art and architectural evidence from the New Kingdom indicates that these practices cannot have been general. It is interesting, however, to speculate that the subgroup represented by these children from Abydos and perhaps the Meroitic and later Nubians, in whom profound cortical bone loss in females has been attributed to lactation (Dewey, Armelagos and Bartley 1969), may have practised customs similar to purdah.

Lesions in a Ptolemaic skull similar to those of the Abydos subadults are reported by Pahl and co-workers (1986-1987), who present the results of careful macroscopic description, radiography, histology and chemical analysis. Their broadly-based differential diagnosis identifies plasma cell tumours, osteolytic metastases and eosinophilic granuloma (or histiocytosis) as the most likely conditions (Pahl et al. 1986-1987: 160). Emphasising the difficulty of further refining their diagnosis, they tentatively suggest that osteolytic metastases are the most likely, eosinophilic granuloma, the least. Unfortunately, this intensive exercise is severely limited by the absence of post-cranial remains.

Concluding Statement

The waning years of the twentieth century bear witness to renewed scholarly interest in the health status of ancient Egyptians, thus reopening many of the debates initiated nearly one hundred years ago. This bimodal distribution of activity, doubtless influenced by the extensive rescue archaeology programmes associated with the dams at Aswan, also follows a historical pattern generally observed in palaeopathology (Jarcho 1965a; 1965b, 1966; Brothwell 1969; Brothwell and Sandison 1967; Buikstra and Cook 1980; but see Sandison 1968 for a disclaimer).

As noted by Brothwell, 'Right from the beginning one or two diseases were destined to be controversial, particularly syphilis' (Brothwell 1969: 338). Indeed, the presence of syphilis in ancient Egypt figured heavily in turn of the century debates over the Columbian theory: '. . .il est erroné d'accuser Christophe Colomb d'avoir infecté le vieux continent par le retour, en Europe, de son équipage contaminé pendant son séjour dans le nouveau monde' (Zambaco-Pacha 1900: 62). In contrast, today's perspective - that there are as yet no unequivocal examples of treponematosis in ancient Egypt - supports the Columbian hypothesis.

The earliest purported cases of tuberculosis (Fouquet 1897) were later identified as spondylitis deformans - a form of degenerative arthritis considered exceptionally frequent in Egyptian materials by Wood Jones (1910) and Ruffer (1918; Ruffer and Rietti 1912). From the beginning, the study of this condition has been plagued by a morass of terminological problems (Bourke 1967, 1971, 1973). Although it is clear that Elliot Smith, Wood Jones and Ruffer were describing osteophytosis rather than ankylosing spondylitis, the condition has suffered a rather unfortunate definitional history, which was not clarified by statements such as that by Wood Jones (1910: 270-271; see also Elliot Smith and Dawson 1924: 159), who commented that 'The various pathological changes in bones and joints which are grouped collectively under the title of "rheumatoid arthritis" are so common in bodies of all periods that it is true to say that "rheumatoid arthritis" is par excellence the bone disease of the ancient Egyptian and Nubian.' The term 'rheumatoid' is clearly not being used in the modern sense and would be labelled degenerative joint disease today.

The assertion that ankylosing spondylitis was exceptionally frequent in ancient Egypt has been re-examined (Zorab 1961, Wells 1964, Bourke 1967, 1971, 1973, Sandison 1968, Armelagos 1968). Although a few convincing cases have been described (Bourke 1967), most ancient Egyptian and Nubian examples present spinal pathology appropriately considered degenerative osteoarthritis and osteophytosis. Bourke (1967, 1971, 1973), in an extensive analysis of osteophytosis in remains excavated by Petrie in 1898 and 1899 concluded that the ancient Egyptian pattern resembled that of Iron Age British and Pueblo Indian skeletons, while contrasting with modern British, modern American whites and Eskimo remains. Age-related expression has been emphasised by Armelagos (1968). Given that these commonly observed degenerative changes are now thought to reflect activity-related stress, the reconstruction of occupational or role-related behaviors based upon the study of arthritis patterns emerges as a potentially productive option for palaeopathologists.

While the presence of tuberculosis-like pathology remains convincing, its origin and impact upon community health persist as unresolved issues. Population-based approaches and differential diagnoses refreshed by epidemiological perspectives should encourage new insights concerning this ancient malady. Emphasis upon such factors as brucellosis and the expected differences between bovine and human tuberculosis, coupled with considerations of social class and occupation, may lead to useful perspectives on the epidemiological relationships between ancient Egyptians and their herd animals. Care must be taken to examine skeletal series prior to and during all phases of cattle domestication.

This paper has examined the course taken by palaeopathology in the study of ancient Egyptians. By focusing upon selected conditions identified at the onset of the twentieth century, we have illustrated how perspectives on disease in the past have changed, as have the ways in which palaeopathologists engage in study. Pervasive, however, is the need for a rigorous scientific method, informed by medical, biological and social (including archaeological) sciences. Advances in the characterisation of community health are developed through epidemiological perspectives on disease; diagnostic models are similarly strengthened. Essential to this enterprise is the skilful sampling of the archaeological record and the long-term curation of human remains. Researchers should retain all excavated remains, the young and the old, the cranial and the post-cranial, the healthy and the ill, the male and the female. In addition, such valuable collections need to be regularly accessible to scholars and to state-of-the-art analytical equipment. Until this occurs, interpretations will continue to confuse important distinctions, including the most basic: mistaking post-mortem changes for ante-mortem lesions.

Clearly, it does not make sense for each expedition to create expensive curation and medical facilities, but if there could be a centralised repository for remains, with laboratory equipment accessible, biological studies could be both cost-effective and efficient. An alternative might be to make portable equipment regularly available for field projects on a loan or rental basis. Unless such steps are taken, scholars will continue to dispute diagnoses due to incomplete or non-standard information, just as did Elliot Smith and Lortet nearly a century ago. The Egyptian archaeological record - its splendour and unique richness - deserves better.

Our goal in this exercise is not to focus upon who has been right - or wrong - in identifying infectious diseases in ancient Egypt, but rather to illustrate the dynamic nature of our knowledge about the past. The quarreling epigrams that begin this paper are not, as they might seem, the erring missteps of an infant field. Rather, they are the vital signs of a healthy science in historical context. The experiments of palaeopathology are the specimens, complete with archaeological documentation, that the past provides. Re-study of remains in palaeopathology is the analogue of repeated, confirmatory experiments in the laboratory sciences. Modern controversies over the interpretation of lesions in ancient Egyptian tissues should be viewed in the same light. They are positive contributions to the ongoing life of our discipline, a most welcome sign of scientific vitality.

Acknowledgments

The Nubian materials viewed at the British Museum of Natural History were generously made available to the senior author by Theya Molleson, Curator of Palaeontology, who also provided photographs and access to diagnoses made by Dr John Price. Dr Price's contributions to this paper are also gratefully acknowledged.

Roxie Walker has graciously shared information concerning skeletal pathology from the Tomb of Iurudef at Saqqara, as well as slides and photographs. Access to her field and laboratory notes, as well as unpublished texts, were also important to the development of this manuscript. The skeletal sample reported here was recovered during the 1985 excavations at Saqqara, co-sponsored by the Egypt Exploration Society and the Rijksmuseum van Oudheden, Leiden.

Helpful comments on the Abydos section of this paper were provided by Dr David O'Connor and Janet Richards. Photographs of skeletal pathology from Abydos are reproduced through their courtesy and by permission of the University Museum and Pennsylvania-Yale Expedition to Abydos. Funding for the 1988 Pennsylvania-Yale Expedition to Abydos (including support for BJB) was provided by the National Science Foundation, National Geographic Society, American Research Center in Egypt/USIA, the University Museum and Department of Anthropology of the University of Pennsylvania, Yale University and private donors.

Figures 8 and 9 from Morse, Brothwell and Ucko (1964) are reprinted courtesy of the *American Review of Respiratory Diseases*. Dan F Morse graciously supplied copies of the original photographs.

The authors also wish to thank the following individuals for their constructive criticism: Shelley Burgess, W Vivian Davies, Lisa Hoshower, Don Ortner and Roxie Walker. Nicole Couture and David Link assisted with translations.

References

Abadir, F, 1953. 'Skeleton found in shaft no. 42 of the uninscribed mastaba "B"', in *Excavations at Giza 1949-1950*, A-M Abu-Bakr (ed.), Cairo, Government Press, 85.

Allison, M J, D Mendoza and A Pezzia, 1973. 'Documentation of a case of tuberculosis in pre-Columbian America', *Am. Rev. Respiratory Diseases*

107, 985-99.
Armelagos, G J, 1968, *Paleopathology of Three Archeological Populations from Sudanese Nubia*. PhD Dissertation, University of Colorado, Boulder.
Baker, B J, 1990. *Differential Diagnosis at Abydos, Egypt*. Unpublished ms on file at the University of Pennsylvania Museum, Egyptian Section, Philadelphia.
Baker, B J and G J Armelagos, 1988. 'The origin and antiquity of syphilis: paleopathological diagnosis and interpretation', *Current Anthropology* 29, 703-737.
Baker, B J, L K Kealhofer and J E Richards, 1989, 'Death and disease in ancient Egypt: The Northern Cemetery at Abydos'. Poster presented at the 58th annual meeting of the American Association of Physical Anthropologists, San Diego, April 1989. Abstract published in the *Am. J. Phys. Anthropol.* 78, 187.
Bartels, P, 1907. 'Tuberkulose in der Jüngeren Steinzeit', *Archiv für Anthropologie* 6, 243-255.
Bass, W M, 1987. *Human Osteology*, 3rd edition. Columbia, Missouri, Missouri Archaeological Society, Special Publication No. 2.
Bourke, J B, 1967. 'A review of the palaeopathology of the arthritic diseases', in *Diseases in Antiquity*, D R Brothwell and A T Sandison (eds.), Springfield, Illinois, Charles C Thomas, 352-370.
Bourke, J B, 1971. 'The palaeopathology of the vertebral column in ancient Egypt and Nubia, *Med. Hist.* 5, 363-375.
Bourke, J B, 1973. 'Trauma and degenerative diseases in ancient Egypt and Nubia', in *The Population Biology of the Ancient Egyptians*, D R Brothwell and B Chiarelli (eds.), New York, Academic Press, 225-232.
Braunstein, E M, S J White, W Russell and J E Harris, 1988. 'Paleoradiologic evaluation of the Egyptian royal mummies', *Skeletal Radiol.* 17, 348-352.
Brothwell, D R, 1969. 'Africa's contribution to paleopathology: from the past to the future', in *Man in Africa,* M Douglas and P M Kaberry (eds.), Tavistock Publications Limited, 337-348.
Brothwell, D R and A T Sandison, 1967. 'Editorial prolegomenon: the present and future', in *Diseases in Antiquity,* D R Brothwell and A T Sandison (eds.), Springfield, Illinois, Charles C Thomas, xi-xiv.
Buikstra, J E, 1976. 'The Caribou Eskimo: general and specific disease', *Am. J. Phys. Anthropol.* 45, 351-368.
Buikstra, J E (ed.), 1981. *Prehistoric Tuberculosis in the Americas*. Northwestern University Archeological Program, Scientific Paper No. 5, Evanston, Illinois.
Buikstra, Jane E and Della C Cook, 1980. 'Paleopathology: an American Account', *Ann. Rev. Anthropol.* 9, 433-470.
Buikstra, J E and S R Williams, 1991. 'Tuberculosis in the Americas: current perspectives', in *Human Paleopathology: Current Syntheses and Future Options,* D J Ortner and A C Aufderheide (eds.), Washington, DC, Smithsonian Press.
Cave, A J E, 1939. 'The evidence for the incidence of tuberculosis in ancient Egypt', *The British Journal of Tuberculosis* 33, 142-152.
Cockburn, A, R Barraco, T A Reyman and W H Peck, 1975. 'Autopsy of an Egyptian mummy', *Science* 187, 1155-1160.
Cockburn, A and E Cockburn (eds.), 1980. *Mummies, Disease and Ancient Cultures,* Cambridge, Cambridge University Press.
Cook, D C, 1976. *Pathologic States and Disease Process in Illinois Woodland Populations: an Epidemiologic Approach,* PhD Dissertation, University of Chicago, Chicago, Illinois.
Cook, D C, 1984. 'Subsistence and health in the lower Illinois valley: osteological evidence', in *Paleopathology at the Origins of Agriculture,* M N Cohen and G J Armelagos (eds.), Orlando, Florida, Academic Press, Chapter 10, 235-269.

Cook, D C, 1990. 'Epidemiology of circular caries: a perspective from prehistoric skeletons', in *A Life in Science: Papers in Honor of J. Lawrence Angel,* J E Buikstra (ed.), Center for American Archeology, Scientific Papers N 6, Kampsville, Illinois, Chapter 6, 64-86.

Crosby, A W Jr, 1969. 'The early history of syphilis: a reappraisal', *Am. Anthropol.* 71, 218-227.

Crosby, A W, 1972. *The Columbian Exchange,* Westport, Connecticut, Greenwood Press.

Derry, D E, 1907. 'Notes on predynastic Egyptian tibiae', *J. Anat. & Physiol.* 41, 123-130.

Derry, D E, 1909. 'Anatomical report', in *The Archaeological Survey of Nubia, Bulletin No. 3,* Cairo, National Printing Department, 29-52.

Derry, D E, 1938. 'Pott's disease in ancient Egypt', *The Medical Press and Circular* 197, 196-199.

Dewey, J R, G J Armelagos and M H Bartley, 1969. 'Femoral cortical involution in three Nubian archeological populations', *Hum. Biol.* 41, 13-28.

Elliot Smith, G, 1908a. 'The alleged discovery of syphilis in prehistoric Egyptians', *The Lancet* 2, 521-524.

Elliot Smith, G, 1908b. 'The Anatomical Report', in *The Archaeological Survey of Nubia, Bulletin No 1,* Cairo, National Printing Department, 25-35.

Elliot Smith, G and W R Dawson, 1924. *Egyptian Mummies,* London, Allen and Unwin.

Elliot Smith, G and D E Derry, 1910. 'Anatomical report', in *The Archaeological Survey of Nubia, Bulletin No 5,* Cairo, National Printing Department.

Elliot Smith, G and F Wood Jones, 1908. 'Anatomical report', in *The Archaeological Survey of Nubia, Bulletin No 2,* Cairo, National Printing Department.

Elliot Smith, G and M A Ruffer, 1910. 'Pott'sche Krankheit an einer ägyptischen Mumie aus der Zeit der 21. Dynastie (um 1000 v. Chr.)', in *Zur historischen Biologie der Krankheitserreger,* K Sudhoff (ed.), Leipzig, Vol 3, 9-16.

Fouquet, D, 1897. 'Observations pathologiques', in J de Morgan, *Recherches sur les Origines de l'Égypte,* Paris, 350-373.

Formicola, V, Q Milanesi and C Scarsini, 1987. 'Evidence of spinal tuberculosis at the beginning of the fourth millennium BC from Arene Candide (Liguria, Italy)', *Am. J. Phys. Anthropol.* 72, 1-6.

Frankenberg, S R and D G Albertson, 1988. 'Pathology Descriptions', in *The Archaic and Woodland Cemeteries at the Elizabeth site in the Lower Illinois Valley,* D K Charles, S R Leigh and J E Buikstra (eds.), Center for American Archeology Research Series, Volume 7, Kampsville, Illinois, Appendix 4, 282-295.

Gangolphe, M, 1912. 'Syphilis osseuse préhistorique', *Mém. de l'Acad. des Sciences, Belles-Lettres et Arts de Lyon* 13, 131-144.

García Frías, J E, 1940. 'La tuberculosis en los antiguos peruanos', *Actualid ad Medica Peruana* 5, 274-291.

Gardner, W A and C N Griffin, 1971. 'A paleopathologic exercise - the Charleston Museum mummy', *J. S. Carolina Med. Assoc.* 67, 269-270.

Goldstein, M S, B Arensburg and H Nathan, 1976. 'Pathology of Bedouin skeletal remains from two sites in Israel', *Am. J. Phys. Anthropol.* 45, 621-640.

Gray, P H K, 1967. 'Calcinosis intervertebralis, with special reference to similar changes found in mummies of ancient Egyptians', in *Diseases in Antiquity,* D R Brothwell and A T Sandison (eds.), Springfield, Illinois, Charles C Thomas, 20-30.

Grin, E I, 1953. *Epidemiology and Control of Endemic Syphilis: report on a Mass-Treatment Campaign in Bosnia,* World Health Organization, Monograph 11, Geneva, Switzerland.

Grin, E I, 1956. 'Endemic syphilis and yaws', *Bulletin of the World Health Organization* 15, 959-973.

Grmek, M D, 1989. *Diseases in the Ancient Greek World,* Johns Hopkins, Baltimore.

Guerra, F, 1978. 'The dispute over syphilis: Europe versus America', *Clio Medica* 13, 39-61.

Hackett, C J, 1976. 'Diagnostic criteria of syphilis, yaws and treponarid (Treponaematoses) and of some other diseases of dry bone', *Sitzungberichte der Heidelberger Akademie der Wissenschaften Mathematisch-naturwissenschaftliche Klasse,* Abhandlung 4, Berlin, Springer-Verlag.

Hamada, G and A Rida, 1972. 'Letter to the Editor, orthopaedics and orthopaedic diseases in ancient and modern Egypt', *Clinical Orthopaedics and Related Research* 89, 253-268.

Haneveld, G T, 1980. 'Pott's disease before Pott', *The Netherlands Journal of Surgery* 32, 2-7.

Hare, R, 1967. 'The antiquity of diseases caused by bacteria and viruses', in *Diseases in Antiquity,* D Brothwell and A T Sandison (eds.), Springfield, Illinois, Charles C Thomas, 115-131.

Hudson, E H, 1964. 'Treponematosis and African slavery', *Brit. J. Vener. Dis.* 40, 43-52.

Hussein, M K, 1949. 'Quelques spécimens de pathologie osseuse chez les anciens Égyptiens', *Bulletin de l'Institut d'Egypte* 32, 11-17.

Jacobi, K P, D C Cook, R S Corruccini and J S Handler (nd). 'Congenital syphilis in the past: slaves at Newton Plantation, Barbados, West Indies', *Am. J. Phys. Anthropol.* (submitted for publication).

Jarcho, S, 1965a. 'Paleopathology', *Science* 147, 1160-1163.

Jarcho, S, 1965b. 'Human Paleopathology', *Archives of Pathology* 79, 425-427.

Jarcho, S, 1966. 'The development and present condition of human palaeopathology in the United States', in *Human Palaeopathology,* S Jarcho (ed.), 3-30.

Kilgore, L, 1989. 'Possible case of rheumatoid arthritis from Sudanese Nubia', *Am. J. Phys. Anthropol.* 79, 177-183.

Klebs, A C, 1917. 'Palaeopathology', *Johns Hopkins Hospital Bulletin* 28, 261-266.

Knox, J M, D Musher and N D Guzick, 1976. 'The pathogenesis of syphilis and the related treponematoses', in *The Biology of Parasitic Spirochetes,* R C Johnson (ed.), Academic Press, New York, 249-253.

Lee, S L and F F Stenn, 1978. 'Characterization of mummy bone ochronotic pigment', *JAMA* 240, 136-138.

Leigh, R W, 1934. 'Notes on the stomatology and pathology of ancient Egypt', *U. Calif. Pubs. in Amer. Archaeol. and Ethnol.* 34, 1-54.

Long, A R, 1931. 'Cardiovascular renal disease', *Archives of Pathology* 12, 92-94.

Lortet, L C, 1907. 'Crâne syphilitique et nécropoles préhistoriques de la Haute Égypte', *Bulletin de la Société d'Anthropologie de Lyon* 26, 211-225.

Lortet, L C and C Gaillard, 1909. 'La faune momifiée de l'ancienne Égypte et recherches anthropologiques', *Arch. d. Mus. d'Hist. Nat. de Lyon* 10, 1-336.

Manchester, K, 1983. *The Archaeology of Disease,* Bradford, The University of Bradford Press.

May, W P, 1897. 'Rheumatoid arthritis (osteitis deformans) affecting bones 5500 years old', *Br. Med. J. (Clinical Research)* 2, 1631-1632.

Møller-Christensen, V, 1969. *The History of Syphilis and Leprosy - an Osteo-archaeological Approach,* Abbottempo, Book 1, Abbott Laboratories, Chicago, Illinois.

Moodie, R L, 1923. *Paleopathology: An Introduction to the Study of Ancient Evidences of Disease,* Urbana, University of Illinois Press.

Morse, D, 1961. 'Prehistoric tuberculosis in America', *Am. Rev. Respir. Dis.* 83, 489-504.
Morse, D, 1967. 'Tuberculosis', in *Diseases in Antiquity*, D Brothwell and A T Sandison (eds.), Springfield, Illinois, Charles C Thomas, 249-271.
Morse, D, Brothwell D R and P J Ucko, 1964. 'Tuberculosis in ancient Egypt', *Am. Rev. Respir. Dis.* 90, 524-541.
Musher, D, and J M Knox, 1983. 'Syphilis and yaws', in *Pathogenesis and Immunology of Treponemal Infection*, R F Schell and D M Musher (eds.), Dekker Press, New York, 101-121.
O'Connor, D B, 1967. 'Abydos: a preliminary report of the Pennsylvania-Yale expedition 1967', *Expedition* 10, 10-23.
O'Connor, D B, 1979. 'Abydos: the University Museum-Yale University Expedition', *Expedition* 21, 46-49.
Ortner, D J, 1979. 'Disease and mortality in the early Bronze Age people of Bab edh-Dhra, Jordan', *Am. J. Phys. Anthropol.* 51, 589-598.
Ortner, D J and W G J Putschar, 1981. *Identification of Pathological Conditions in Human Skeletal Remains*, Washington, DC, Smithsonian Institution Press.
Pahl, W M, C A Baud and R Lagier, 1986-1987. 'Untersuchungen an einem Ptolemäerzeitlichen Mumienschädel aus Ägypten: Ein Beitrag zur Differentialdiagnose Osteolytischer Schädelveränderungen', *Ossa* 13, 145-165.
Pales, L, 1930. *Paléopathologie et Pathologie Comparative*, Paris, Masson & Co.
Powell, M L, 1988. *Status and Health in Prehistory, A Case Study of the Moundville Chiefdom*. Smithsonian Institution Press, Washington, DC.
Raven, M J, D A Aston, G T Martin, J Taylor and R Walker, 1991. *The Tomb of Iurudef, a Memphite Official in the Reign of Ramesses II*, EES, Fifty-seventh Excavation Memoir, London.
Resnick, D and G Niwayama, 1988. *Diagnosis of Bone and Joint Disorders*, Second edition, Volume 4, W B Saunders Company, Philadelphia, Pennsylvania.
Richards, J E, 1988. 'Understanding mortuary remains at Abydos', *Newsletter of the American Research Center in Egypt* 142, 5-8.
Richards, J E, 1989. 'The Abydos Northern Cemetery project: Pennsylvania-Yale Expedition to Abydos, 1988'. Paper presented at the annual meeting of the American Research Center in Egypt, Philadelphia.
Rowling, J T, 1960. *Disease in Ancient Egypt: Evidence from Pathological Lesions found in Mummies*, M.D. thesis, University of Cambridge, England.
Rowling, J T, 1961. 'Pathological changes in mummies', *Proc. Roy. Soc. Med.* 54, 409-415.
Rösing, F W, 1990. *Qubbet el Hawa und Elephantine*, Gustav Fischer Verlag, Stuttgart.
Ruffer, M A, 1910. 'Remarks on the histology and pathological anatomy of Egyptian mummies', *Cairo Sci. Jour.* 4, 1-5.
Ruffer, M A, 1914. 'Studies in palaeopathology', *Mitteilungen zur Geschichte der Medizin und der Naturwissenschaften* 13, 453-460.
Ruffer, M A, 1918. 'Studies in Palaeopathology. Arthritis deformans and spondylitis in ancient Egypt', *J. of Pathology and Bacteriology* 22, 152-196.
Ruffer, M A, 1921. *Studies in the Palaeopathology of Egypt*. Chicago, University of Chicago Press.
Ruffer, M A and A Rietti, 1912. 'On osseous lesions in ancient Egyptians', *J. of Pathology and Bacteriology* 16, 439-465.
Sandison, A T, 1968. 'Pathological changes in the skeletons of earlier populations due to acquired disease, and difficulties in their interpretation', in *The Skeletal Biology of Earlier Human Populations*, D R Brothwell (ed.), Pergamon Press, Oxford.
Satinoff, M I, 1972. 'The medical biology of the early Egyptian populations from Asswan, Assyut and Gebelen', *J. Human Evol.* 1, 247-257.

Shaw, A F B, 1938. 'A histological study of the mummy of Har-Mose, the singer of the Eighteenth dynasty (*c.* 1490 BC)', *The Journal of Pathology and Bacteriology* 47, 115-123.
Shoemaker, S A, 1986. 'Mummies, mycobacteria, and molecular biology - the old and the new', *Am. Rev. Respir. Dis.* 134, 642-643.
Shoemaker, S A, J H Fisher, W D Jones Jr and C H Scoggin, 1986. 'Restriction fragment analysis of chromosomal DNA defines different strains of *Mycobacterium tuberculosis*', *Am. Rev. Respir. Dis.* 134, 210-214.
Shore, L R, 1936. 'Some examples of disease of the vertebral column found in skeletons of ancient Egypt; a contribution to palaeopathology', *The British Journal of Surgery* 24, 256-271.
Simon, G and P A Zorab, 1961. 'The radiographic changes in alkaptonuric arthritis: a report of three cases (one an Egyptian mummy)', *Br. J. Radiol.* 34, 384-386.
Steinbock, R T, 1976. *Paleopathological Diagnosis and Interpretation.* Springfield, Illinois, Charles C Thomas.
Stenn, F, J W Milgram, S L Lee, R J Weigand and A Veis, 1977. 'Biochemical identification of homogentisic acid pigment in an ochronotic Egyptian mummy', *Science* 197, 566-568.
Strouhal, E, 1976-77. 'Two cases of polytopic osteolytic lesions in the pyramid age Egyptians', *Ossa* 3/4, 11-52.
Strouhal, E 1987. 'La tuberculose vertébrale en Égyte et Nubie anciennes', *Bull. et Mém. de la Soc. d'Anthrop. de Paris,* série XIV, 4, 261-270.
Strouhal, E, 1989. 'Palaeopathology of the Christian population at Sayala (Egyptian Nubia, 5th-11th cent. A.D.)', in *Advances in Paleopathology*, L Capasso, (ed.), Chieti, Italy, 191-196.
Strouhal, E, 1991. 'Vertebral tuberculosis in ancient Egypt and Nubia', in *Human Paleopathology: Current Syntheses and Future Options,* D J Ortner and A C Aufderheide (eds.), Washington, DC, Smithsonian Press.
Strouhal, E and L Vyhnánek, 1981. 'New cases of malign tumors from Late Period cemeteries at Abusir and Saqqara (Egypt)', *Ossa* 8, 165-189.
Strouhal, E and L Vyhnánek, 1987. 'Nouveaux exemples de tumeurs osseuses malignes provenant de cimetiéres Égyptiens de la basse époque', *Bull. et Mém. de la Soc. d'Anthrop. de Paris,* série XIV, 4, 159-170.
Sudhoff, K, 1910. 'Zur Einführung und Orientierung', in *Zur historischen Biologie der Krankheitserreger,* Leipzig, Vol 3, 3-8.
Suzuki, T, 1984. *Palaeopathological and Palaeoepidemiological Study of Osseous Syphilis in Skulls of the Edo Period,* Tokyo, U. of Tokyo Press.
Tapp, E, A Curry and C Anfield, 1975. 'Sand pneumoconiosis in an Egyptian mummy', *Brit. J. Med.* 2, 276.
Virchow, R, 1858. 'Über die Natur der Constitutionell-syphilitischen Affectionen', *Virchow's Archiv für Pathologische Anatomie und Physiologie* 15, 217-236, 243-253.
Vyhnánek, L and E Strouhal, 1976. 'Radiography of Egyptian mummies', *ZÄS* 103, 118-128.
Wallgren, J E, R Caple and A C Aufderheide, 1986. 'Contributions of nuclear magnetic resonance studies to the question of alkaptonuria (ochronosis) in an Egyptian mummy', in *Science in Egyptology,* R A David (ed.), Manchester, England, University of Manchester Press, 321-327.
Walker, R, 1991. *The People Buried in Iurudef's Tomb.* Manuscript on file with the Bioanthropology Foundation, Lignières, Switzerland.
Walker, R, F Parsche, M Bierbrier and J H McKerrow, 1987. 'Tissue identification and histologic study of six lung specimens from Egyptian mummies', *Am. J. Phys. Anthropol.* 72, 43-48.
Watermann, R, 1960. 'Palaeopathologische Beobachtungen an altägyptischen Skeletten und Mumien', *Homo* 11, 167-179.
Wells, C, 1964. *Bones, Bodies, and Disease,* New York, Frederick A. Praeger.

Wells, C, 1967. 'Pseudopathology', in *Diseases in Antiquity*, D Brothwell and A T Sandison (eds.), Springfield, Illinois, Charles C Thomas, 5-19.

Wells, C and B M Maxwell, 1962. 'Alkaptonuria in an Egyptian mummy', *Brit. J. Radiol.* 35, 679-682.

Williams, H U, 1929. 'Human paleopathology, with some original observations on symmetrical osteoporosis of the skull', *Archives of Pathology* 7, 839-902.

Wood Jones, F, 1908. 'A supplementary anatomical report', in *The Archaeological Survey of Nubia, Bulletin No.1*, Cairo, National Printing Department, 37-39.

Wood Jones, F, 1910. 'Chapter VII. General Pathology (Including Diseases of Teeth)', in *The Archaeological Survey of Nubia Report for 1907-1908*, Vol. II: *Report on the Human Remains*, G Elliot Smith and F Wood Jones (eds.), Cairo, National Printing Department, 263-292.

Young, R A, B R Bloom, C M Grossinsky, J Ivany, D Thomas and R W Davis, 1985a. 'Dissection of *Mycobacterium tuberculosis* antigens using recombinant DNA', *Proc. Natl. Acad. Sci.* 82, 2583-2587.

Young, R A, V Mehra, D Sweester et al., 1985b. 'Genes for the major protein antigens of the leprosy parasite, *Mycobacterium leprae*', *Nature* 316, 450-452.

Zambaco-Pacha, D A, 1897. 'L'antiquité de la syphilis', *Paris Médic.* 52, 372-381.

Zambaco-Pacha, D A, 1900. 'De quelques lésions pathologiques datant des temps des Pharaons', *Bulletin de l'Académie de Médecine* 44, 58-68.

Zimmerman, M R, 1976. *A Paleopathologic and Archeologic Investigation of the Human Remains of the Dra Abu el-Naga site, Egypt, based on an experimental study of mummification*. PhD Dissertation, University of Pennsylvania, Philadelphia.

Zimmerman, M R, 1979. 'Pulmonary and osseous tuberculosis in an Egyptian mummy', *Bulletin of the New York Academy of Medicine* 55, 604-608.

Zorab, P A, 1961. 'The historical and prehistorical background of ankylosing spondylitis', *Proceedings of the Royal Society of Medicine* 54, 415-420.

PREDYNASTIC SCHISTOSOMIASIS

R L Miller, N De Jonge, F W Krijger and A M Deelder

Epidemiology and Parasitology

Schistosomiasis, one of the most widespread human parasitic infections, is caused by trematode worms of the genus *Schistosoma*. Infection occurs during contact with water infested with the parasite larvae. Human schistosomiasis is endemic in the tropics where an estimated 600 million people are exposed to the risk of infection. In rural areas, the socio-economic impact of the disease is great. The recent diagnosis of a Predynastic case (Deelder et al. 1990) suggests that schistosomiasis has been an important environmental and occupational hazard for at least 5000 years (Miller 1991) *Schistosoma* parasites depend on both vertebrate and molluscan hosts at different stages of their development. Adult schistosomes are equipped with suckers which enable them to attach to the blood-vessels of the host. The schistosome species infecting humans in the Nile valley are *S. haemotabium*, and *S. mansoni*. The life cycle of the parasite is shown in Fig. 1. Adult worms do not multiply in the vertebrate host. Eggs which are not caught in the body of the host, leading to bladder calcification, abdominal swelling (ascites) and, in *S. mansoni* infections, portal fibrosis of the liver, are excreted in the urine or faeces.

If eggs reach water, a larva (miracidium) hatches which can enter a *Bulinus* or *Biomphalaria* snail, depending on the schistosome species. Once within a snail, asexual multiplication occurs as several generations of sporocysts develop. These sporocysts produce large numbers of infective larvae (cercariae). Several thousand cercariae leave an infected snail each day, and some of these cercariae succeed in penetrating the skin of the final host who comes into contact with infested water during wading, irrigation, fishing, doing laundry or collecting water from an infested stream. As they penetrate the host, the cercariae shed their tail and enter the bloodstream as immature schistosomes (schistosomula) that migrate to the lungs and then to the liver where the adult worms pair about twenty-six days after infection. The paired adult worms then migrate either to the veins of the blood vessels surrounding the bladder (*S. haematobium*) or to the veins of the intestine (*S. mansoni*), where eggs are produced at the rate of 200-300 per female a day. The usual number of worms in an infected individual is less than twenty, although in extreme cases up to 2000 worms may be found. Adult worms usually live three to seven years although survival of more than thirty years has been reported.

In modern times, the schistosome parasite was first reported by Theodor Bilharz on 1 May, 1851, while performing an autopsy on a young male. In his letter describing his discovery, Bilharz wrote: 'After my attention had been drawn to the liver, I soon found a white, long helminth in the blood of the portal vein in quantity, which I assumed to be a nematode, but which I immediately recognised as something new. The microscope revealed a splendid distomum, with a flat body and a curving tail, which exceeded the body about ten times in length' (Bilharz and Siebold 1853: 59-62, quoted in Abdel-Wahab 1982: 1). The female is 2.0-2.6 cm in length and Bilharz did not need a microscope to identify the parasite, which is about the same size as a small scarab or ring bezel (Fig. 2).

Although ancient Egyptians certainly did not perform autopsies, evisceration did occur before burial, and when livers were removed from individuals who had died while immature worms were in circulation, the circumstances under which Bilharz made his first observation of the parasite

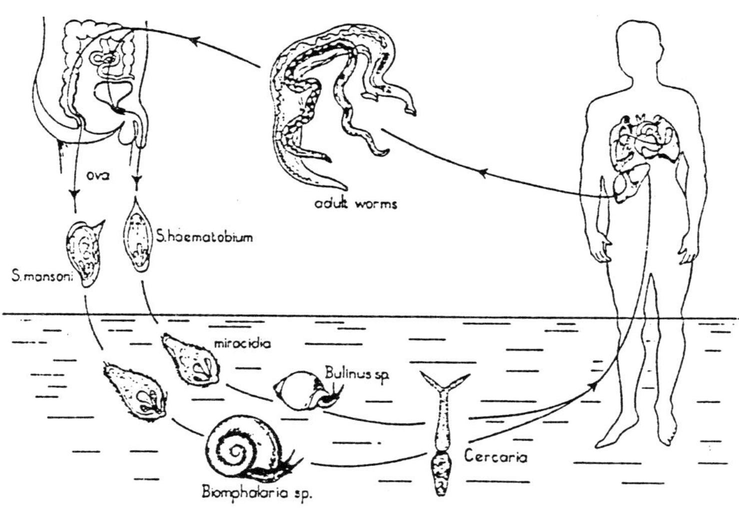

Fig. 1. Life cycle of *S. mansoni and S. haematobium.*

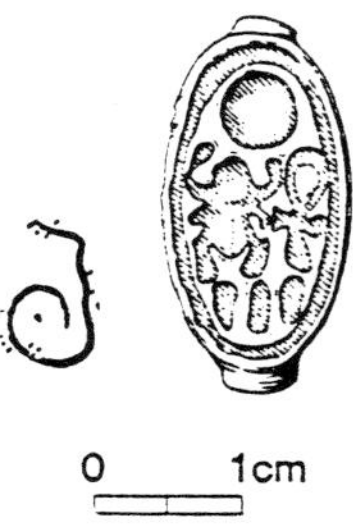

Fig. 2. Adult schistosome worm and New Kingdom ring ornament. Adapted from Manson-Bahr and Apted, 1982 (worm) and A Boyce in Kemp, 1989 (ring bezel).

would have also occurred. The presence of worms in the viscera of children and adolescents who died from severe schistosome infections associated with bloody urine is unlikely to have passed unnoticed, even if in P Ebers 62 the aetiology of the condition was reversed with the worms being attributed to the urinary bleeding (Abdel-Wahab 1982: 2), the sympton that would have been observed before the parasite became apparent.

Current methods of diagnosing cases of schistosomiasis rely on three methods: (1) parasitological diagnosis, i.e., determining the presence of eggs in the urine, faeces or visceral tissue, the technique used eighty years ago by Ruffer to diagnose the first cases of schistosomiasis in New Kingdom Egypt and also used subsequently (Ruffer 1910; Millet et al. 1980); (2) immunological diagnosis either by detection of specific antibodies or by the detection of circulating antigens, a technique recently used to demonstrate the presence of an active schistosomiasis infection in a Predynastic adolescent (Deelder et al. 1990); (3) by looking for pathological changes, e.g., ultrasound demonstration of periportal fibrosis in the liver or radiological detection of bladder wall calcification, a technique that has been successfully applied to mummy populations by the Manchester Museum team (Isherwood et al. 1979).

Diagnosing Ancient Cases of Schistosomiasis by Detection of Schistosome Adult Worm Antigen in Mummy Tissue

Pathological features such as the presence of eggs in visceral tissue of infected individuals or bladder wall calcification associated with *S. haematobium* infections do not necessarily indicate active infections. Adults who have survived childhood encounters with schistosomiasis and retain an acquired partial immunity that protects them from further massive superinfection (Butterworth et al. 1985) could still have bladder wall calcification and retain some eggs from earlier infections from which they have recovered, although these features are gradually reabsorbed over the years and disappear. In contrast, schistosome circulating anodic antigen (CAA), a proteoglycan from a portion of the worm gut is regurgitated by a healthy worm along with undigested host blood components. When CAA is regurgitated and released into the blood stream of the host it can be detected in circulation during active infections before part is degraded and cleared in the liver and part is excreted with the urine. CAA is only present during active infections, and following the destruction of the parasites, it disappears from serum rapidly, so that the level in circulation is reduced by half in about 2.5 days and is completely eliminated after a time-span of about ten days.

Enzyme-linked immunosorbent assay (ELISA) has been shown to have a useful role in determining the presence of circulating antigens during infections with schistosome worms. A recently developed sandwich ELISA technique is sensitive to 0.2 ng/ml concentrations of schistosome circulating anodic antigen with high precision, high analytical recovery and 100% specificity (Deelder et al. 1989). This ELISA was carried out using anti-CAA monoclonal antibody 120-1B10-A both as capture antibody and as alkaline phosphatase conjugated second antibody (Fig. 3). This IgG_1 monoclonal antibody was selected from a panel of cell lines because of its excellent performance in sandwich ELISA. Before the samples were assayed, all specimens were homogenised in an all-glass homogeniser in 7.5% trichloroacetic acid (TCA) solution to remove proteins and dissociate immune complexes (De Jonge et al. 1987) and then spun down (25,000 g, 15 min); the supernates were dialysed 24 h against distilled water (4° C) and lyophilised by freeze-drying.

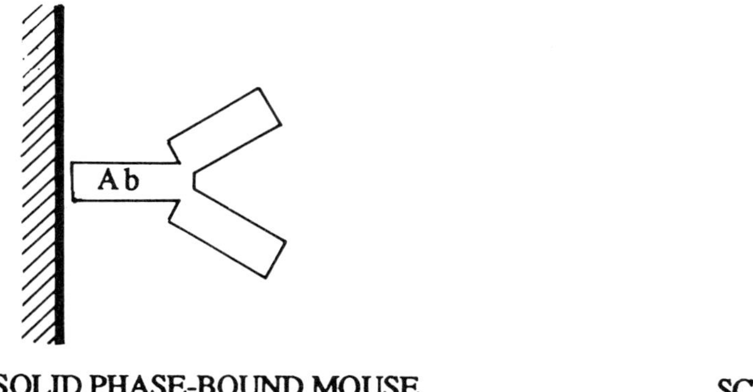

SOLID PHASE-BOUND MOUSE
MONOCLONAL ANTIBODY

SCHISTOSOME
CIRCULATING ANTIGEN

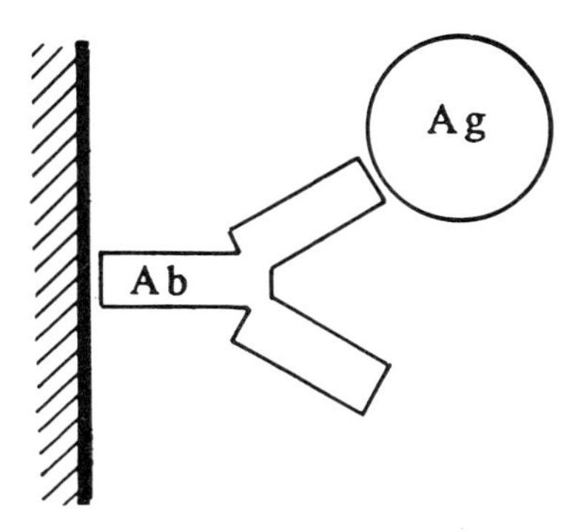

SOLID PHASE-BOUND
Ab-Ag COMPLEX

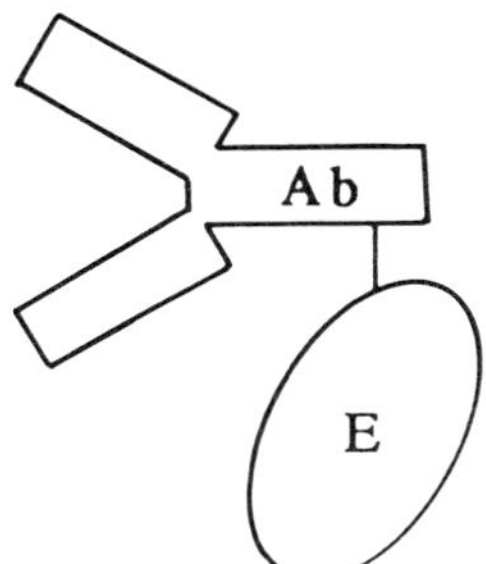

ENZYME-MOUSE MONOCLONAL
ANTIBODY CONJUGATE

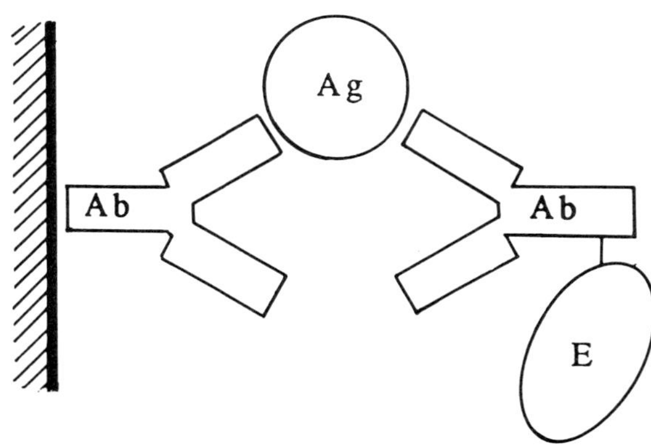

ENZYME-LABELLED Ab-Ag-Ab SANDWICH COMPLEX

Fig.3. Stages of enzyme-linked immunosorbent assay (ELISA) of schistosome circulating anodic antigen (CAA).

Three positive samples from unembalmed, desiccated Egyptian mummies were tested (Deelder et al. 1990). Concentrations of respectively 123 and 13 x 10^{-12} g CAA/mg of mummy skin were found in both cheek and colon tissue from an adolscent male Theban temple weaver, Nakht (early 12th century BC), whose schistosomiasis had been previously diagnosed from *S. haematobium* eggs in gut, kidney and liver (Millet et al. 1980). Although this adolescent also suffered from cirrhosis of the liver, it should be noted that hepatitis or alcohol abuse, rather than schistosomiasis, was probably responsible for the liver cirrhosis (Miller 1990). 8 x 10^{-12} g CAA/mg was also detected in skin tissue from a Predynastic adolescent, British Museum EA 32753 (Plate 6).

Negative controls included uninfected hamsters, blood samples from a modern Dutch population without exposure to schistosomiasis or travel to areas of endemic infection, and gut tissue from five unembalmed, desiccated mummies from the Atacama desert near Arica, northern Chile, dated c. 2000 BC or earlier-AD 600 (Arriaza et al. 1988). As schistosomiasis did not reach the New World until carried there by slaves following European contact, these Pre-Columbian individuals would not have been exposed to the disease. With the negative controls no CAA could be detected even at very high sample concentrations, i.e. at a cut-off value of 0.5 pg/mg no positive results were detected, results in line with our observation that with serum samples of uninfected individuals no false-positive results were ever found (Deelder et al. 1989).

The results of this study show the feasibility of detecting CAA after 3200-5000 years in unembalmed, desiccated tissue preserved under optimum conditions in a desert environment not subject to postmortem leaching from weathering or embalming. Allowing for a ten-fold reduction in the weight of dried tissue and assuming 10% blood content of cheek tissue, the CAA level determined in Nakht's cheek would correspond to a serum level of approximately 25 ng per ml in circulation. This is remarkably well in accordance with CAA levels in contemporary *S. haematobium*-infected individuals where levels of 10-40 ng CAA per ml serum were demonstrated in patients with heavy infection (De Jonge et al. 1989), a degree of infection already indicated by the pathological studies of Nakht (Millet et al. 1980). It also demonstrates that the complex proteoglycan produced by the adult schistosome worm is stable over a period of several thousand years, survives at levels consistent with a diagnosis of moderate to severe active infections, and can be assayed in non-visceral tissue, providing a new technique of palaeoseroepidemiology in the diagnosis of infection via detection of a genus-specific proteoglycan antigen.

Thanks to the generosity of the British Museum in allowing samples to be taken from other Predynastic individuals from Gebelein, we were able to begin to fill in the outline of the palaeoepidemiology of schistosomiasis in Predynastic times. Although tissue from the shin of the Predynastic adolescent BM EA 32753 tested positive for the presence of schistosome circulating anodic antigen (CAA), tissue from two adults, BM EA 32754 (left cheek) and BM EA 32755 (right shin), did not reveal the presence of CAA or indication of active infection. While this is only a preliminary result needing confirmation from further testing of these samples to determine the degree to which other polysaccharides such as blood group serotypes have survived, shin and cheek skin has previously been shown to be suitable for detecting ancient schistosome infections using the new antigen assay.

Conclusions and Prospects

Antigen-antibody complexes from smallpox infection have been previously reported from a sixteenth century AD Italian mummy (Fornaciari and Marchetti 1986); the identification of schistosome worm antigens from Nakht and BM EA 32753 shows that in some environments complex polysaccharides are capable of surviving for considerable periods of time at least an order of magnitudue greater. Comparably long-term survival of short sequences of DNA from desiccated Egyptian mummy tissue after 5000-2400 years has been observed in samples which also included tissue from BM EA 32753 (Pääbo 1985; 1989).

Use of ELISA to identify active infections from samples of dried, unembalmed tissue opens up the possibility of determining the prevalence of schistosomiasis within ancient populations without the need for invasive techniques. Although *Bulinus truncatus* shells which indicate the presence of potential intermediate vectors for schistosome worms have been recorded on Palaeolithic (Leakey 1971; Gautier 1976) and Neolithic sites in Africa, and their possible relation to schistosomiasis noted (Gautier 1980; Wright 1977), the earliest history of the disease is significantly extended by the discovery of an active infection in a Predynastic adolescent, BM EA 32753, the earliest case of schistosomiasis diagnosed to date. A number of questions remain to be answered. Did the prevalence of schistosomiasis change between Predynastic and later times with innovations in hydraulic technology (Butzer 1976) which might be expected to have an effect on the degree of exposure to schistosomiasis? What was the prevalence of the infection within different age groups, professions and social classes? Did the use of wells for water supply and the siting of workmen's villages at distances of over 1 km from the Nile have a positive effect on the health of New Kingdom royal tomb workers with water deliveries from wells (Kemp 1989) and laundry services provided by washermen from outside their village (Janssen 1979)?

It is now feasible to investigate the presence of other pathogen and parasite products in mummy tissues using modern immunological and molecular biological techniques to determine the diagnosis of ancient disease directly from nucleic acids and antigens rather than from the indirect effects of diseases on the skeleton and soft tissue. New, relatively non-destructive techniques are now available which could be used on fragments of mummy tissue where viscera are either not preserved or not accessible, perhaps in connection with radiocarbon tests to determine dating of tissue where only fragmentary samples were preserved.

Acknowledgments

This research was supported by the Wellcome Trust and by the Science and Technology for Development Programme of the European Community. We thank F Nowell-Smith, Academy of Medicine, Toronto; W V Davies and M L Bierbrier, British Museum, London; T Holden, Institute of Archaeology, London; and A C Aufderheide, University of Minnesota, Duluth, for providing samples.

References (medical journal titles in standard *Index Medicus* format)

Abdel-Wahab, M F, 1982. *Schistosomiasis in Egypt*, Boca Raton, Florida, CRC.

Arriaza, B, Allison, M and Gerszten, E, 1988. 'Maternal mortality in pre-Columbian Indians of Arica, Chile', *Am. J. Phys. Anthropol.* 77, 35-41.

Bilharz, T and von Siebold, C T, 1853. 'Ein Beitrag zur Helminthographia humana', *Zeitschrift für Wissenschaftliche Zoologie* 4, 53-76.

Butterworth, A E, Capron, M, Cordingley, J S, Dalton, P R, Dunne, D W, Kariuki, H C, Kimani, G, Koech, D, Mugambi, M, Ouma, J H, Prentice, M A, Richardson, B A, Arap Siongok, T K, Sturrock, R F and Taylor, D W, 1985. 'Immunity after treatment of human schistosomiasis mansoni. II. Identification of resistant individuals, and analysis of their immune responses', *Trans. R. Soc. Trop. Med. Hyg.* 79, 393-408.

Butzer, K W, 1976. *Early Hydraulic Civilization in Egypt*, Chicago, University of Chicago Press.

Deelder, A M, De Jonge, N, Boerman, O C, Fillié, Y E, Hilberath, G W, Rotmans, J P, Gerritse, M J and Schut, D W O A, 1989. 'Sensitive determination of circulating anodic antigen in *Schistosoma mansoni* infected individuals by an enzyme-linked immunosorbent assay using monoclonal antibodies', *Am. J. Trop. Med. Hyg.* 40, 268-272.

Deelder, A M, Miller, R L, De Jonge, N, Krijger, F W, 1990. 'Detection of schistosome antigen in mummies', *Lancet* Vol. 335, No. 8691, 724-725.

De Jonge, N, Fillié, Y E, Deelder, A M, 1987. 'A simple and rapid treatment (trichloroacetic acid precipitation) of serum samples to prevent non-specific reactions in the immunoassay of a preoteoglycan', *J. Immunol. Methods*. 99, 195-197.

De Jonge, N, Fillié, Y E, Hilberath, G W, Krijger, F W, Lengeler, C, De Savigny, D H, Van Vliet, N G, Deelder, A M, 1989. 'Presence of the schistosome circulating antigen (CAA) in urine of patients with *Schistosoma mansoni* or *S haematobium* infections', *Am. J. Trop. Med. Hyg.* 41, 563-569.

Fornaciari, G and Marchetti, A, 1986. 'Italian smallpox of the sixteenth century', *Lancet* Vol. II, No. 8521/22, 1469-1470.

Gautier, A, 1976. 'Assemblages of fossil fresh-water molluscs from the Omo group and related fossils in the Lake Rudolf Basin', in *Earliest Man and Environments in the Lake Rudolf Basin* (eds. Coppens, M Y, Howell, F C, Isaac, G Ll. and Leakey, R E F) Chicago, University of Chicago Press, 379-382.

Gautier, A, 1980. 'Terminal paleolithic and neolithic materials from Nabta', in *Prehistory of the Eastern Sahara* (eds. Wendorf, F and Schild, R), New York, Academic Press, 349-364.

Isherwood, I, Jarvis, H and Fawcitt, R A, 1979. 'Radiology of the Manchester Mummies', in *Manchester Museum Mummy Project* (ed. David, A R), Manchester, Manchester University Press, 25-64.

Janssen, J J, 1979. 'The water supply of a desert village', *Bulletin Medelhavsmuseet* 4, 9-15.

Kemp, B J, 1989. *Amarna Reports*, V, London, Egypt Exploration Society.

Leakey, M D, 1971. *Olduvai Gorge*, Vol. 3, Cambridge, Cambridge University Press, 253, 290.

Manson-Bahr, P E C and Apted, F I C, 1982. *Manson's Tropical Diseases,* 18th ed., London, Balliere Tindall.

Miller, R, 1990. '*Ds*-vessels, beer mugs, cirrhosis and casting slag', *Göttinger Miszellen* 115, 63-82.

Miller, R, 1991. 'Palaeoepidemiology, literacy, and medical tradition among necropolis workmen in New Kingdom Egypt', *Medical History* 35, 1-24.

Millet, N B, Hart, G D, Reyman, T A, Zimmerman, M R, and Lewin, P K, 1980. 'ROM I: mummification for the common people', in *Mummies, Disease and Ancient Cultures* (eds. Cockburn, A and Cockburn, E), Cambridge, Cambridge University Press, 71-84.

Pääbo, S, 1985. 'Molecular cloning of ancient Egyptian mummy DNA', *Nature* 314, 644-645.

Pääbo, S, 1989. 'Ancient DNA: extraction, characterisation, molecular cloning, and enzymatic amplification', *Proc. Natl. Acad. Sci. USA* 86, 1939-1943.

Ruffer, M A, 1910. 'Note on the presence of "Bilharzia haematobia" in Egyptian mummies of the Twentieth Dynasty', *Brit. Med. J.* 1, 16.

Wright, C A, 1977. 'The ecology of African schistosomiasis', in *Human Ecology in the Tropics* (eds. Garlick, J P and Keay, R W J) London, Taylor and Francis, 127-143.

DENTAL ANTHROPOLOGY OF THE NILE VALLEY

Jerome C Rose, George J Armelagos and L Stephen Perry

Introduction

In his 1972 article titled 'Dental Anthropology of Early Egypt and Nubia' David Greene states that 'the study of Egyptian and Nubian dental anthropology is at present based upon a few published reports ...' (p 315). He lists nineteen references concerned with Egyptian and Nubian teeth and skeletal remains in the bibliography. His stated goal is 'to review this meagre data and present a synthesis of present knowledge' (Greene 1972, 315). Our goals for this paper are to present a synthesis of Nile Valley dental anthropology and to assess the changes which have taken place during the intervening eighteen years.

The historical literature review is organised into three interpretative themes: 1) dental disease, disease process, and dietary reconstruction; 2) childhood stress; and 3) genetics, dental morphology, metrics, and adaptation. The literature synthesis indicates several areas for methodological improvement and these suggestions are organised into the section titled methodological critique. In the conclusion the present state of Nile Valley dental anthropology is compared to the state of affairs in 1972.

Dental Disease, Disease Process, and Dietary Reconstruction

Dental Caries

Since Mummery, in his classic work of 1870, established the relationship between dental caries and diet, numerous researchers have argued about temporal trends in Egyptian and Nubian caries frequencies. Ruffer (1920; 1921) states that the caries rates remained relatively stable from predynastic to Christian times before rising suddenly during the Ptolemaic period. Other more recent investigators provide data demonstrating an increase from predynastic to dynastic periods (Derry 1933; Hillson 1978; 1979; Leek 1986). In contrast, Grilletto (1964; 1973; 1977; Grilletto and Davide 1967), employing the Marro collection at Turin demonstrates a decline in caries from 33% to 28% affected individuals between the predynastic and dynastic Egyptian samples. Egyptian caries rates reach their highest levels from Roman times onward (David 1985; Gaballah et al. 1980; Strouhal 1984).

In Nubia caries are infrequent during the Mesolithic at 1% (Green 1972; Greene et al. 1967), increase slightly but remain low during predynastic times (Smith and Jones 1910), and reach a high of 18 to 23% during the Christian occupation (Beck 1988; Beck and Greene 1989; Martin et al. 1984). As the frequency of caries is related to the physical consistency and composition of the diet, these trends have implications for the reconstruction of Egyptian and Nubian diets.

Our review of the literature clearly demonstrates that any trend established for the long time spans of the Nile archaeological sequences is dependent upon the samples being used. For example, dynastic caries rates per individual range from a low of 10% affected individuals reported by Hillson (1979) to a high of 42% reported by Mummery (1870), while rates by tooth range from a low of 0.4% (Leek 1966) to highs of 2.3% (Brothwell 1963) and 4.65% (Grilletto 1977) affected teeth.

Hillson (1979) establishes the importance of this variation within time periods when he points out that the increase he found between the predynastic and Middle Kingdom samples is overshadowed by the much greater differences between contemporaneous individual cemeteries. Grilletto (1977) demonstrates differences associated with settlement and presumably social class by reporting a per-tooth caries rate of 4.03% from a village settlement and a higher rate of 5.42% from the urban centre at Assiut.

Similarly in Nubia, increases over time and variation between contemporary populations are attributed to changing socio-economic factors which affect the diet (Armelagos 1968; Beck and Greene 1989). For example Beck (1988) demonstrates that caries increase over time among the Medieval Christians of Kulubnarti as increasing village autonomy provides greater local access to resources and improved health. Strouhal (1984, 164) articulates the importance of this variation by suggesting that 'it seems necessary to continue the research by studying clear-cut samples from single localities and well-defined periods, taking into account their socio-economic level.'

Dental Attrition
Ruffer (1921) notes the extreme dental attrition found during all periods in Egypt from the predynastic to the Coptic. Leek (1967; 1972) attributes the extensive wear found throughout the Nile Valley to the abundant sand in the environment, use of stone grinding utensils and the presence of whole grains of seed in the bread. Ruffer (1921) observed a temporal trend of wear decreasing with time and a geographic variation with wear greatest in Upper Egypt. Grilletto's (1977) study of dental attrition also demonstrates a decrease in dental wear between his predynastic and dynastic samples. He attributes this decline to development of more elaborate food preparation techniques. Ibrahim (1987) shows a decline in attrition over time and attributes variation in degree and pattern of wear to dietary differences between cultures, localities and socio-economic levels. These studies establish that anlaysis of dental attrition can provide insight into both dietary and socio-economic change.

Dental Microwear
Puech et al. (1983) employ nitrocellular impressions and a light microscope at 100 magnifications and, more recently, a scanning electron microscope (Puech and Leek 1986) to examine the microwear of predynastic and dynastic tooth surfaces. They describe the 'blunt' character of the few striations visible and note the virtual absence of large striations. Puech and co-workers (1983; 1986) attribute the blunt striations to the rarity of large abrasive particles and the presence of particles too small to leave striations. They suggest that the polishing observed on the tooth surfaces is associated with the frequent chewing of papyrus masticatories (wads of papyrus fibre chewed to stimulate saliva flow) as suggested by Dixon (1972). Dixon (1972) notes the similarities of mastication and expectoration between Egyptian papyrus chewing and the chewing of fibrous desert plants in prehistoric southwest Texas, USA. The high dietary fibre content of these prehistoric Texans has been established by coprolite analysis and the presence of fibre quids in the dry caves, while scanning electron microscopy of the molars has demonstrated polished tooth surfaces similar to those reported for Egypt (Marks et al. 1988).

Dental Abscess and Tooth Loss
Ruffer (1921) is one of many to note that dental abscessing can be found in every skeletal collection from Egypt and Nubia. One of the consequences of abscessing is antemortem tooth loss which has increased over time in both Nubia and Egypt (Ibrahim 1987). Both caries and extensive dental attrition have been identified by Satinoff (1972) as the major causes of abscessing among Nile Valley peoples at various times and locations. The extensive attrition of Mesolithic teeth from Sudanese Nubia is clearly identified as the cause of the 28% affected individual abscess rate (Greene et al. 1967). If attrition declines with time and abscesses increase, then caries must have played an increasingly important role in antemortem tooth loss (Ibrahim 1987).

Burns (1979; 1982) shows that the interrelationship between caries and attrition is a complex one with various degrees of attrition being associated with both increased and decreased caries frequencies. Brothwell (1963) establishes a direct relationship between severe dental attrition and abscessing from the predynastic period to the Thirteenth Dynasty, but an inverse relationship in the Twenty-Sixth through Thirtieth Dynasty. Presumably this increase in abscessing in the face of decreased attrition can be attributed to an increase in dental caries.

Elemental and Microscopic Analysis
Elemental or isotopic analyses have to date not been employed in dietary reconstruction in Egypt. Grandjean et al. (1979) and Nielson et al. (1986) use teeth to show that Nubian lead levels are thirty times less than those in modern Denmark, but small increases in Nubian lead levels appear to be associated with increased use of lead. Stack (1986) employed atomic absorption spectrometry of root tips to compare amounts of cadmium, copper, iron, lead, strontium, zinc and magnesium from the Bristol mummy with four Egyptian and four Nubian mummies.

One interesting discovery by Bassett et al. (1980; see also Armelagos and Mills in this volume) is tetracycline staining in Nubian bone. The ingestion of this broad spectrum antibiotic in the food provides an explanation for the low skeletal infection rates found in some groups (Martin et al. 1984; Armelagos et al. 1981). Recently two of three teeth from the late Roman occupation of the Dakhleh Oasis showed tetracycline labelling of the enamel matrix (Cook et al. 1989). These discoveries not only explain the low observed skeletal infection rates, but also have implications for variation in dental caries rates. The presence of tetracycline in the food could have had a cariostatic effect and may explain some of the anomalously low caries rates. Comparison of caries rates between groups with and without tetracycline consumption should prove of interest.

Tooth Loss in Sudanese Nubia
A detailed examination of dental caries, attrition, tooth loss and cultural variation in Sudanese Nubia will illustrate the intricate interrelationship of the disease processes involved in antemortem tooth loss. The data derived from five cemeteries in the Wadi Halfa area (see Adams 1977; Armelagos 1968; Martin et al. 1984). Cemetery 6B16 is Meroitic, NAX and 2413 are X-Group, and 6B13 and 6G8 are Christian. In Fig. 1 the percentages of teeth with caries and resorbed sockets in the mandible and maxilla for the combined samples are plotted by tooth type. The decreasing rates of caries from the molars to the incisors are a normal agricultural pattern. The concordance of the caries and resorbed socket rates indicates that caries played a significant role in antemortem tooth loss. High correlation coefficients indicate a close relationship between caries and antemortem tooth loss in the Meroitic (6B16 = 0.86) and X-Group (NAX = 0.86, 2413 = 0.85) samples. Much lower correlation coefficients for the two Christian cemeteries (6B13 = 0.59, 6G8 = 0.40) suggest that other factors are involved in tooth loss during this later period.

Fig. 2 shows little relationship between the rates of high attrition scores and resorbed sockets indicating that wear is not the prime causal agent for tooth loss. In other words, extreme wear which exposes the pulp chamber to infection and subsequent abscessing is not the cause of high antemortem tooth loss. Fig. 3 displays a direct relationship between extreme attrition and caries rates at all but the Christian site of 6G8. The data suggest that items in the diet which resulted in more frequent high wear scores also contributed to higher caries rates. At 6G8 the physical consistency of the diet remains the same as at 6B13, but its cariogenicity increases.

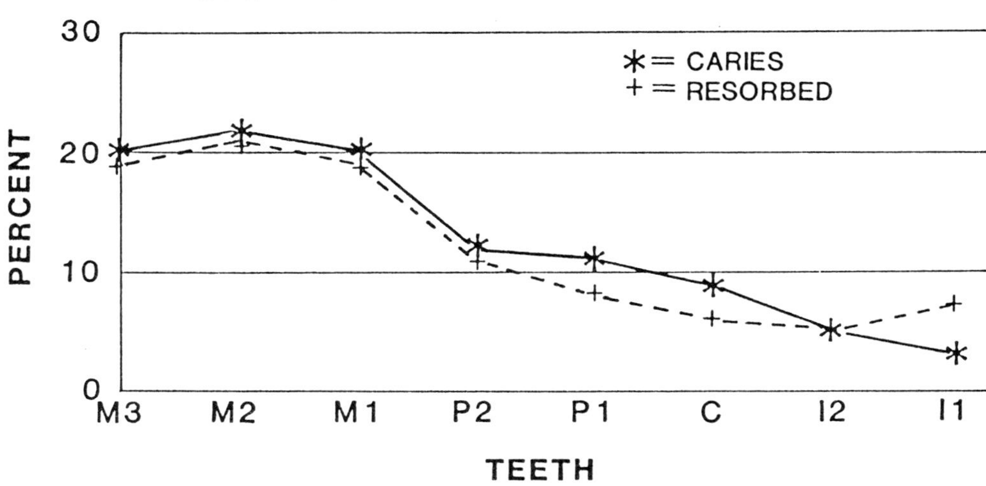

Fig. 1. Percentages of teeth with caries and resorbed sockets by tooth type with maxillary and mandibular teeth combined for all samples.

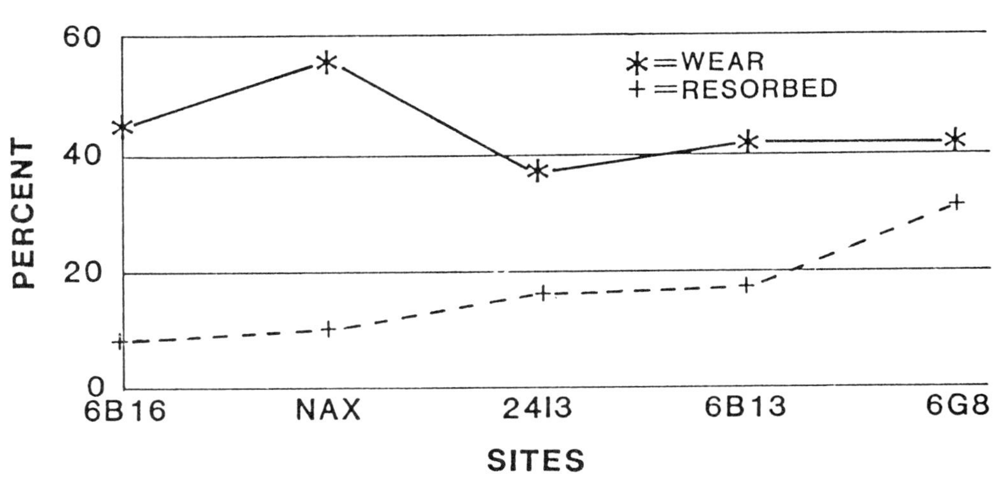

Fig. 2. Percentages of extreme wear and resorbed sockets by cemetery sites.

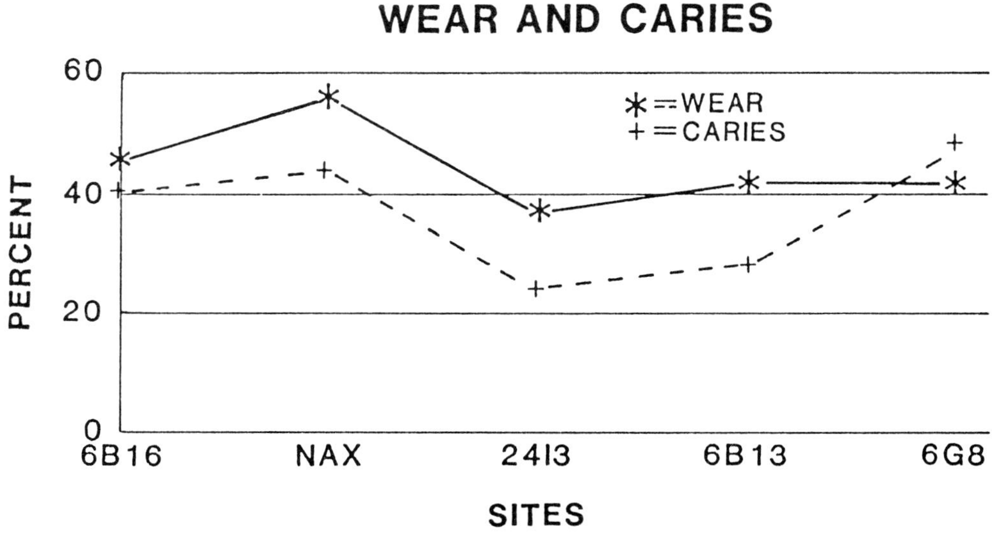

Fig. 3. Percentages of extreme wear and caries by cemetery sites.

This study illustrates the importance of analysing variation between skeletal samples belonging to a single culture. Both the caries and attrition rates are different between the two X-Group sites of NAX and 2413. In fact the diet of NAX appears more similar to the Meroitic site of 6B16. Of special note is the fact that the Christian occupants of 6G8 show the only reversal of wear and caries rates, which suggests that this is the first major dietary change in this sequence of sites. Palaeodemographic analysis by Armelagos et al. (1981) shows that the Christian occupations experienced the first major improvement in life expectancy, and dietary changes intimated by these dental data may have been responsible. Socio-economic stratification, demonstrated by Green et al. (1974) for the site of Meinarti, suggests that the dietary difference between the two Christian sites of 6B13 and 6G8 may be due to social stratification and increased access to more or different cariogenic foods. A similar change in dental pathology over time has been demonstrated for the Christian population of Kulubnarti and has been attributed to increased local autonomy and economic development (Beck 1988; Beck and Greene 1989).

This brief examination of Nubian dental data illustrates the necessity of comparing homogeneous samples and testing for socio-economic differences before samples are combined for further rate comparisons of caries, antemortem tooth loss, abscessing, and dental attrition.

Childhood Stress

Enamel Hypoplasia

Enamel hypoplasias, linear depressions in the enamel surface, are frequently employed in the reconstruction of childhood stress. Increased hypoplasia frequencies are associated with increased childhood mortality and morbidity, and the age distribution of the defects is used to define the chronological pattern of childhood stress.

Hillson (1978; 1979) found fairly constant 40% hypoplasia rates in skeletal samples from predynastic and dynastic Egypt and Nubia. Three Egyptian samples from Badari, Sedment and Hawara have higher rates. Hillson (1979) concludes that childhood stress was higher for these populations. Variations in the peak frequencies are attributed to variation in the ages of weaning between temporal and social groups (Hillson 1979). In contrast Ibrahim (1987) notes an increase in hypoplasias over time in his Egyptian samples with the implication that childhood stress also increased over time.

Blakey et al. (1989) find no difference in the frequency of hypoplasias and hypocalcifications in incisors and canines between the Meroitic and Christian samples from El Geili located along the Middle Nile. However, when only canines are tabulated, enamel defects increase over time (Blakey et al. 1989). These results imply that stress increased only in the later ages of the individuals, when the canines were still forming and the incisor crowns were complete.

Although trends in hypoplasia frequencies over long spans of time and between wide spread locations may be informative, it is changes or lack of change within short time spans and restricted geographic regions which are most informative. For example Karhu (1990) reports that hypoplasias are ubiquitous among the Medieval Christian Nubians from Kulubnarti with some indication of increased stress over time. Van Gerven et al. (1990) report a 4.2 mean hypoplasia rate for the early Kulubnarti Christian sample and a 3.7 mean rate for the later sample. The earlier sample shows a prolonged period of hypoplastic occurrences indicating a longer period of childhood stress associated with shorter time spans between hypoplastic episodes. Taken as a whole, these data demonstrate that childhood stress declined over time. This decline in childhood stress is connected, as previously mentioned, with decreased morbidity and mortality associated with increased local autonomy (Van Gerven et al. 1990).

Microdefects of Enamel

In 1956 Sognnaes sectioned four teeth from a predynastic skeletal sample from Keneh in Upper Egypt. Microscopic examination showed that two of the teeth had pronounced incremental lines in the enamel and three of the teeth exhibited interglobular dentine (a defect of calcification). This was the first histological study of ancient Egyptian teeth. Accentuated striae of Retzius, now termed Wilson bands, extend from the enamel surface to the enamel adjacent to the dentine and are the result of physiological disturbance of the ameloblasts or enamel forming cells. This physiological response to stress leaves a permanent record in the enamel. Because these microdefects record milder stress episodes than hypoplasias, they are more frequent and provide a more sensitive measure of childhood stress.

Rudney (1981; 1983a; 1983b; Rudney and Greene 1982) examined the frequency of Wilson bands in first molars from two Nubian samples to determine the relative differences in childhood stress. He demonstrates that, contrary to expectation, there was a decline in Wilson band frequency between the Meroitic and X-Group samples. This decline in stress is attributed to the development of local autonomy, adaptation to helminthic disease and differential ingestion of tetracycline contaminated foods (Rudney 1983b). The age peaks in the defect frequencies between fourteen and seventeen months are attributed to weaning (Rudney 1983a).

Genetics, Dental Morphology, Metrics and Adaptation

Cranial changes over time and variation between contemporaneous skeletal samples in both Egypt and Nubia are frequently explained by various combinations of population migration, replacement and admixture (see Greene 1981; Van Gerven et al. 1979 for discussions). Dental measurements, anomalies and morphology have also been used to argue for the presence or absence of migration and admixture. The use of dental data to establish genetic stability over time has made it possible to explain cranial variation in terms of in-situ evolutionary change and adaptation.

Dental Anomalies

Ruffer (1921) was one of the first to use dental data to imply genetic differences between Nubians and Lower Egyptians, when he reported that reduced tooth number was rare in Nubia, while agenesis of the central mandibular incisor was more common in Upper Egypt. Despite Ruffer's early reference to dental anomalies, others, such as Leek (1972) and Smith (1986), have reported the presence of traits such as supernumerary teeth and agenesis (i.e., failure of teeth to form), but none has used them in populational studies. These rapidly scored traits are highly heritable and could be profitably employed to establish populational differences and similarities.

Dental Morphology

Dental morphology has been extensively used to establish genetic stability in the Wadi Halfa region of Sudanese Nubia. Greene (1967a; 1967b; 1972; 1973) employs sixteen morphological traits to establish genetic continuity between the Meroitic, X-Group and Christian occupants of Wadi Halfa. This conclusion contrasts with those drawn by others from cranial analyses. His 1982 work not only establishes the validity of this conclusion using multivariate statistics, but extends the genetic continuity to the more recently excavated Christian cemeteries from Kulubnarti. Establishing populational continuity has made it possible to attribute changes in health and mortality to changes in socio-economic conditions rather than population migration and/or replacement.

Dental Metrics

These dental morphological similarities have been strengthened by metrical studies using bucco-lingual and mesio-distal tooth diameters. Calcagno (1986a) demonstrates close relationships between the earlier Nubian A-Group, C-Group and Pharaonic samples and the later Meroitic, X-Group, and Christian occupants of the Wadi Halfa region.

Calcagno (1984; 1986b; 1989) Chamla (1980), Burrell (1988) and Smith (1979) have all used dental measurements to establish a temporal trend for reduction in tooth size using Nile Valley data. Greene (1984) demonstrates that available sample sizes are too small to use fluctuating asymmetry of tooth diameters to assess the severity of childhood stress.

Nubian Cranio-dental Change

Having established genetic continuity in Sudanese Nubia using dental morphology and metrics, the observed craniometric variation begged an evolutionary explanation. Three hypotheses have been proposed to account for morphological changes. The 'masticatory function hypothesis' (Carlson 1976; Carlson and Van Gerven 1977), the 'caries-resistance-dental reduction hypothesis' (Armelagos 1968; Calcagno 1984; Greene 1970; 1972; Goodman et al. 1986; Van Gerven et al. 1977) and the 'increasing population density hypothesis' (Macchiarelli and Bondioli 1984; 1986). The masticatory function hypothesis suggests that the chewing complex of the face will reduce and there will be associated cranial remodelling with changes in subsistence patterns. Consumption of processed foods will lead to an alteration in the growth of the chewing complex. In response to this reduction there is a secondary reorientation of the face and vault. While this hypothesis does not exclude the possibility of genetic changes in facial reduction, it does not consider genetic evolution to be the major source of change.

The dental reduction hypothesis places greater emphasis on the genetic evolution of tooth size. The dietary change from highly abrasive foods shifted the selective advantage from large morphologically complex teeth. The large, complex teeth are capable of resisting wear, while smaller, simpler teeth are more resistant to dental caries. The selection for smaller, less complex teeth occurred within the overall masticatory apparatus and cranial remodelling.

Macchiarelli and Bondioli (1984; 1986) have argued against a need to consider selective factors in understanding the reduction in tooth size or differential reduction in the craniofacial complex. They interpret the decrease in dental and facial size observed in post-Pleistocene populations as a result of an overall decrease in body size related to stresses associated with increased population density. They argue that there is no evidence for selection favouring either smaller teeth or alterations in craniofacial architecture. Specifically, they do not believe that a shift in diet could have caused the rapid facial reduction and the 15% reduction in occlusal surface area (from 2319 to 1971 square millimetres) reported by Frayer (1978). The magnitude of this change in the shift from the upper Paleolithic to Mesolithic periods cannot be due to a 'completely different survival capacity' over the last ten thousand years. According to Macchiarelli and Bondioli, the reliance on a selectionist interpretation is due to sampling error. We have an abundance of dental remains (that do show size reduction) but lack the postcranial material necessary to interpret systematic changes.

Armelagos and co-workers (1984) have countered the position of Macchiarelli and Bondioli. They note that over the last eleven thousand years Sudanese Nubian populations underwent precisely the transition from a low density dispersed pattern to a higher density central place pattern suggested by Macchiarelli and Bondioli in their hypothesis, but experienced a more rapid change in the craniofacial complex than in the postcranial skeleton. The Nubian series is especially appropriate for an assessment of the body size hypothesis, since there is evidence for biological continuity over this time period and there are postcranial skeletal remains.

The analysis of changes in femur length shows that there is a four percent decrease in stature from the Mesolithic through Christian periods (Armelagos et al. 1989). Changes in cranio-facial morphology of these same populations would have been far more profound and far more complex, if they were just a reflection of the reduction in body size. Examination of percent changes in craniofacial measurements from Mesolithic through Christian times reveals three clusters of change. Facial height, cranial height, parietal chord and frontal chord (complex one) show an increase in size. Facial length, cranial length, symphyseal height and mandibular ramus height (complex two) demonstrate a slight decrease of from 0.8% to 4.8%. In cluster three (length of masseter origin, length of mandibular corpus and symphyseal thickness) there is reduction of from 13% to 26% between Mesolithic and Christian times (Carlson and Van Gerven 1977; Armelagos et al. 1989).

In this transformation of the Nubian skull there is a reduction in the attachment of masticatory muscles suggesting a reduction in their size. The slight increase in cranial height and a decrease in cranial length are secondary responses to facial reduction. These changes in the chewing musculature and bony attachments resulted in a smaller face that is more inferior and posterior in position. The cranium is transformed into a higher and more spheroid form. This pattern of transformation, while not explicable in terms of simple body size reduction, can be understood in terms of alternative hypotheses (masticatory function hypothesis and caries resistance-dental reduction hypothesis).

Calcagno (1984) presents dental data from ancient Nubia that make it possible to evaluate the relative roles of facial reduction and dental reduction from Mesolithic through Christian times. By analysing the diachronic order of facial reduction and dental reduction, it is possible to understand their respective roles in cranio-facial remodelling.

Tooth breadth, which is a reliable indicator of crown size, decreases at the rate of one percent per thousand years between Mesolithic and A-Group-C-Group (early agriculturalist) periods. This rate of dental reduction would suggest a strong selective pressure favouring smaller teeth (Calcagno 1984). The reduction is seen in anterior and posterior, mandibular and maxillary teeth in males and females.

There is a deceleration in the rate of reduction between early agriculturalists (A-Group-C-Group) and later intensive agriculturalists (Meroitic, X-Group and Christian periods). During this time, anterior tooth size approaches stability, while posterior teeth continue to reduce, but at a slower rate.

This pattern of dental reduction closely follows the pattern of masticatory reduction reflected in the third cluster of cranio-facial measurements. As previously discussed, these dimensions show an initial rapid reduction between Mesolithic and early agricultural periods of 12.5% to 21.6%. This, however, is followed by a slower reduction of 6% from early to later agricultural periods.

The correspondence between the rate and timing of dental and masticatory features makes it difficult to accept either of the selectionist hypotheses as the primary factor in explaining the change. The fact that there is a continuing reduction of the posterior dentition suggests that there may be a continuing advantage of smaller caries-resistant teeth.

Our inability to separate changes in the dental system from the masticatory features of the skull may just be a function of biological reality. The dentition, the bony system that supports the teeth, the musculature that moves the jaws are all a single functional complex that has evolved through time in response to dietary changes. More highly processed cariogenic foods of the Neolithic may have simultaneously shifted the selective advantage in favour of smaller teeth, while at the same time contributing to a new nutritional environment in which reduced neuromuscular stimulation results in changed patterns of bone deposition and craniofacial growth.

Methodological Critique

The literature review provided above makes it obvious that dental caries are the most consistently reported and most comparable of the dental data sets. The use of caries for documenting dietary changes could be enhanced by employing a tooth surface recording system such as the one used by Moore and Corbett (1971). Comparisons between studies would be enhanced if the reported caries data also included the numbers of teeth examined and the age structure of the samples. At present few data sets from the literature can be aggregated for analysis, because the basic sample size data are lacking.

In contrast, the large number of dental attrition schemes makes it virtually impossible to compare data between studies. We recommend the adoption of a single scoring system by all researchers. One, developed by Scott (1979), is already used by some Egyptologists and discriminates best between samples when tooth wear is slight to moderate. Another, proposed by Smith (1983; 1984), has the advantage of scoring both anterior and posterior teeth and discriminates well, when attrition is moderate to extreme.

The study of dental microwear patterns has proven useful in many parts of the world for identifying dietary differences and changes in food preparation technology (see Teaford 1988 for a review). The application of this technique to Nile Valley dental samples should be expanded and the actual teeth or high quality replicas should be examined with a scanning electron microscope. Alternatively, if scanning microcopy is too costly, teeth can be directly observed with a Nomarski reflected light differential interference microscope, if the teeth are coated with an opaque material such as gold.

Enamel hypoplasias are frequently used in palaeopathological studies, but systems for scoring and determining the age of formation have not been standardised. Comparability between analyses requires that the same teeth be recorded because the different tooth classes (e.g., incisors, canines) have variable susceptibilities to the stressors causing hypoplasias. The recommended method is to record hypoplasias on the maxillary central incisors and mandibular canines. It has been suggested that magnitude of the stress can be estimated from the width and/or depth measurements of hypoplasias. Hillson and Jones (1989) offer a technique for obtaining these measurements.

The analysis of enamel microdefects can add significantly to the study of childhood stress. Because Wilson bands represent less severe episodes of stress, hypoplasia and microdefect frequencies interpreted together provide a more detailed picture of the chronology and magnitude of childhood health than does either defect alone. Preliminary research by the senior author suggests that diseases with cyclical episodes, such as malaria, can be identified with these data. However, Wilson band analysis suffers from the same problems as hypoplasias. Defect definitions and the tooth types chosen for analysis must be standardised. Although Rudney (1981) used first molars, the recommended teeth are the same as for hypoplasia anlaysis, maxillary central incisors and mandibular canines.

Despite the significant contributions which dental morphology analysis has made to the study of Nile skeletal samples, comparability of data between studies is virtually impossible. This is due to lack of standardisation, which is attributed to failing to publish detailed morphological descriptions and photographs of the traits being scored or to publishing in relatively obscure and hard to get vehicles. This situation could be alleviated by adopting a standardised widely available system. Our experience recommends the thirty-nine trait dental morphology system established by Christy Turner at Arizona State University, USA. The advantage of this system is that plaster models and descriptions of the morphological variants are available.

In summary, the synthesis of Nile Valley dental anthropology recommends two changes in the overall research strategy. First, there must be standardisation of data collection procedures, which would provide comparability of data between studies. Preparing this review has made it evident to us that anyone desiring to synthesise the data will be compelled to re-collect data from already studied collections. Second, many of the problem domains chosen for study are too broad in both time and socio-economic stratification. Meaningful interpretations require focusing on narrower time frames and teasing out the rank and occupational differences which may exist (see Saffirio 1972). Some of the most recent studies have done this to the great benefit of understanding the ancient populations and cultures of the Nile Valley.

Conclusions

Greene's 1972 paper cites only nineteen skeletal or dental publications, while we have been able to include seventy-three contributions to Egyptian and Nubian dental anthropology. This represents a significant increase in our knowledge during the intervening eighteen years. It also clearly indicates an increased interest in dental anthropology. The identification of seven completed dissertations and one thesis concerning dental anthropology since 1972 documents an increase in the number of new researchers entering the field, while many of the pre-1972 people are still active.

Greene (1972) concentrates his discussion on three topics: 1) using dental morphology to establish population stability; 2) a general discussion of caries and dental attrition; and 3) a consideration of the dental reduction hypothesis. The present synthesis also considers these same topics, but differs in introducing new methodologies, new research domains and bringing the previous areas to new levels of understanding. New methodologies include microwear, hypoplasia, Wilson band and elemental analyses. New research domains include a focus upon variation within limited temporal and geographic areas. Differences in caries, dental attrition, and dental defects are demonstrating the complexity of Egyptian socio-economic organisation and adaptation to local conditions. This goes significantly beyond documenting differences between the predynastic and dynastic periods. Where Greene discussed genetic stability and dental reduction, these two concepts have been united into complex models of evolutionary change and adaptation.

In summary, the future of Nile Valley dental anthropology looks bright with new researchers, new methodologies and the introduction of a comprehensive bio-cultural research paradigm.

References

Adams, W Y, 1977. *Nubia. Corridor to Africa*, Princeton University Press, Princeton.

Armelagos, George J, 1968. *The Paleopathology of Three Archaeological Populations from Sudanese Nubia*. PhD, dissertation, Department of Anthropology, University of Colorado.

Armelagos, George J, Jacobs, Kenneth H and Martin, Debra L, 1981. 'Death and Demography in Prehistoric Sudanese Nubia', in Humphreys, S C and King, H (eds.), *Mortality and Immortality: The Anthropology and Archaeology of Death*, 33-57, Academic Press, London.

Armelagos, George J, Van Gerven, D P, Goodman, A H and Calcagno J M, 1989. 'Post-Pleistocene Facial Reduction, Rejoinder to Macchiarelli and Bondioli', *Journ. Human Evolution* 4, 1-7.

Bassett, E J, Keith, M S, Armelagos, G J, Martin, D K and Villanueva, A R, 1980. 'Tetracycline-Labelled Human-Bone from Ancient Sudanese Nubia', *Science* 209, 1532-34.
Beck, Rosemary, 1988. *The Dental Pathology of Medieval Christian Sudanese Nubians from the Batn-el-Hajar*, MA thesis, Department of Anthropology, University of Colorado.
Beck, Rosemary and Greene, D L, 1989. 'Dental Disease Among Medieval Christian Sudanese Nubians from the Batn-el-Hajar', *Am. Journal Physical Anthrop.* 78, 190.
Blakey, M L, Coppa, A, Damadio, S and Vargiu, R, 1989. 'A Comparison of Dental Enamel Defects in Christian and Meroitic Populations at Geili, Central Sudan', *Am. Journ. Physical Anthrop.* 78, 193.
Brothwell, Don R, 1963. 'The Macroscopic Dental Pathology of Some Earlier Human Populations', in Brothwell, D R (ed.), *Dental Anthropology*, 271-88, Pergamon Press, Oxford.
Burns, Peter E, 1979. 'Log-Linear Analysis of Dental Caries Occurrence in Four Skeletal Series', *Am. Journ. Physical Anthrop.* 51, 637-48.
Burns, Peter E, 1982. *A Study of Sexual Dimorphism in the Dental Pathology of Ancient Peoples*, PhD dissertation, Department of Anthropology, Arizona State University.
Burrell, Lydia Lambe, 1988. A *Diachronic Study of Sexual Dimorphism in a Series of Human Skeletal Populations from Ancient Nubia*, PhD dissertation, Department of Anthropology, University of Colorado.
Calcagno, James M, 1984. *Human Dental Evolution in Post-Pleistocene Nubia*, PhD dissertation, Department of Anthropology, University of Kansas, Lawrence, Kansas.
Calcagno, James M, 1986a. 'Odontometrics and Biological Continuity in the Meroitic, X-Group, and Christian Phases of Nubia', *Current Anthrop.* 27, 66-9.
Calcagno, James M, 1986b. 'Dental Reduction in Post-Pleistocene Nubia', *Am. Journ. Physical Anthrop.* 70, 349-63.
Calcagno, James M, 1989. *Mechanisms of Human Dental Reduction: A Case Study from Post-Pleistocene Nubia*. University of Kansas Publication in Anthropology, No 18, Lawrence, Kansas.
Carlson, David S, 1977. 'Temporal Variation in Prehistoric Nubian Crania', *Am. Journ. Physical Anthrop.* 45, 467-84.
Carlson, David S and Van Gerven, Dennis P, 1977. 'Masticatory Function and Post-Pleistocene Evolution in Nubia', *Am. Journ. Physical Anthrop.* 46, 495-506.
Chamla, Marie-Claude, 1980. 'Etude des Variations Metriques des Couronnes Dentaires des Nort-Africains, De L'Epipaleolithique à L'Epoque Actuelle', *L'Anthropologie* 84, 254-71.
Cook, Megan, El Molto and Anderson, C, 1989. 'Fluorochrome Labelling in Roman Period Skeletons from Dakhleh Oasis, Egypt', *Am. Journ. Physical Anthrop.* 80, 137-43.
David, A R, 1985. 'The Manchester Mummy Project', *Archaeology* (Nov/Dec), 40-7.
Derry, D E, 1933. 'Incidence of Dental Disease', *British Medical Journal* 1933, Part I, 112.
Dixon, D M, 1972. 'Masticatories in Ancient Egypt', *Journ. Human Evol.* 1, 433-49.
Gaballah, M F, Abd El-Malek, Ratiba, G and Badaway, Zaizafon H, 1980. 'A Study of Dental Attrition in Relation to Suture Closure in Ancient Coptic Skulls from Upper Egypt', *Egypt Journ. Anatomy* 3, 2, 37-45.
Goodman, A H, Armelagos, G J, Van Gerven, D P and Calcagno, J P, 1986. 'Diet and Post Mesolithic Craniofacial and Dental Evolution in Sudanese Nubia', in David, A R, (ed.), *Science in Egyptology*, 201-11, Manchester University Press, Manchester.

Grandjean, P, Neilsen, O V, Shapiro, I M, 1979. 'Lead Retention in Ancient Nubian and Contemporary Populations', *Journal Environmental Pathol. Toxico.* 2, 3, 781-7.

Green, S, Green, S and Armelagos, G J, 1974. 'Settlement and Mortality of the Christian Site (1050 AD-1300 AD) of Meinarti, Sudan', *Journ. Human Evol.* 3, 297-311.

Greene, David L, 1967a. *Genetics, Dentition and Taxonomy*, University of Wyoming Publications, Vol 33, No 2, Laramie, Wyoming.

Greene, David L, 1967b. *Dentition of Meroitic, X-Group and Christian Populations from Wadi Halfa, Sudan*, University of Utah Anthropological Papers, No 85, Salt Lake City, Utah.

Greene, David L, 1970. 'Environmental Influences on Pleistocene Hominid Dental Evolution', *Bioscience* 20, 276-9.

Greene, David L, 1972. 'Dental Anthropology of Early Egypt and Nubia', *Journ. Human Evol.* 1, 315-24.

Greene, David L, 1973. 'Dental Anthropology of Early Egypt and Nubia', in Brothwell, D R and Chiarelli, B A (eds.), *Population Biology of the Ancient Egyptians*, 315-24, Academic Press, New York.

Greene, David L, 1981. 'A Critique of Methods Used to Reconstruct Racial and Population Affinities in the Nile Valley', *Bulletins et Memoires de la Sociète d'Anthropolgie de Paris* 8, 3, 357-65.

Greene, David L, 1982. 'Discrete Dental Variations and Biological Distances of Nubian Populations', *Am. Journ. Physical Anthrop.* 58, 75-9.

Greene, David L, 1984. 'Fluctuating Dental Asymmetry and Measurement Error', *Am. Journ. Physical Anthrop.* 65, 283-9.

Greene, David L, Ewing, George H and Armelagos, George J, 1967. 'Dentition of a Mesolithic Population from Wadi Halfa, Sudan', *Am. Journ. Physical Anthrop.* 27, 41-56.

Grilletto, Renato R, 1964. 'Osservazioni Preliminari Sulla Carie Dentaria in una Serie di Crani Egiziani Predinastici', *VII Congres International des Sciences Anthropologiques et Ethnologiques*, Moscou 2, 491-7.

Grilletto, Renato R, 1973. 'Caries and Dental Attrition in the Early Egyptians as Seen in the Turin Collections', in Brothwell, D R and Chiarelli, B A (eds.), *Population Biology of the Ancient Egyptians*, 325-31. Academic Press, New York.

Grilletto, Renato R, 1977. 'Carie et Usure Dentaire chez les Egyptiens Predynastiques et Dynastiques de la Collection de Turin (Italie)', *L'Anthropologie* 81, 459-72.

Grilletto, Renato R and Davide, D, 1967. 'Dati Preliminari Sulla Carie e Sull' usura Dei Denti Negli Egiziani Antichi', *Archivio per l'Antropologia e l'Etnologia* 98, 177-80.

Hillson, Simon W, 1978. *Human Biological Variation in the Nile Valley, in Relation to Environmental Factors*. PhD thesis, University of London.

Hillson, Simon W, 1979. 'Diet and Dental Disease', *World Archaeology* 11, 147-62.

Hillson, Simon W and Jones, B K, 1989. 'Instruments for Measuring Surface Profiles - An Application in the Study of Ancient Human Tooth Crown Surfaces', *Journ. Archaeol. Science* 16, 95-105.

Ibrahim, Moustafa Awad, 1987. *A Study of Dental Attrition and Diet in Some Ancient Egyptian Populations*, PhD dissertation, University of Durham (United Kingdom).

Karhu, S L, 1990. 'Inter-tooth Distribution of Dental Enamel Hypoplasias among Medieval Christian Nubians from Kulubnarti', *Am. Journ. Physical Anthrop.* 81, 247.

Leek, F Filce, 1966. 'Observations on the Dental Pathology Seen in Ancient Egyptian Skulls', *Journ. Egyptian Archaeol.* 52, 59-64.

Leek, F Filce, 1967. 'Observations on the Dental Pathology Seen in Ancient Egyptian Skulls', *Journ. Dental Assoc. South Africa* 22, 6, 187-95.

Leek, F Filce, 1972. 'Bite, Attrition and Associated Oral Conditions as Seen in Ancient Egyptian Skulls', *Journ. Human Evol.* 1, 289-95.

Leek, F Filce, 1986. 'Cheops' Courtiers: Their Skeletal Remains', in David, A R (ed.), *Science in Egyptology*, 183-99, Manchester University Press, Manchester.

Macchiarelli, R and Bondioli, L, 1984. 'Time and Dental Structure in Man: a Reassessment of Mechanism', *Anthro. Contemporanea* 7, 102-39.

Macchiarelli, R and Bondioli, L, 1986. 'Post-Pleistocene Reduction in Human Dental Structure: a Reappraisal in Terms of Increasing Population Density', *Journ. Human Evol.* 1, 405-18.

Marks, Murray, Rose, J C and Buie, E L, 1988. 'Bioarchaeology of Seminole Sink', *Plains Anthropologist* 33, No. 122, Part 2, 75-118.

Martin, Debra L, Armelagos, G J, Goodman, A H and Van Gerven, D P, 1984. 'Chapter 8: The Effects of Socio-economic Change in Prehistoric Africa: Sudanese Nubia as a Case Study', in Cohen, M N and Armelagos, G J (eds.), *Paleopathology and the Origins of Agriculture*, 193-214, Academic Press, Orlando.

Moore, W J and Corbett, M E, 1971. 'The Distribution of Dental Caries in Ancient British Populations 1. Anglo-Saxon Period', *Dental Caries Res.* 5, 151-68.

Mummery, J, 1870. 'On the relations which dental caries, as discovered among the ancient inhabitants of Britain, and amongst existing aboriginal races, may be supposed to hold to their food and social conditions', *Trans Odontol. Soc. Great Britain* 2, 7-24, 27-90, 95-102.

Nielsen, O V, Grandjean P and Shapiro, I M, 1986. 'Lead Retention in Ancient Nubian Bones, Teeth and Mummified Brains', in David, A R (ed.), *Science in Egyptology,* 25-33, Manchester University Press, Manchester.

Puech, P-F and Leek, F F, 1986. 'Dental Microwear as an Indication of Plant Food in Early Man', in David, A R (ed.), *Science in Egyptology*, 239-41. Manchester University Press, Manchester.

Puech, P-F, Serratrice, C and Leek, F F, 1983. 'Tooth Wear as Observed in Ancient Egyptian Skulls', *Journ. Human Evol.* 12, 617-29.

Rudney, Joel D, 1981. *The Paleoepidemiology of Early Childhood Stress in Two Ancient Nubian Populations*, PhD dissertation, Department of Anthropology, University of Colorado.

Rudney, Joel D and Greene, D L, 1982. 'Interpopulation differences in the Severity of Early Childhood Stress in Ancient Lower Nubia: Implications for Hypotheses of X-Group Origins', *Journ. Human Evol.* 11, 559-65.

Rudney, Joel D, 1983a. 'The Age-related Distribution of Dental Indicators of Growth Disturbance in Ancient Lower Nubia: an Etiological Model from the Ethnographic Record', *Journ. Human Evol.* 12, 535-43.

Rudney, Joel D, 1983b. 'Dental Indicators of Growth Disturbance in a Series of Ancient Lower Nubian Populations: Changes Over Time', *Am. Journ. Physical Anthrop.* 60, 463-70.

Ruffer, Marc Armand, 1920. 'Study of Abnormalities and Pathology of Ancient Egyptian Teeth', *Am. Journ. Physical Anthrop.* 3, 335-82.

Ruffer, Marc Armand, 1921. *Studies in the Palaeopathology of Egypt* (edited by Roy L Moodie), University of Chicago Press, Chicago.

Saffirio, L, 1972. 'Food and Dietary Habits in Ancient Egypt', *Journ. Human Evol.* 1, 297-305.

Satinoff, Merton I, 1972. 'The Medical Biology of the Early Egyptian Populations from Asswan, Assyut and Gebelen', *Journ. Human Evol.* 1, 247-57.

Scott, E C, 1979. 'Dental Wear Scoring Techniques', *Am. Journ. Physical Anthrop.* 51, 213-8.

Smith, Bennett Holly, 1983. *Dental Attrition in Hunter-Gatherers and Agriculturalists,* PhD dissertation, Department of Anthropology, University of Michigan.

Smith, Bennett Holly, 1984. 'Patterns of Molar Wear in Hunter-Gatherers and Agriculturalists', *Am. Journ. Physical Anthrop.* 63, 39-56.

Smith, G Elliot and Jones, F Wood, 1910. 'Report of the Human Remains', in *Archaeological Survey of Nubia, 1907-8*, Vol II, Cairo.

Smith, N J D, 1986. 'Dental Pathology in an Ancient Egyptian Population', in David, A R (ed.), *Science in Egyptology*, 43-8. Manchester University Press, Manchester.

Smith, Patricia, 1979. 'Regional Diversity in Epipaleolithic Populations', *OSSA* 6, 243-50.

Sognnaes, Reidar F, 1956. 'Histological Evidence of Developmental Lesions in Teeth Originating from Paleolithic, Prehistoric, and Ancient Man', *Am. Journ. Path.* 32, 547-77.

Stack, M V, 1986. 'Trace Elements in Teeth of Egyptians and Nubians', in David, A R (ed.), *Science in Egyptology*, 219-22. Manchester University Press, Manchester.

Strouhal, Eugen, 1984. 'Paleopathology of Dentition of the Ancient Egyptians from Abusir', *Garcia de Orta, Ser Antropobiol.* 3, 163-72.

Teaford, Mark F, 1988. 'A Review of Dental Microwear and Diet in Modern Mammals', *Scanning Microscopy* 2, 2, 1149-66.

Van Gerven, Dennis P, Armelagos, G J and Rohr, A, 1977. 'Continuity and Change in Cranial Morphology of Three Nubian Archaeological Populations', *Man (NS)* 12, 270-7.

Van Gerven, Dennis P, Carlson, D S and Armelagos, G J, 1979. 'Racial History and Biocultural Adaptation of Nubian Archaeological Populations', *Journ. African History* 14, 555-64.

Van Gerven, Dennis P, Beck, R and Hummert, J R, 1990. 'Enamel Hypoplasia in Two Medieval Populations from Nubia's Batn el Hajar', *Am. Journ. Physical Anthrop.* 82, 413-20.

HISTOLOGICAL STUDIES OF ANCIENT TOOTH CROWN SURFACES

Simon W Hillson

Abstract
Microscopical study of tooth crown surfaces offers a unique opportunity to reconstruct a detailed record of growth during childhood for individuals in a collection of human remains. This paper outlines the physiological basis of the layering observed in the enamel of tooth crowns and summarises the methods by which it may be studied. The possibilities for producing growth layer chronologies are discussed using archaeological examples which include ancient Egyptian and Nubian material.

Growth of Human Tooth Crowns

Growth is an important process to consider when making a biological study of ancient human communities. The rate at which children grow, and any disruptions to their progress, strongly reflect the conditions in which they grew up. This includes particularly their overall health and nutritional plane. If archaeologists are to understand the ways in which ancient communities lived, then these two factors are fundamental. Growth can be investigated in various parts of the skeleton and dentition, but the tooth crowns offer what is probably the only real possibility at present for making a detailed study of growth. This is partly because the enamel which jackets the tooth crowns is the best preserved element of most human burials and so yields most detail under the microscope. It is also to do with the way in which the enamel grows - one of the most complex processes in the living world. All this complexity is woven during childhood into the mineral structure of the enamel and is preserved throughout life and after burial so long as dental wear, disease, or erosion in the soil do not destroy the enamel. Most important for the study of growth are the various types of regular layering which are visible in the crystalline enamel structure. The most prominent of these are the *brown striae of Retzius*, named after the Swedish anatomist who provided the most well-known early description of them (Retzius 1837). They are the dominant feature of any thin section of enamel, seen under low magnification (Pl.7,1), and they mark out the pattern of growth followed during the construction of the tooth crown. This invariably starts with the deposition of dome-like layers, one on top of another, underneath where the cusps and biting edges of the teeth are eventually to be formed. When the full height of the cusps has been achieved, this is followed by the deposition of sleeve-like layers around the outside of the domes, again one on top of another and overlapping down the sides of the tooth crown. A pattern like this was clearly described by Retzius in his original paper and has been recognised many times since then. The brown striae show a progression in spacing which is so regular that they have almost universally been interpreted as representing a regular growth rhythm. They are thought to mark out the boundaries between layers of enamel which each took a constant time to form. The American dental anatomist Isaac Schour and his colleagues (Massler et al. 1941) likened the layering of enamel to tree rings and, although the geometry of crown growth is very different to that of a tree, there are indeed parallels between dendrochronology and the study of tooth crown layers.

Brown striae of Retzius are best studied in polished thin sections of tooth crowns, but these are both time consuming and destructive to prepare from archaeological specimens. Fortunately, the layering which the brown striae represent leaves its mark on the crown surface as well. When a little worn crown is examined under moderate magnification, alternate shallow ridges and furrows can be seen running around the circumference (Pl. 7,2). These surface features as a whole are usually knows as *perikymata* and are split into *perikyma furrows* and *perikyma ridges* (Risnes 1984). Each perikyma furrow represents one brown stria of Retzius (Risnes 1985a; 1985b) and so the perikymata can be examined in the place of brown striae, although this task is not so simple as it might seem. First of all, a tooth crown must be relatively unworn for the perikyma furrows to be seen easily and their examination is therefore restricted, for most practical purposes, to the teeth of children and young adults. Secondly, perikymata are not usually prominent (Boyde 1975) on deciduous tooth crown surfaces and so their study is normally confined to the permanent dentition. Thirdly, the perikyma furrows represent only the sleeve-like region of enamel layering. Brown striae in the dome-like central part of the crown do not reach the surface and are not represented in the perikymata. Assuming that a regular growth rhythm lies behind the brown striae, their count in the different parts of the growth sequence gives an idea of the relative amounts of time taken to form the crown surface, as opposed to the enamel which is hidden under the cusps (Bromage and Dean 1985; Bullion 1987; Hillson in press). In an incisor or canine tooth, the hidden region of the crown formed by dome-like layering accounts for between 10 and 20% of the total count of brown striae in the sequence. This suggests that the perikymata seen on the surface represent the latter 80-90% of the total time taken to form the crown. In molars, dome-like layering typically accounts for 30-50% of the brown striae, so that the perikymata on the surface must represent only the last 50-70% of crown formation time. This idea, again, is not new and Isaac Schour and his colleagues clearly represented this type of division in their diagram of crown formation chronology (Figure 6 in Massler et al. 1941). In spite of this limitation, however, the perikymata still represent a useful slice of time in childhood growth, particularly when it is understood that different classes of teeth are formed between different ages and that each tooth crown surface can contribute a small part of a rather longer growth history (below).

Methods of Examination

If tooth crowns are viewed in strong, oblique illumination, the most widely spaced of the perikyma furrows can be seen with the naked eye (below). Even the smaller perikyma furrows may be made out with modest magnification under a low power stereomicroscope, if the illumination can be arranged effectively. Original tooth surfaces are, however, too pale and too glossy to see small details of structure clearly and most work is done with replicas of the crown surface. The initial impression of the crown surface is normally taken using a proprietary dental material and the replica is cast in this impression using an epoxy resin. Many impression materials and many resin systems have been tried, and the best of them have yielded a resolution of better than 0.5μm (e.g. Beynon 1987). This is more than adequate to see perikymata clearly. Transparent replicas can be examined effectively by ordinary transmitted light microscopes, but the difficulty with this technique is that the small depth of focus which it allows makes it difficult to view large areas of the bulging crown sides at one time.

The scanning electron microscope has a very much greater depth of focus and for this reason perikymata are most commonly examined by this means. Counting of perikyma furrows is usually carried out on microphotographs, but a computer interfaced system has been developed for counting them and measuring their spacing directly from light microscopic examination of replicas (Hillson and Jones 1989). The perikyma furrow spacing curves of Figs. 1 and 2 were produced by this method.

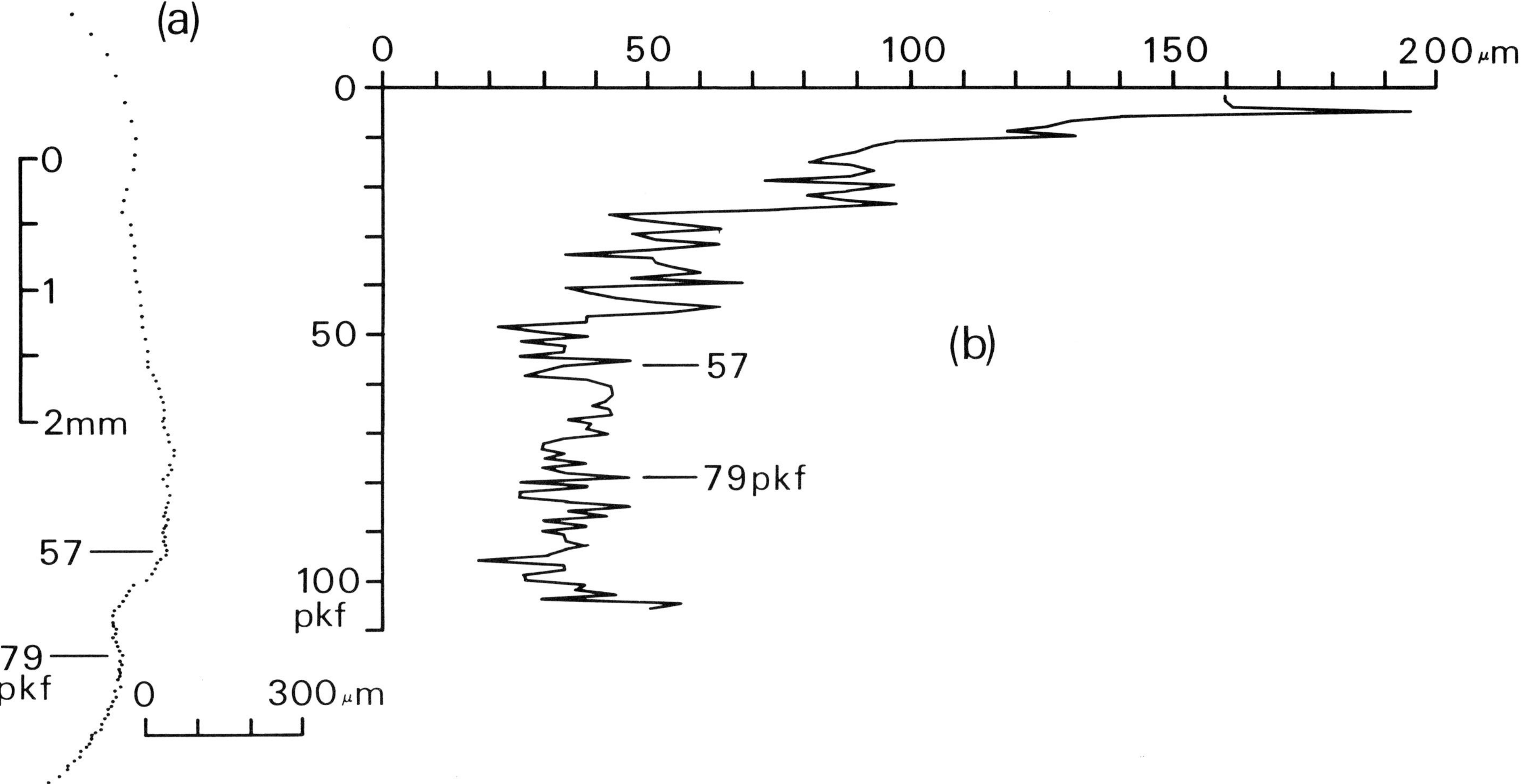

Fig. 1. Perikyma furrow profile (a) and spacing graph (b) for the tooth shown in Pl. 7,2. Both plots represent the buccal crown surface, with the occlusal limit at the top and the cervical margin at the bottom. Each dot in Fig. 1a shows the point at which a perikyma furrow was encountered in a straight transect down the crown replica. Fig. 1b shows the information recorded in the profile converted into a graph of perikyma furrow spacing. The enamel hypoplasia defect mentioned in the text and shown in Pl. 7,2 is located between perikyma furrows (pkf) 57 and 79.

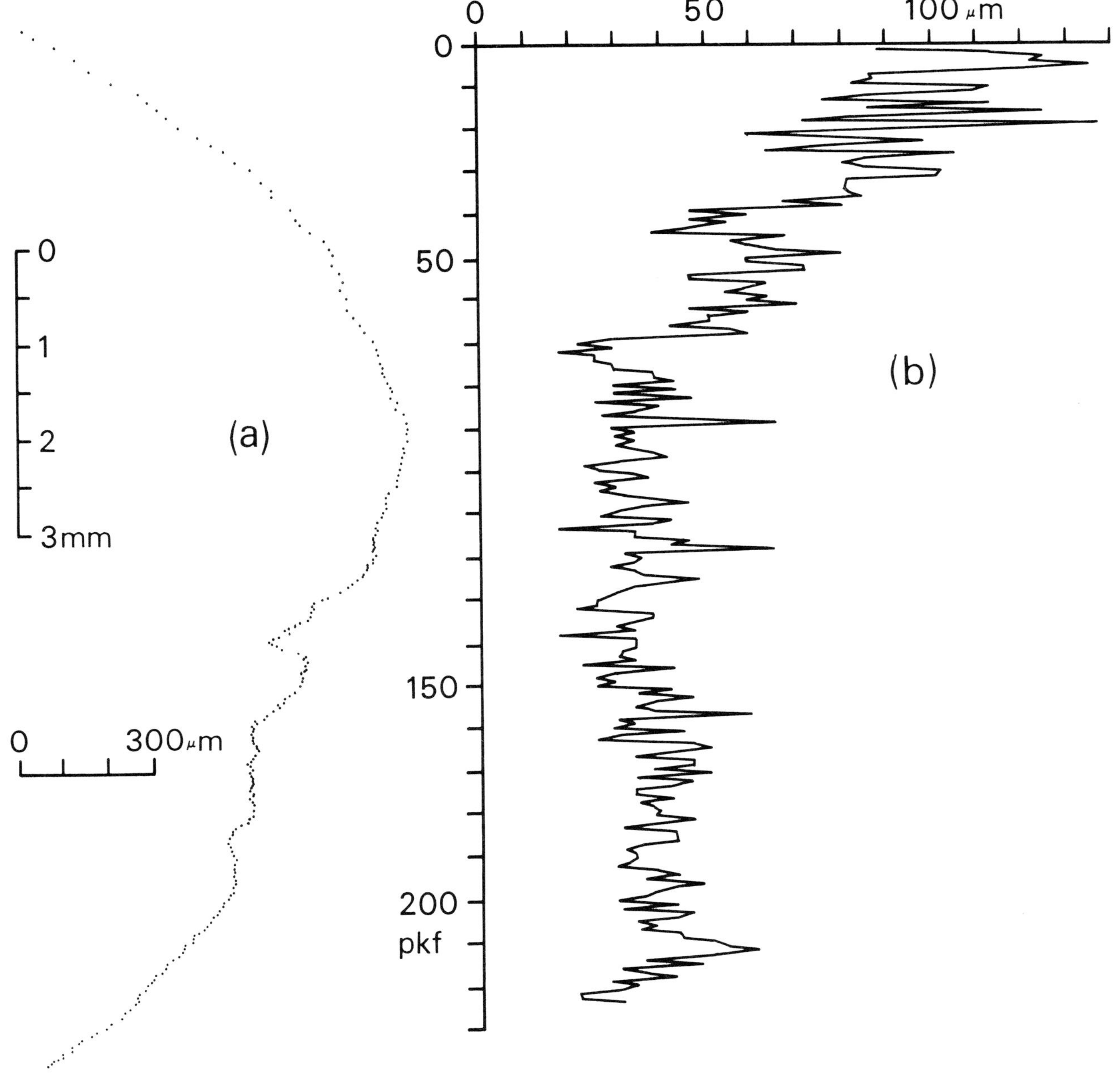

Fig. 2. Perikyma furrow profile (a) and spacing graph (b) for a lower canine from the same individual as the tooth in Pl. 7,2 and Fig. 1. Both plots represent the labial crown surface, with the occlusal tip at the top and cervical margin at the bottom.

Paradoxically, archaeological teeth are rather better than fresh teeth for examining perikymata. It is true that the heavy wear of the grinding surfaces seen in ancient Egyptian and Nubian specimens causes the loss of considerable areas of crown surface at an early age, but it is the areas of lowest perikyma furrow density that are lost first in this way (below). Far more damaging in modern teeth is toothbrush abrasion which tends to concentrate around the neck of the crown instead - the area of highest perikyma furrow density.

Identification of perikyma furrows and ridges is not straight-forward. Other normal features of the crown surface may confuse matters and all crown surfaces are abraded to some extent (except in those cases where teeth have not yet erupted into the mouth). Considerable experience is needed to follow a perikyma furrow across an abraded surface.

The Pattern of Growth at the Crown Surface

The surface of a tooth crown forms over a period of months and years by the accumulation of perikyma ridges. The first of these appears around the very tip of the earliest formed cusp of the crown and, following on from this, there is a regular progression of perikyma ridges and their accompanying furrows over the maximum bulge of the crown side and down to the neck of the crown, where it joins the root. The perikyma ridges are broadest (i.e. the furrows are furthest apart) near the cusps and are narrowest towards the crown neck. This trend is most pronounced in molars, where the perikyma furrows have a spacing of more than 150µm at the cusps but, just after the maximum bulge of the crown sides, very rapidly crowd together to a spacing of less than 40µm (Fig. 1). A canine, by contrast, has perikyma furrows spaced less widely (100-130µm) near the tip of its single cusp, followed by a more gradual decrease in spacing to a long 'plateau' of 40-50µm and a final sharp decrease to 20µm (Fig. 2) at the crown neck. These trends are due to the geometry of crown growth, and the differences between tooth classes reflect differences in the way in which they grow. They do not in themselves appear to be due to dietary or epidemiological factors that might disturb growth.

That is not to say, however, that growth at the tooth surface is unaffected by health and diet. The growth curves of Fig. 1 show a large number of minor fluctuations in the spacing of perikyma furrows. They are complex and at first sight appear random but, when they are analysed in other ways, are seen to be related to defects which appear in the surface profile. Such defects are very common, but show a large range in size. Some involve just one or two perikyma furrows and are, in effect, simply accentuated perikyma furrows. Others involve 10-20 or more perikyma furrows and are large enough to see with the naked eye, or to be felt with a fine probe. At this level, they are clinically detectable defects of a type known as *enamel hypoplasia.* Many studies have been carried out on the epidemiology of enamel hypoplasia (summary in Hillson 1990), and such defects are clearly related to dietary deficiencies and fever-generating diseases. One of the problems encountered in such studies is diagnosis. How large does a defect have to be before it is recorded as enamel hypoplasia? The smaller defects would not normally be detected clinically and so presumably do not appear in the results, but the author's experience of many crown surfaces has suggested that they are part of a continuous range of defects. It therefore seems logical to suppose that they too are the result of systemic disturbances to growth of a similar kind, but of a shorter duration.

Perikyma Furrow Counts

If the brown striae of Retzius and thus the perikymata represent a regular growth rhythm, then counts of perikyma furrows within and between surface defects allow those defects to be fitted into the growth sequence of the crown surface with some precision. On the second permanent molar profile shown in Pl.7,2 and Fig. 1, there are 106 perikyma furrows from cusp tip to neck. The first of these is probably very near to the first perikyma ridge formed and, even if the very slight wear on this particular tooth crown has obscured the relationship, the wide spacing of perikyma furrows in this region of the crown means that very little error would be involved in assuming that furrow number 1 represents the start of crown surface formation. A defect involving 22 perikyma furrows is visible on the crown side. It is of a size detectable clinically and had been diagnosed as enamel hypoplasia even before a replica was made of the tooth. The defect contains 21% of the perikyma furrows in the profile, and so could be said to represent 21% of the time taken to form the crown surface. The perikyma furrow nearest to the start of the defect is the 57th from the cusp tip and therefore the defect could be said to have been initiated only after 54% of the total time taken to form the crown surface had elapsed. Similarly, 27 perikyma furrows follow on after the last furrow of the defect, implying that 25% of total crown formation time had still to elapse before the crown was completed.

In addition to these calculations, it is possible to extend Schour's tree-ring analogy (above) by matching defect/perikyma furrow sequences between different teeth from one individual's dentition in the same kind of way that dendrochronology matches tree-ring sequences between different trees. The tooth surfaces shown in Pl.7,2 and Figs. 1 and 2 cannot be used to demonstrate this, because the second molar defect appears to have been formed at an age when the canine had already been completed. A non-Egyptian example is given in Pl.8,1-2. The lower first molar in this pair of photographs shows a pitted defect on the cusp tip, with a narrow smooth band of normal crown surface below it, followed by another pronounced defect, a wider band of normal crown and finally a less pronounced groove-like defect. The lower second incisor from the same individual shows a similar sequence of defects. Counts of perikyma furrows within and between defects match well between the two tooth crowns (Fig. 3). This is an exceptionally clear example, demonstrating that all tooth crowns being formed at the time of a growth disturbance can potentially show a defect. The difficulty with such matches is that one particular growth disturbance may be represented by different types and levels of defect in the different crowns of one individual dentition. This is particularly true in comparisons between molars and incisors and is another reason why Pl.8,1-2 is an abnormally clear example.

In spite of such difficulties, which can largely be surmounted by experience and careful selection of specimens, there is considerable potential for the technique. Where counts of perikyma furrows within and between defects match closely, it is possible to suggest with some confidence that isolated teeth come from one individual. The author has found this to be of use in multiple fragmentary burials where the number of individuals involved has been unknown. Where several teeth of different class can be matched for one individual, there is the further possibility of producing an extended defect-and-perikyma-furrow-count sequence. In this way, the growth sequence is not confined just to the limited period covered by one tooth crown surface, but over a longer period represented by the surfaces of all the crowns in the dentition. Matches between tooth crowns are also important in confirming the sequences produced for each individual crown surface. Indeed, it is never wise to assess a defect on the evidence of one class of tooth crown alone.

Before such matches can take place, it is first necessary to know where to look for them, that is, which zones of the crown surfaces of different teeth are likely to show matching defects. Timing of crown formation varies

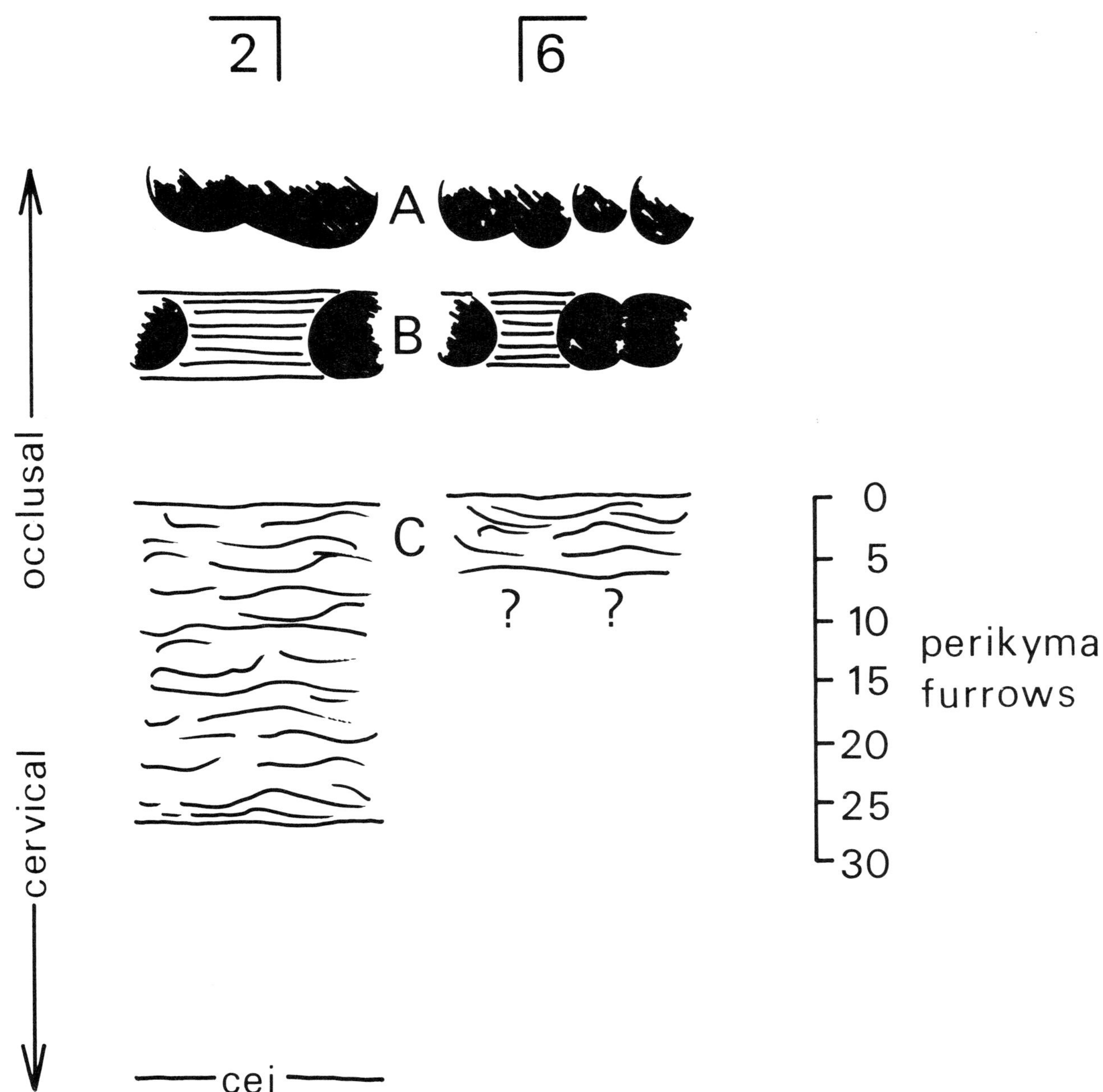

Fig. 3. A diagrammatic representation of enamel hypoplasia defects and perikyma furrow counts from the teeth in Pl. 8, 1-2. The three defects described in the text are labelled as in Pl. 8, 1-2. Groove-like defects are shaded horizontally, with zones of irregular perikymata represented by irregular shading. The cement-enamel junction (cej) is marked for the second incisor but not for the first molar, where the perikymata in the cervical region are too indistinct for counting.

somewhat between individuals, is different between males and females, and varies consistently between populations from different parts of the world. The true timing is, in fact, very difficult to establish because the crowns are growing deep in the bone of the jaws where only a programme of regular radiography can monitor them in living children. Apart from the ethical problems such a study poses, there are real difficulties in relating the radiograph image to histological features of the crown such as brown striae of Retzius and perikyma furrows. These features are produced in the enamel matrix some time before it mineralises enough to be detectable by radiograph. Add to all this the fact that crown surfaces do not grow linearly, but extend much faster in the earlier stages than the later, and it becomes clear that it would be difficult to construct a reliable table of age versus crown surface formed for all teeth. In view of this, the author now follows an approach that was first suggested by Logan and Kronfeld (1933). This is to use observations of defects in many individuals to build up a table of the most common matches, without considering the elements of time and age. Logan and Kronfeld produced a table of this type (their Table 3), which the author has extended (Table 1) using observations on several hundred Egyptian and Nubian dentitions (Hillson 1979). It is used for preliminary assessments of dentitions, before a microscopic examination is carried out.

Perikyma Furrow Counts and Time

The regular growth rhythm represented by brown striae of Retzius and perikymata is now thought to be relatively constant between different individuals. Dean (1987; pers. comm.) summarised a number of studies which have suggested that, on average, the enamel enclosed between each pair of neighbouring brown striae of Retzius represents almost exactly 8 days of crown growth. There is a small amount of variation around this figure, but the great majority of people do not seem to depart from it. If the brown striae are 8 days apart, then so are the perikyma furrows and it becomes possible to apply a chronology to the defect sequences described above. For example, in the second molar shown in Pl.7,2 and Fig. 1, the large groove-like defect could be said to have occurred some 16 months after the crown surface first started forming at the cusp tip (57 perikyma furrows multiplied by 8 days each). By the same means, the disruption causing the defect itself could be said to have lasted about 6 months, and it can be estimated that a further 7 or 8 months was required to complete the crown afterwards.

How reliable are such estimates? They are extremely difficult to test as the tooth germs grow deep in childrens' jaws. A comparison can, however, be made between total brown striae of Retzius and perikyma furrow counts and the figures for crown formation timing obtained by radiographic studies. As outlined above, the comparability of these figures is open to question, but they represent the only chance of an independent test and one would not expect the discrepancy to amount to more than a month or two in any case.

In the second molar from Pl.7,2 and Fig. 1, 106 perikyma furrows were counted down the crown surface from cusp tip to neck. Counts in a number of second molar microscope sections (Bullion 1987) suggest that on average some 70 brown striae of Retzius are hidden under the cusps and do not register as perikyma grooves on the surface. The number of such dome-like layers seems to vary relatively little within each class of tooth crown. This makes it possible to estimate a total of 176 brown striae to form the crown of the second molar from the very first dome-like layers under its cusps to the last sleeve-like layer at its neck. Multiplying 176 up by 8 days yields an estiamte of 1408 days, or just over 50 months for the duration of crown formation. Gustafson and Koch (1974), which tabulates crown formation

TABLE 1. DEFECT MATCHES IN PERMANENT TOOTH CROWN SURFACES.

	I	II	III	IV	V	VI	VII	VIII	IX
1st incisors		<	*	*	>				
Upper 2nd incisors					<*	*>			
Lower 2nd incisors			<	*	*	>			
Canines				<	*	*	>		
1st & 2nd premolars					<	*	*	>	
1st molars	<	*	*	>					
2nd molars					<	*	*	>	
3rd molars									<*>

Key to symbols:

< = the occlusal or incisal part of the crown, just under the cusps.

* = in the middle part of the crown profile.

> = near the cervical, or neck, part of the crown.

The nine matching categories (I to IX) are in order of tooth crown development, but do not relate to chronological age and represent very unequal periods of time. In addition, they do not represent an equal division of the crown surface of any one tooth. Instead, they merely give a rough indication of where a defect on one crown is likely to be matched by another. For example, a defect near the incisal edge of an upper second incisor could be expected to match with corresponding defects near the neck of the neighbouring upper first incisor, somewhere towards the middle of the canine crown and around the cusps of premolars and second molars.

The original idea for the table comes from Logan and Kronfeld (1933), but the table has been extended and the numbering of categories differs from theirs.

timing from a large number of radiographic studies, suggests that permanent lower second molars start to form between 2.5 and 3 years, and are completed sometime between 6.25 and 8 years. The shortest duration that these figures could represent is 3.25 years (39 months) and the longest is 5.5 years (66 months). The estimate of 50 months for the tooth in Pl. 7,2 and Fig. 1 fits comfortably within this limit. Other studies, this time on incisor teeth, have shown similar agreement (Bromage and Dean, 1985; Hillson, in press). The relationship has been good enough to allow attempts to determine age at death from perikyma furrow counts in the remains of children whose teeth were still being formed at the point of death (Bromage and Dean 1985; Dean et al. 1986; Molleson and Cox, in press).

The same principles can be applied to estimating the age at which defect-causing growth disruptions occurred. If the defect in the second molar of Fig. 1 started 57 perikyma furrows from the cusp tip, then this can be added to the average 70 brown striae of Retzius hidden under the cusps (above) to produce a total count of 127, which would represent 1016 days or just over 36 months. If the permanent lower second molar starts to form between 2.5 and 3 years (Gustafson and Koch 1974; above), or 30-36 months after birth, then the disruption causing the defect can be estimated to have started at 66-72 months or 5.5-6 years of age. The disruption seems to have been continuous in its effects for another 22 perikyma furrows and the defect can thus be estimated to have finished at 73-79 months or 6-6.5 years of age.

Several assumptions underlie such estimates, but they seem at present to be reasonable assumptions. Archaeological specimens do offer the possibility (as yet untried) of checking the results if a permanent first molar crown can be sectioned. Of all the permanent tooth crowns, this is the earliest to start forming and in most individuals the first enamel is deposited around the time of birth. Clear brown striae of Retzius are only formed in the enamel after birth and the first brown stria, representing the point of birth itself, is often very prominent - the *neonatal line* (Schour 1936). The neonatal line can normally be detected under at least one of the four main cusps of the first molar and, even if this cannot be done, very little error is involved in assuming that the tooth started to form at birth. If a replica of the crown surface has been made before sectioning, it is usually possible to find a prominent feature of the perikymata that can be recognised at the edge of the enamel in the section. Brown striae could be counted from the neonatal line to this feature, giving a firm starting point to the sequence of perikyma furrows and defects.

Conclusions

Histological studies of the crown surface offer a number of possibilities for studying dental material from archaeological sites. They can act like fingerprints and help to identify the teeth belonging to particular individuals in mixed and fragmentary burials. They have already been used to arrive at precise estimates of age-at-death in the remains of children whose tooth crowns were still in formation at the point of death. Finally, they can be used to estimate with some precision the timing of growth disruptions during childhood, which cause the common and widely studied crown surface defects known as enamel hypoplasia. Such studies add detail to the information which can be extracted from human remains in archaeology and have the particular advantage that they can largely be carried out on replicas which are made rapidly and non-destructively. Collections remote from the main laboratory can thus be visited briefly, impressions can be taken for large collections in field conditions, these can provide a record of material which must be reburied after examination, and the light and tough epoxy replicas can conveniently be sent through the post for others to work upon.

References

Beynon, A D, 1987. 'Replication technique for studying microstructure in fossil enamel', *Scanning Microscopy* 1, 663-669.

Bromage, T G and Dean, M C, 1985. 'Re-evaluation of the age at death of Plio-Pleistocene fossil hominids', *Nature* 317, 525-528.

Bullion, S K, 1987. *Incremental structures of enamel and their applications to archaeology*, unpublished PhD thesis, University of Lancaster.

Dean, M C, 1987. 'Growth layers and incremental markings in hard tissues; a review of the literature and some preliminary observations about enamel structure in *Paranthropus boisei*', *Journal of Human Evolution* 16, 157-172.

Dean, M C, Stringer, C B and Bromage, T G, 1987. 'A new age at death for the Neanderthal child from Devil's Tower, Gibraltar and the implications for studies of general growth and development in Neanderthals', *American Journal of Physical Anthropology* 70, 301-309.

Gustafson, G and Koch, G, 1974. 'Age estimation up to 16 years of age based on dental development', *Odontologisk Revy* 25, 297-306.

Hillson, S W, 1979. 'Diet and dental disease', *World Archaeology* 11, 147-162.

Hillson, S W, 1990. *Teeth*, 1st paperback edn (revised), Cambridge University Press.

Hillson, S W, in press. 'Studies of growth in dental tissues' in Lukacs, J R (ed.), *Culture, Ecology and Dental Anthropology* (*Journal of Human Ecology Special Issue*).

Logan, W H G and Kronfeld, R, 1933. 'Development of the human jaws and surrounding structures from birth to the age of fifteen years', *Journal of the American Dental Association* 20, 379-427.

Massler, M, Schour, I and Poncher, H, 1941. 'Developmental pattern of the child as reflected in the calcification pattern of the teeth', *American Journal of Diseases of Children* 62, 33-67.

Molleson, T I and Cox, M, in press. *Spitalfields: the middling sort*, Council for British Archaeology, London.

Retzius, A, 1837. 'Bemerkungen über den inneren Bau der Zähne', *Archiv. Anat. Physiol.* 8, 486-566.

Risnes, S, 1984. 'Rationale for consistency in the use of enamel surface terms: perikymata and imbrications', *Scandinavian Journal of Dental Research* 92, 1-5.

Risnes, S, 1985a. 'A scanning electron microscope study of the three dimensional extent of Retzius lines in human dental enamel', *Scandinavian Journal of Dental Research* 93, 145-152.

Risnes, S, 1985b. 'Circumferential continuity of perikymata in human dental enamel investigated by scanning electron microscopy', *Scandinavian Journal of Dental Research* 93, 185-191.

A MOLECULAR APPROACH TO THE STUDY OF EGYPTIAN HISTORY

Svante Pääbo and Anna Di Rienzo

An interest common to molecular anthropologists and to many archaeologists is the understanding of the history of human populations. However, the approach to this common goal is very different in the two disciplines. Molecular anthropologists study modern populations and accumulate genetic data, which can be used to relate present-day populations to each other and also to infer the historic processes that have resulted in the genetic variability that is observed. However, these inferences are based on assumptions about the accumulation of genetic changes that may or may not be true. Archaeologists and physical anthropologists, on the other hand, are in the enviable situation of being able to study directly the ancient populations that once were present in an area. However, the archaeological data leave the student of population history in the constant dilemma of distinguishing between the two alternatives of acculturation and population replacement. A case in point is the population history of Nubia. When Reisner discovered and first described the X-group he interpreted it as a new people that moved into Lower Nubia. Since then, many investigators have reinterpreted the X-group as a continuity from the Meroitic Kingdom. Similarly, physical anthropologists have come to different conclusions concerning the population history of Nubia. Whereas continuity has been claimed between Late Pleistocene populations on the one hand and Meroitic and later Nubian populations on the other (Carlson and van Gerven 1977), this is contradicted by other investigators (Turner and Markowitz 1990).

Thus, there is a fundamental dichotomy in the types of data that are produced by archaeologists and geneticists, where the former study the ancient populations directly and the latter infer the history from data collected in modern populations. However, both have problems inherent in their reconstruction of history. It is our belief that some of these problems in the future may be relieved and that the two disciplines will be able to interact more fruitfully than in the past. The reason is recent technical developments that allow molecular genetic data to be accumulated on a large scale from present populations as well as retrieved from archaeological finds.

The genetic system that has been most successfully exploited in the molecular study of human evolution is the DNA molecule that exists in the mitochondria of most cells of our bodies. The mitochondrial DNA is uniquely suited for anthropological studies, since it evolves much more rapidly than nuclear genes and, therefore, accumulates changes fast enough to allow differences to be observed over the short time period during which modern humans have differentiated. Also, mitochondrial DNA is maternally inherited, i.e. all the mitochondrial DNA molecules in an individual appear to stem from the mother of that individual. Therefore, only one type of mitochondrial DNA exists in a person and no recombination between different types of molecules will be able to confuse the results. In consequence, the evolutionary history of the mitochondrial DNA molecule can be reconstructed by comparing the nucleotide sequences of the fastest evolving part of the genome, the so-called control region. Such sequences from different individuals are aligned with each other; the differences are scored and used to construct a phylogenetic tree. Such a tree represents an estimate of the evolutionary history of the molecules. It reconstructs the structure of the common ancestors of the present-day molecules, generally under the assumption of maximum parsimony, i.e. the differences observed are explained by the minimum number of evolutionary changes. It is worth noting that this way of reconstructing history is fundamentally different from the gene frequency approach common in physical and molecular anthropology. For both approaches the occurrence of several genetic markers is determined in each population under study and used to infer relationships among them. However, while the tree-building approach reconstructs the histories of populations by inferring the history of each single lineage in the population, the gene frequency approach treats each population as a unit and summarises all the information for a population in a parameter that is the gene frequency. Thus, while in the tree-building approach the unit of study is the lineage, in the gene frequency approach the unit of study is the entire population. When it comes to studying ancient populations, the tree-building approach is superior to the gene frequency approach, since it is clearly difficult or impossible to obtain a representative population sample of an ancient population.

The first demonstration that DNA is preserved in ancient human remains and can be retrieved and replicated in bacteria (Pääbo 1985) sparked a lot of enthusiasm but was followed by relatively little further progress. The reason is to be found in the vast amount of damage that occurs in the old DNA molecules. Many processes, some acting rapidly after death in the decaying tissues, others acting slowly over the millennia, contribute to erasing the information present in the DNA. This causes the short pieces that can be replicated in bacteria to be so few that very little information can be obtained. This quite depressing situation has changed during the past four years with the invention of a technique that is able to multiply the few intact copies that are present in the extract from an ancient tissue. Thus, the intact copies of a mitochondrial gene segment that one wishes to study can be specifically amplified and sequenced. This technique, the polymerase chain reaction (abbreviated PCR), can retrieve mitochondrial DNA sequences from archaeological remains of soft tissues that are several thousands of years old (Pääbo et al. 1987). There is also recent evidence that this may work for bones as well as for soft tissues (Hagelberg et al. 1989; Hänni et al. 1990; Horai et al. 1990; Thuesen and Engberg 1990; Hedges and Sykes in this volume). Thus, PCR has created an entirely new field of 'molecular archaeology' (Pääbo et al. 1989). In addition, it is revolutionising molecular genetics and will in the near future contribute substantially to our understanding of the population history of humans. The reason for this is not only the fact that old sequences can be retrieved but also, and equally or more importantly, that the PCR and sequence determination of PCR products are highly amenable to automation. Therefore, DNA sequences can be generated on a large scale from populations. This allows a reconstruction of the history of the populations with the highest resolution attainable for molecular data (Wilson et al. 1989).

The vast sets of DNA sequences that will be generated for modern populations will be the starting-points for the formulation of hypotheses that can then be directly tested by diachronic studies, where ancient populations are investigated for the same DNA regions. This will particularly be the case in areas such as the Nile Valley, where copious amounts of ancient humans are preserved. Such studies, that follow the molecular evolutionary history of a population by going back in time, can provide information on a number of questions which are only indirectly approachable by studies that are limited to modern populations. Examples of such questions are: (1) Different modern subpopulations, whether geographically, culturally or socially defined, can be related to previous populations that are retrieved archaeologically. (2) The demographic history of the population can be directly assessed in terms of phenomena such as expansions and contractions in size (Di Rienzo and Wilson 1991). (3) If a population can be studied before and after a migration event, a quantitative estimate of the gene-flow caused by the migration can be achieved.

The first studies involving the sequencing of mitochondrial DNA from modern populations have been conducted in Sub-Saharan Africa (Vigilant et al. 1989), Japan (Horai and Hayasaka 1990), Sardinia (Di Rienzo et al. 1989), and Western North America (Ward et al. 1991). In Egypt, an extensive survey of the mtDNA variability in the Nile Delta population is in progress and preliminary results (Di Rienzo and Wilson in preparation) allow one to formulate hypotheses on the origin and history of this population. For example, one interesting finding is that a small subset of modern Egyptian mitochondrial DNA lineages are closely related to Sub-Saharan African lineages. Fig. 1 shows the phylogenetic relationships among such lineages. Two different patterns are observed. In one case (Fig. 1A) two Egyptian lineages branch off from different African lineages. A reconstruction of migration events, indicated by arrows, suggests that the two Egyptian lineages originated independently from an ancestral African population. The application of a time scale to the tree could allow an estimation of the time interval when these migrations occurred. In the case of the diagram in Fig. 1A, one migration has occurred in the interval between t2 in the past and the present, while the other migration took place in the interval between t3 in the past and the present. In the case depicted in Fig. 1B, two Egyptians are also closely related to Sub-Saharan lineages. However, in this case the two Egyptian lineages are each other's closest relatives. This allows one to explain the occurrence of these lineages in Egypt with only one migration event involving an ancestor of the two Egyptian lineages (arrow). In this case, an older (t3) as well as a younger time limit (t1) in the past can be inferred for the migration. We envision that when we gain a better knowledge of the mitochondrial DNA variability of the Egyptian population, we will be able to arrive at a quantitative estimate of the numbers of such migration events that have occurred as well as of their timing. These inferences can then be tested by going back in time to mummies and skeletal remains. An illustration of the feasibility of this approach is provided by the mummy of Nekht-Ankh, a priest of the Middle Kingdom. Short mitochondrial DNA sequences (45 and 81 nucleotides long) have been determined by PCR and direct sequencing from the remains of his liver found in a canopic jar (Pääbo, 1989). When this sequence is compared to the sequences determined from the Delta population, it is found that it is identical to four of the modern Egyptian mitochondrial lineages. However, in order to achieve comparisons that are statistically meaningful, sequences that are as long as the modern ones (approximately 400 nucleotides) will have to be determined from the ancient populations. This is a laborious task because the ancient DNA is degraded to such an extent that a 400 nucleotide sequence needs to be determined in approximately eight shorter, overlapping pieces. However, we believe that this work is well worth the effort, since it will give us the first molecular view ever of an ancient human population and of its modern descendants.

This fascinating task will require close cooperation between molecular geneticists and archaeologists. Particularly, it is necessary for the field archaeologist and museum curators to be aware of the greatest problem that mars the use of PCR to study ancient human remains - i.e. contamination by contemporary human DNA. This may stem from minute fragments of skin or droplets of sputum deposited on the specimens. Since the contaminating DNA is relatively undamaged, even extremely small amounts of mitochondrial DNA will be easily amplified by PCR and can be mistaken for ancient sequences. It is therefore necessary to excavate and store the material in a way that minimises the handling of the specimens by humans and to use plastic gloves when touching specimens (see e.g. Pääbo 1990). If bodies are reburied, several samples of one or a few grams of soft tissues as well as bones should be removed from different parts of each body using clean surgical instruments and should be stored in clean containers under dry conditions.

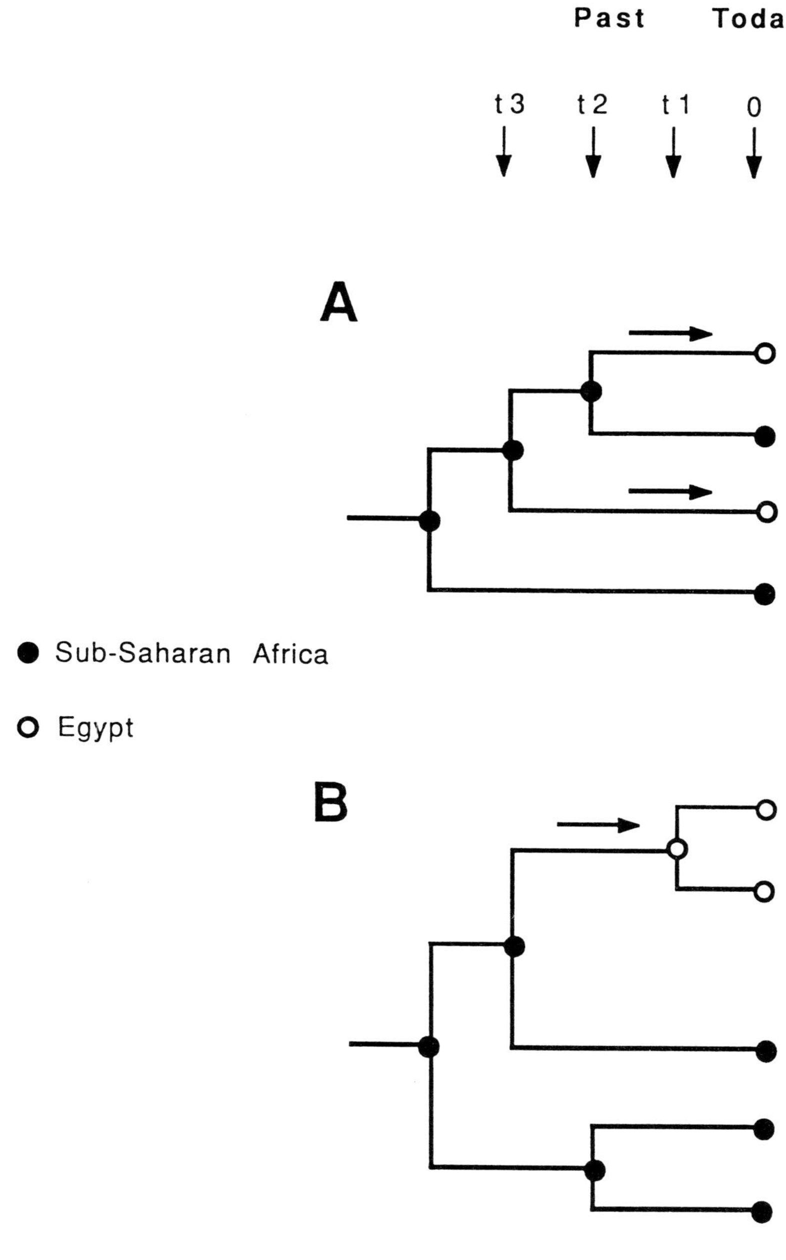

Fig. 1. Two reconstructions of partial phylogenetic trees relating mitochondrial lineages present in the Nile Delta (open circles) to Sub-Saharan (filled circles) lineages. The trees are estimated using the maximum parsimony criterion that infers past evolutionary events (illustrated by the scale below) and reconstructs the DNA sequences of putative ancestral mitochondrial types that can be either Egyptian or Sub-Saharan. Migration events (arrows) are inferred to explain the occurrence of these lineages in Egypt. Above, various time points with relevance to the migration events (see text) are indicated.

Obviously, remains of domestic as well as wild animals are as valuable as human remains and should also be collected. Such a collaborative strategy involving molecular biologists and archaeologists will eventually shed new light on the population history of the Nile Valley as well as other regions of the world.

References

Carlson, D S and van Gerven, D P, 1977. *Am. J. Phys. Anthrop.* 46, 495-506.

Di Rienzo, A and Wilson, A C, 1991. *Proc. Natl. Acad. Sci. USA* 88, 1597-1601.

Hagelberg, E, Sykes, B and Hedges, R, 1989. *Nature* 342, 485.

Hänni, C, Laudet, V, Sakka, M, Bègue, A and Stéhelin, D, 1990. *C R Acad. Sci. Paris*, t. 310, Série III, 365-370.

Horai, S and Hayasaka, K, 1990. *Am. J. Hum. Genet.* 46, 828-842.

Horai, S, Hayasaka, K, Murayama, K, Wate, N, Koike, H and Nakai, N, 1990. *Proc. Japan Acad.* 65, 229-233, ser B.

Pääbo, S, 1985. *Nature* 314, 644-645.

Pääbo, S, Gifford, J A and Wilson, A C, 1988. *Nucl. Acids Res.* 16(20), 9775-9787.

Pääbo, S, 1989. *Proc. Natl. Acad. Sci. USA* 86, 1939-1943.

Pääbo, S, Higuchi, R G and Wilson, A C, 1989. *J. Biol. Chem.* 264, 9709-9712.

Pääbo, S, 1990. In *PCR-Protocols and Applications - A Laboratory Manual*, Innis, M A, Gelfand, D H, Sninsky, J J and White, T J (eds.). Academic Press, San Diego.

Theusen, I and Engberg, J, 1990. *J. Archaeol. Sci.* 17, 679-689.

Turner II, C G and Markowitz, M A, 1990. *Homo* 41, 32-41.

Wilson, A C, Zimmer, E A, Prager, E M and Kocher, T A, 1989. In *The Hierarchy of Life*, Fernholm, B, Bremer, K and Jörnwall, H (eds.). Nobel Symposia 70.

Ward, R H, Frazier, B S, Dew, K and Pääbo, S, 1991. *Proc. Natl. Acad. Sci. USA* 88.

ANALYSIS OF RETROVIRUS SEQUENCES IN *c.* 5300 YEARS OLD EGYPTIAN MUMMY DNA OBTAINED VIA PCR AMPLIFICATION AND MOLECULAR CLONING

Jaap Goudsmit, John Dekker, Lia Smit,
Carla Kuiken, Jan Geelen and Rutger Perizonius

1. Introduction

1.1 Origin of Genomic Information

Several lines of evidence indicate that the origin of genetic information, the basis for life, is in RNA and not in DNA (Eigen et al. 1981). Recently Eigen and co-workers presented convincing evidence that the genetic code is not older than, but almost as old as, our planet (Eigen et al. 1989). Although RNA-dependent polymerisation processes encoded by host cell genes are rare among contemporary higher organisms, this may not have been so in the primordial DNA world. Sequelae of this enzymatic pathway can still be found. Two main types of enzymes perform such an RNA-dependent task: *the RNA-dependent RNA polymerases* (Poch et al. 1989, 1990), involved in the replication of plus-, minus- and double stranded RNA viruses, and *the RNA-dependent DNA polymerases*, involved in the replication of retroid elements. RNA-dependent DNA polymerase catalyses DNA replication from an RNA template and is, therefore, called reverse transcriptase (RT). RT was independently discovered by Baltimore (Baltimore 1970) and Temin and Mizutani (Temin and Mizutani 1970). Genetic elements with similarity in sequence to RT have been discovered in a wide variety of organisms. RTs are present in retroviruses such as the AIDS virus, and in DNA viruses like the hepadna viruses that cause hepatitis B in humans. RT-like sequences are found in so-called transposable elements, first discovered in yeast and Drosophila melanogaster, in fungal mitochondrial introns and in a mitochondrial plasmid. Xiong and Eickbush (1990) argue that retroposons are the most likely progenitor of all current retroid elements as they come with and without long terminal repeats (LTRs). Because neither RNA viruses (except the retroviruses) nor hepadna viruses contain LTRs, they assume that the progenitor element did not contain LTRs. Both LTR containing and non-LTR containing retroposons are found in animals, plants and either protozoans or fungi. In contrast, viral RTs are confined to particular taxa: hepadna viruses and retroviruses to vertebrates and caulimo viruses to plants. Xiong and Eickbush explain the current distribution of retroid elements by horizontal transfer of retroposons across major taxonomic groups of organisms. New viruses might have evolved either by the capture of RT sequences from retroposons by pre-existing viruses or by transposons acquiring additional genes and becoming a virus.

1.2 Endogenous Versus Exogenous Retroviruses

Retroviruses can be characterised as *endogenous* and *exogenous*. Exogenous retroviruses are transmitted horizontally, i.e. by infection, and are continuously replicating to produce copies of themselves. They replicate only in cells bearing a receptor for the viral envelope. Because an extra selective pressure is exerted on the virus by the host immune system, the mutation rate of the virus is extremely high. Endogenous retroviruses are transmitted in the genome in a Mendelian fashion. Generally, all members of a species carry multiple copies of an endogenous retrovirus in the genome. Based on phylogenetic tree analysis of RT sequences there is no clear dichotomy between endogenous and exogenous retroviruses (Doolittle et al. 1989; Xiong and Eickbush 1990).

1.3 Human Endogenous Retroviruses

Two major families of human endogenous retroviruses have been described (Xiong and Eickbush 1990). One family is designated Human Endogenous Retrovirus type C (HERV-C or HERV-E; clone 4-1; HURRS-E), most closely related to the retrovirus family of Baboon Endogenous Retrovirus (BAEV) and Gibbon Ape Leukemia Virus (GALV). The other family is designated Human Endogenous Retrovirus type K (HERV-K), most closely related to the retrovirus family of Squirrel Monkey Retrovirus (SMRV-H), Mason-Pfizer Monkey Virus (MPMV) and the Simian Retroviruses types 1 and 2 (SRV-1 and SRV-2).

1.4 Molecular Archaeology

The increasing possibilities of investigating DNA from ancient human, animal and vegetable remains, hold great promise for the future of palaeoanthropological, archaeozoological, archaeobotanical and molecular palaeontological research (Pääbo 1987; Perizonius and Goudsmit 1989). Up till now DNA has been extracted from a wide variety of remains ranging in age up to 45,000 years (Pääbo et al. 1989). Success has been reported with dry, frozen or preserved remains (Pääbo et al. 1989) as well as with both soft and bony tissue (Hagelberg et al. 1989). Although cloning and sequencing of such DNA is possible, the DNA is heavily modified, making cloning inefficient and error prone. The size is reduced to a few hundred base pairs (bp) at the most and the DNA itself is mutilated, showing baseless sites, oxidized pyrimidines and cross-links. Pääbo and co-workers (1989) expect only 1% of the DNA to be undamaged. Venanzie and co-workers (1990) suggest that this is not the reason for inefficient analysis of ancient DNA but is caused by the fact that the majority of the preserved nucleic acids are RNA. This notion, however, is still not confirmed.

The use of the PCR has eliminated some of the problems caused by the low cloning efficiency and cloning artefacts arising from modifications in ancient DNA. Pääbo and co-workers (1989) argue that the PCR is the ideal tool to amplify a small number of intact ancient DNA molecules present in a vast excess of damaged molecules. The PCR has the advantage of being able to detect minute amounts of a particular DNA stretch and has no capacity for repair or misrepair, being an *in vitro* system.

PCR amplification has been performed on several ancient DNAs (Pääbo et al. 1989). Mitochondrial DNA fragments 140bp long have been amplified from DNA isolated from a seven thousand year old brain (Pääbo et al. 1988). Longer fragments could not be amplified. This appears to be a general rule: 150bp long fragments can regularly be obtained, up to 500bp long fragments rarely and longer fragments have not been reported thus far. Virus genes have not been amplified from any ancient DNA to date.

1.5 The Present Study

In this paper we describe the comparison of HERV-C sequences in ancient and contemporary human DNA. The discovery of HERV-C is only a few years old. Originally, Martin (Martin et al. 1981) isolated clones from an African green monkey genetic library that showed hybridisation with a murine retrovirus probe. Subsequently the presence of homologous sequences in human DNA was demonstrated (Repaske et al. 1983). The full length HERV-C clone 4-1 was sequenced in 1985 (Repaske et al. 1985). This 8.8 Kb clone contained nine stop codons, indicating that no functional retrovirus is encoded by this sequence. All retroviral RTs contain two highly conserved coding regions: LPQ (R,G) and Y (V, M) DD about 30 amino acids apart (Mach 1988; Shih et al. 1989). Our experiments were focused on this RT region.

2. Materials and Methods

The ancient Egyptian mummified skin (BM EA 57353) used in the present experiments was kindly offered to us by the Department of Egyptian Antiquities (W V Davies, Keeper) of the British Museum. The skin sample belongs to the body of a male lying in a flexed position which was found in Gebelein, Egypt, and dates from the late predynastic period (approximately 3300 BC). This naturally mummified body is still present in the Egyptian Department and is described in the *Catalogue of Egyptian Antiquities in the British Museum* (Dawson and Gray 1968). The skin originates from the area covering the right tibia and the sample was taken on 20 July 1989 by Jeffrey Spencer (Egyptian Antiquities, British Museum) and one of the authors (Rutger Perizonius). The choice of this particular mummy, EA 57353, was based not only on the fact that it was a natural mummy (i.e. no ancient mummification techniques were applied) but also on the fact that Svante Pääbo had previously been able to extract DNA from samples of it (Pääbo 1987). As control, we used contemporary human T-cell line Molt-3.

DNA from the ancient tissue was isolated by a combination of the methods described by Pääbo and co-workers (Pääbo 1989; Pääbo et al. 1988) and by Boom and co-workers (1990). In short, the Pääbo extraction was performed on approximately 1g of mummified skin which was cut into small pieces. Subsequently, an extraction according to Boom and co-workers (1990) was performed on 200µl supernatant obtained by the Pääbo isolation.

The PCR was performed on 10µl of the ancient DNA in a 100µl reaction mixture containing 10pmol of the primers TA GGT ACC CAG CTT CCC CAA AGG TTC AAG (5'Asp718) and AT GGA TCC AAG GTC ATC AAC GTA CTG GAG (3'BamH1), 50mM KCl, 10mM Tris (pH=8.3), 0.01% Gelatin, 2.2mM $MgCl_2$, 125µM of each dNTP and 12.5 U Taq polymerase (Cetus, CA, USA) (Repaske et al. 1985). The PCR schedule consisted of an initial denaturation of 5 minutes at 95°C, followed by 40 cycles of 1 minute at 95°C, 1 minute at 37°C, 2 minutes at 72°C and a final extension of 10 minutes at 72°C in a thermal cycler (Perkin Elmer, CA, USA).

Amplified fragments were electrophoresed on a preparative low-melting agarose gel and purified on Spin-X columns (Costar, Cambridge, MA, USA), followed by a phenol/chloroform extraction and ethanol precipitation. Subsequently, the fragments were digested with BamH1 and Asp718 in a total volume of 150µl. Digested fragments were purified as described by Boom and co-workers (1990), cloned into a BamH1 and Asp718 digested plasmid pGEM7 (Promega Biotec, Madison, WI, USA) and transformed into *Escherichia coli* strain HB101.

Plasmid DNA from 50ml cultures was extracted. Double stranded sequencing was performed using the Sequenase 2 kit (United States Biochemical, Cleveland, OH, USA) with $[^{35}S]$ dATP following the manufacturers' recommendations. DNA from the Molt-3 cell line was isolated directly as described by Boom and co-workers (1990).

3. Results

After the DNA extraction procedure using approximately one gram of naturally mummified tissue, only a faint smear of very small (<50bp) DNA was visible on the gel. After PCR, two bands of 120-150bp and 100-120bp were visible on gel, while in contemporary DNA (Molt-3 cells) only one band of approximately 120bp was detected. Molecular cloning of the amplification products of ancient DNA resulted in three clones of the small size (100-120bp) fragments (clones 38-1, 38-2 and 38-3). Based on PvuII restriction enzyme analysis, the three clones of the small size fragment were positive (clones 38-1, 38-2 and 38-3). The sequences of these ancient DNAs were compared to four clones (11-1, 11-7, 19-3 and 19-4) amplified from contemporary DNA, using the same experimental conditions as for ancient DNA. In addition the sequen-

```
HERV-C    TGGACCCAGC.TTCCCCAAAGGTTCAAGAACTCCCCCACCAT..CTTTGGGGAGGCGTTGGCTCG
38-1      ......----.----------------------------T--..---C------A-A---C----
38-2      ......----G-------------------------------TC---C-A---AA-A----T---
38-3      ......----.----------------------------ATA...-----------A------G-
11-1      G-T-------.-------------------------------..---C--------A-G------
11-7      G-T-------.----------------------------T--..---C------A-A---C----
19-3      G-T-------.-------------------------------..---C--------A-A------
19-4      G-T-------.-------------------------------..A-----------A------G-
BAEV      -----T-GA-.-------GG-------A-----T-----TC-..---C-AT-----TC-CCACA-
MMTV      ----AAGTTT.-G-----GG-TA-G--A--TAG---T--TT-..A-G-CAAA-ATTTG---ACAA
HTLV-1    ----AAGTA-.-A------G----T--A--TAGT------C-..G--C-AAAT-CA-C----C-A
HIV-1Z2   -AC-ATGT--.----A--GG-A-GG--AGGA--A--GG-A--..A--CCAAAGTAGCA--A-AAA
MMLV      ------AGA-.-C--A--GG-T-----A---AGT------C-..G----AT-----AC--CACA-
FELV      -----A-GC-.----T--GG-----------AG-------C-..A----AT-----TC--CACTC
HIV-1MN   -AC-ATGT--.----A--GG-A-GG--AGGA--A--AG-A--..A--CCAAAGTAGCA--A-AAA
SRV-1     ----AAGTTT.-A--G---C-TA-GGCC---AG---T---T-..A-GCCAAA-ATATG----CAC
MPMV      ----AGGTTT.-A--A---G-TA-GGCC---AGT--T---T-..A-G-CAAA-ATATG----CAC
HIV-2     -AC-AAGTCT.-A--A--GG-A-GG---GGG--A--AG-A--..T---CAAT-CA-AA--AGG-A
HERV-K    ----AAGT-T.-A--T--GG-AA-GCTT--TAGT--A--T--..T-G-CA-ACTTTTG-A-G---
SIV-MAC   -AT-AGGTT-.-G--T--GG-A-GG---GGG--A--AG----..---CCAAT-CA-TA--AGA-A

HERV-C    AGACCTCCAGAAGTTTCCCACC.AGAGACCTAGGCTGCGTGTTGCTCCAGTACGTTGATGACCTT
38-1      ----------------------.--------G-A-------------------------------
38-2      G---T-------------TG--.--------C---G-----------------------------
38-3      ----------------------.--------------A.-A------------------------
11-1      ------------------T-G-.---------------T--------------------------
11-7      ----------------------.-----------A------------------------------
19-3      ---------------T-T-G-C----------------T--------------------------
19-4      ----------------------.--------------A-A-------------------------
BAEV      G------ACCG-C--C-GG---.CAGC-T-C--AAGTGACCC-----------T--A-------C
MMTV      --CTA-ATT--CTG-AAGGGAT.-A-T---A--A--CATATA-TG-G--T---A-G------A--
HTLV-1    TAT---G---CCCA---GGCAA.GCTTT--CCCAA---ACTA-T--T------A-G------A--
HIV-1Z2   -AT-T-AG--CCC---AGA-AA.CA-A-T-C--AAATA--TA-CTAT--A---A-G-----TT-G
MMLV      ------AGCAG-C--C-GG-T-.CAGC---C--A--TGA-CC----A--------G------T-A
FELV      ------GGCCG-T--CAGGGTA.--GT---CG-CTCTA--CC-C--A--A--T--A--------C
HIV-1MN   -AT-T-AG--CCT---AGA-AA.CA-A-T-C--A-ATA--TA-CTAT--A---A-G-----TT-G
SRV-1     --C-A-A--T--AG--AGACAT.GCCTGGAA-CAAATGTATA-TA-A--T---A-G-----TA-C
MPMV      --C-A-A--T---G--AGACAT.GCCTGGAA-CAAATGTATA-TA-A--T---A-G------A-C
HIV-2     --T-T-AG-ACCA--CAGA-AA.GC-A---C--ATGT-A-TA-CG-T------A-G-----TA-C
HERV-K    --CT--T--ACCAG-GAGAGAA.-AGTTTTC--A---TTATA-TA-T--T--TA-------TA--
SIV-MAC   T-TG--AG-ACCC--CAGG-AG.GC-A-T-C--ATGTGACC--AG-------TA-G------A-C
```

░ insertion or deletion specific to mummy material

Fig. 1. Alignment of *pol* gene sequences encoding a portion of the reverse transcriptase. The sequence of the Human Endogenous Retrovirus type C (HERV-C) is used as reference, because the PCR primers were derived from this sequence. The clones 38-1, 38-2 and 38-3 were obtained from ancient Egyptian human DNA. The clones 11-1, 11-7, 19-3 and 19-4 were obtained from human contemporary DNA. The other sequences were obtained from the genebank database.

Abbreviations

HERV-C.....Human Endogenous Retrovirus-C
BAEV.........Baboon Endogenous Retrovirus
MMTV.......Mouse Mammary Tumor Virus
HTLV.........Human T-cell Leukemia Virus
HIV............Human Immunodeficiency Virus
MMLV......Moloney Mouse Leukemia Virus
FELV.........Feline Leukemia Virus
SRV...........Simian Retrovirus
MPMV.....Mason-Pfizer Monkey Virus
HERV-K..Human Endogenous Retrovirus-K
SIV..........Simian Immunodeficiency Virus

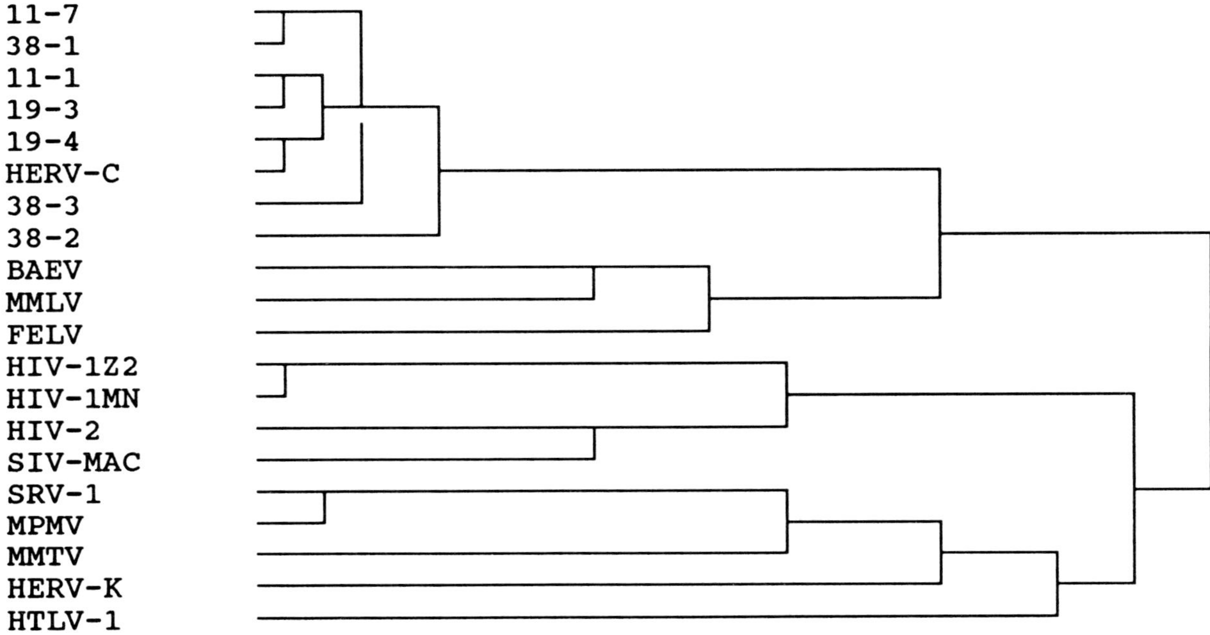

Fig. 2. Dendrogram of the RT sequences in ancient and contemporary human DNA, obtained with the primer set derived from the HERV-C sequence. In addition, the relationship between these new sequences with known RT sequences is indicated.

ces were compared to other members of the retrovirus family (Xiong and Eickbush 1990).

Fig. 1 shows the sequences obtained using either ancient DNA, clones 38-1, 38-2 and 38-3 or contemporary DNA, clones 11-1, 11-7, 19-3 and 19-4. Three of the four clones obtained from contemporary DNA showed only point mutations relative to known RT sequences, leaving the reading frame intact. One clone (19-3) showed an insertion of a C, resulting in a frame shift. Only one clone obtained from ancient DNA (38-1) lacked insertions or deletions, compared to known RT sequences. One clone (38-2) showed two insertions. In the 5' end a G was inserted, causing a frame shift in the reading frame. A subsequent insertion of two nucleotides, TC, restored the original reading frame. The third clone (38-3) showed two deletions: one C at the 5' end and one G at the 3' end of the sequence. After the first deletion a frame shift occurred that was not restored by the second deletion. Subsequently a dendrogram of the aligned nucleotide sequences was composed (Fig. 2). All sequences obtained from contemporary and ancient DNA clustered with the HERV-C sequence. None of the sequences was identical. The HERV-C sequence was most closely related to the 11-1, 19-3 and 19-4 clones, obtained from contemporary DNA. All clones from ancient DNA were more distinct from HERV-C. All these human endogenous retrovirus sequences were only distantly related to the BAEV family, although this was the closest relative.

4. Discussion

The main question remains whether the retroviral sequences obtained from ancient DNA are genuine. Contamination still has to be considered. Some of our data plead against it. The PCR reaction yielded multiple bands in ancient DNA and only a single band in contemporary DNA. However, we intend to extend our studies to the following controls: PCR amplification of longer (250-500bp) HERV-C sequences from both the contemporary and ancient DNA. Pääbo and co-workers (1989) identify several lines of evidence for authenticity of the sequences obtained, among which is the notion that amplification products beyond a size of 150bp for ancient DNA points to contamination with contemporary DNA. In addition, we will repeat this analysis with mtDNA or HERV-K primer pairs. One has to compare independent amplifications of sequences of 500/250 and 120bp in length with a nested or semi-nested approach. Nested PCR might yield longer fragments from heavily damaged DNA, even with the 'jumping' PCR operational. We will also use non-human primate ancient DNA and cat ancient DNA with several primer pairs to obtain species-specific endogenous retrovirus sequences.

Ancient DNA has been shown to be damaged. Pääbo and co-workers (1990) indicate that illegitimate adenosine and thymidine insertions may result from a 'jumping' PCR. However, we saw G, T and C insertions in our template (clone 38-2) as well as a C and G deletion (clone 38-3). In addition, we had seen a C insertion in one of the four clones obtained from contemporary DNA (clone 19-3). Therefore, more experiments are needed to ensure that the insertions and deletions observed in the 38-2 and 38-3 clones are evidence for the clones being chimeric molecules. Although we do not think our 38-1, 38-2 and 38-3 clones from ancient DNA are the result of PCR contamination, firm confirmation awaits the formerly proposed control experiments.

Our preliminary data indicate that endogenous retrovirus sequences that occur in a frequency of approximately 50 copies per cellular genome can be amplified by PCR, molecularly cloned and sequenced. The next step will be to demonstrate species specificity by using monkey and cat ancient DNA. In addition, contamination will be excluded. Subsequently, our goal will be to analyse exogenous retrovirus sequences in the appropriate tissues (spleen etc.) obtained from mummified primates and cats, in order to gain insight into retrovirus evolution.

References

Baltimore D, 1970. 'Viral RNA-dependent DNA polymerase', *Nature* 226, 1209-1211.

Boom R, Sol C J A, Salimans M M M, Jansen C L, Wertheim-van Dillen P M E, Noordaa van der J, 1990. 'A rapid and simple method for purification of nucleic acids', *J. Clin. Microbiol.* 28, 3, 495-503.

Dawson W R, Gray P H K, 1968. *Catalogue of Egyptian Antiquities in the British Museum* I, *Mummies and Human Remains*, London.

Doolittle R F, Feng D-F, Johnson M S, McClure M A, 1989. 'Origins and evolutionary relationships of retroviruses', *Q. Rev. Biol.* 64, 1-30.

Eigen M, Gardiner W, Schuster P, Winkler-Oswatitsch R, 1981. 'The origin of genetic information', *Sc. Am.* 24, 673-679.

Eigen M, Lindemann B F, Tietze M, Winkler-Oswatitsch R, Dress A, Haeseler von A, 1989. 'How old is the genetic code? Statistical geometry of tRNA provides an answer', *Science* 244, 673-679.

Hagelberg E, Sykes B, Hedges R E M, 1989. 'Ancient bone DNA amplified', *Nature* 342, 485.

Mach D H, Sninsky J J, 1988. 'A sensitive method for the identification of uncharacterised viruses related to known virus groups: hepadnavirus model system', *Proc. Natl. Acad. Sci. USA* 85, 6977-6981.

Martin M A, Bryan T, Rasheed S, Khan A S, 1981. 'Identification and cloning of endogenous retroviral sequences present in human DNA', *Proc. Natl. Acad. Sci. USA* 78, 4892-4896.

Pääbo S, 1987. 'Molecular genetic methods in archaeology, a prospect', *Anthrop. Anz.* 45, 9-17.

Pääbo S, 1989. 'Ancient DNA: extraction, characterisation, molecular cloning, and enzymatic amplification', *Proc. Natl. Acad. Sci. USA* 86, 1939-1943.

Pääbo S, Gifford J A, Wilson A C, 1988. 'Mitochondrial DNA sequences from a 7000-year old brain', *Nucleic Acids Res.* 16, 20, 9775-9787.

Pääbo S, Higuchi R G, Wilson A C, 1989. 'Ancient DNA and the Polymerase Chain Reaction, the emerging field of molecular archaeology', *J. Biol. Chem.* 264, 17, 9709-9712.

Pääbo S, Irwin D M, Wilson A C, 1990. 'DNA damage promotes jumping between templates during enzymatic amplification', *J. Biol. Chem.* 265, 8 , 4718-4721.

Perizonius W R K, Goudsmit J, 1989. 'Ancient DNA and archaeovirology: a question of samples', *Göttinger Miszellen*, 110, 47-53.

Poch O, Blumberg B M, Bougueleret L, Tordo N, 1990. 'Sequence comparison of five polymerases (L proteins) of unsegmented negative-strand RNA viruses: theoretical assignment of functional domains', *J. Gen. Virol.* 71, 1153-1162.

Poch O, Sauvaget I, Delarue M, Tordo N, 1989. 'Identification of four conserved motifs among the RNA-dependent polymerase encoding elements', *EMBO J.* 8, 12, 3867-3874.

Repaske R, O'Neill R R, Steele P E, Martin M A, 1983. 'Characterization and partial nucleotide sequence of endogenous type C retrovirus segments in human chromosomal DNA', *Proc. Natl. Acad. Sci. USA* 80, 678-682.

Repaske R, Steele P E, O'Neill R R, Rabson A B, Martin M A, 1985. 'Nucleotide sequence of a full-length human endogenous retroviral segment', *J. Virol.* 54, 764-772.

Shih A, Misra R, Rush M G, 1989. 'Detection of multiple, novel reverse transcriptase coding sequences in human nucleic acids: relation to primate retroviruses', *J. Virol.* 63, 1, 64-75.

Temin H M, Mizutani S, 1970. 'RNA-dependent DNA polymerase in virions of Rous Sarcoma Virus', *Nature* 226, 1211-1213.

Venanzie F M, Rollo F, 1990. 'Mummy RNA lasts longer', *Nature* 343, 25-26.

Xiong Y, Eickbush T H, 1990. 'Origin and evolution of retroelements based upon their reverse transcriptase sequences', *EMBO J.* 9, 3353-3362.

THE EXTRACTION AND ISOLATION OF DNA FROM ARCHAEOLOGICAL BONE

R E M Hedges and B A Sykes

Introduction

This contribution is a condensed version of the paper presented at the Colloquium on Biological Anthropology and the Study of Ancient Egypt. The study and analysis of DNA from archaeological material is still in its very early stages. Work on, and current understanding of, surviving DNA in soft tissue has been admirably summarised by Pääbo (in this volume). As far as bone itself is concerned, we have no further technical developments to add beyond Hagelberg et al. (1989), although some of the implications are also explored in Hedges (in press). Similar work to that described in Hagelberg et al. (1989) has now also been reported elsewhere (Hänni 1990; Horai et al. 1989 and Herrman).

Here we describe the particular features (archaeological and biochemical) of using bone for analysis of surviving genetic information; we summarise the present limited state of understanding of the occurrence and recognition of DNA in ancient bone and we outline some of the foreseeable applications.

Bone as a Repository for Molecular Genetic Information

Although bone undergoes chemical alteration during burial (see, for example, Schwarcz et al. 1989), it survives over a much wider range of environmental conditions, and for a much longer time, than most other biological tissues and is therefore common on very many archaeological sites. For this reason, it is the preferred material with which to conduct genetic studies, particularly at the population level, and with which to make comparisons from one site to another. An additional and important advantage is that the same material also carries phenotypic as well as potential pathological and dietary information. For these reasons, it is worth making a special effort to determine if original, indigenous DNA can survive in archaeological bone.

Numerous studies have been made on other surviving biomolecules in bone. These tend to divide into immunological methods for the detection of genetically polymorphic antigens (for example, the blood group antigens, Lengyel 1975), or more direct methods on the major proteins in bone (collagen being by far the most abundant) (Hedges and Wallace 1978; Ambrose 1990; Weiner and Bar-Yosef 1990). Such work has mainly served to demonstrate the complexity of studying the degradation of proteins in a non-isolated environment. Abundant data exist on the loss of collagen from bone with time (Hedges and Law 1989). These show that environmental parameters, especially temperature and water availability, are especially important influences on survival. Recently, immunochemical work has improved to the point where proteins exhibiting interesting polymorphisms (such as the immunoglobulins) can reasonably clearly be detected (Tuross 1989; Smith and Wilson 1990; Ulrich et al. 1987; Cattaneo et al. 1990). However, there is still much to be done before ancient protein from bone can be analysed in order to obtain useful genetic information. Although DNA is, potentially, a much richer source of genetic information, it could be that protein sequences might offer a technically simpler approach for some applications. In any case, knowledge of the survival of such material is also relevant to the question of survival of DNA. For example, the protein osteocalcin apparently survives particularly well in bone, probably because of its affinity with the hydroxyapatite matrix (Ajie 1990). The same may turn out to be true for DNA fragments. It is worth pointing out that most bones from climates as hot as Egypt generally have lost most of their collagen; whether this is also true of the DNA remains to be seen.

DNA is less abundant in bone than some of the molecules considered above. However, techniques for the recognition and analysis of DNA are now extremely sensitive. Under ideal conditions the latest methods are capable of detecting even single molecules. DNA is present in the nuclei and mitochondria of cells, so the survival of cells and the survival of DNA are to some extent linked. Dense bone contains many different cell types which, for our purposes, can be divided into those which exist on the outside of the calcified matrix and those which are entombed within it. The former category includes endothelial cells lining blood capillaries and the blood cells within them, and osteoblasts and osteoclasts involved in bone synthesis and degradation at active fronts on the bone surface. These cells are less protected than the osteocytes deep inside the calcified matrix whose only contact with the outside world are thin cellular processes tunnelling through to their neighbours. There are of the order of 10^7 osteocytes per gram of modern dense bone and these are likely to be the source of DNA surviving over the very long periods necessary for archaeological investigation. Finally, the contribution of exogenous DNA, from infiltrating fungi and bacteria, and also from contamination from the soil and handling after excavation and in the laboratory, should be borne in mind.

Survival of DNA

The effects of time and the burial environment are likely to lead to a loss of DNA and subsequent leaching of degraded products, so that there is less available for analysis. The remaining DNA will also be modified by breakage and cross-linking, making long sequences difficult or impossible to recover, and by chemical changes to the bases themselves which might corrupt the sequence. The extent and nature of some of these changes have been documented by Päabo (1989). How detrimental to the recovery of genetic information this will be depends on the particular bone, and also on the ability to reconstruct the sequence from a consensus of numerous damaged copies. Crucial genetic information can be obtained from very short valid sequences of only a few base pairs in length and a great deal of work can be done with surviving sequences of 100-500 bp. (To put this in context, the human genome is about 3 x 10^9 bp long; mitochondrial DNA amounts to 16000 bp).

The Technology of DNA Retrieval, Recognition and Analysis

DNA may be extracted from bone by first chemically dissolving the inorganic matrix, then selectively breaking down the protein component and extracting the DNA fragments with phenol/chloroform. The presence of DNA and its molecular weight distribution can be seen using electrophoresis to separate the fragments and ethidium bromide to stain them. Typically a microgram or so of DNA can be extracted from about a gram of bone (which, say, is less than 5000 years old and in reasonable preservation). This DNA is of low molecular weight (e.g. 500 bp lengths) and may or may not be mainly indigenous. Further details, with references, are briefly given in Hagelberg et al. (1989).

DNA recovered from bone can be analysed in a number of ways, all of which have been developed over the past 10-15 years during the revolution in molecular biology. Most simply, indigenous sequences in the bone extract can be detected by their ability to form very specific and stable hybrids with probes made from their modern equivalent which are labelled with the radioactive isotope ^{32}P. Stable hybrids can be detected by auto-radiography using X-ray film. This general approach has not been particularly successful in the context of ancient DNA, possibly because of contamination with non-indigenous material and problems of ensuring specificity with damaged sequences. Damage also makes DNA difficult to clone into bacteria compared to its modern counterpart. The technique which has really made recovery and analysis a realistic possibility is the polymerase chain reaction (PCR). In the simplest terms, this technique mimics in the test-tube the natural duplication process which doubles the DNA content of a cell every time it divides. Consecutive rounds of duplication amplify the starting material without limit. For instance twenty rounds at full efficiency will produce 2^{20} or approximately a million copies. In practice, efficiencies are around 80% but this is not a serious problem - the solution is to do a few more cycles. Usually the sequence to be amplified must be known in advance. The target is defined by synthetic oligonucleotides, about 20 bases long, which are exact matches of the sequence at either end of the stretch to be amplified. Once amplified, its nucleotide sequence can be deduced by standard methods. Since it is the variation in this sequence which will form the basis for genetic comparisons, the targets are picked on the basis of their known variability in modern populations. The obvious point of attack is the mitochondrial genome, which as well as being very polymorphic has the additional advantage of far higher abundance than nuclear sequences. Potential targets in the nuclear genome are at the histocompatability loci (HLA) and the very variable and hence useful families of tandemly repeated sequences (VNTR).

Though the theoretical sensitivity of PCR is equivalent to one molecule, in practice, and especially on degraded material, there are many complications. Some of these relate to the ability to amplify badly degraded material (Päabo 1989). Also PCR may fail to amplify material which is present, for example because the enzymatic reaction is being inhibited by an unknown contaminant. Much further work is necessary in order to understand the particular problems involved in amplifying ancient DNA. One of the most important of these arises from the extreme sensitivity of the method itself, so that contamination by a single molecule from a previous amplification may be sufficient to give a positive amplification. The use of PCR to investigate ancient DNA may well become common in the future and it is essential that workers have a keen regard for the likelihood of generating spurious results from contamination. To some extent it is possible to employ laboratory controls to show that, for example, no amplification takes place when DNA not containing the sequence in question is used. Other forms of contamination are harder to prove to be absent; for example, in amplifying sequences from human bone, it is not easy to show that such sequences are indigenous rather than contamination from subsequent human handling (although, if the latter, it is likely that the contaminating DNA will be less degraded and will amplify as longer sequences).

Summary of Results

Our results, and the work of others (Hagelberg et al. 1989; Hänni et al. 1990; Horai et al. 1989; Herrman), show that mitochondrial sequences (which are the most abundant forms of DNA in cells) can be amplified from archaeological bone up to at least a few thousand years old. Nearly all this work has been done on human bone and it is not always possible to prove that human DNA contamination has been eliminated, but the weight and reproducibility of the evidence strongly suggest this is generally the case. In our work, which mainly concentrated on bone only a few hundred years old, a useful proportion of bones studied, though by no means all, yielded amplifiable mitochondrial DNA. The one bone from the Middle East that yielded positive results contained exceptionally well preserved collagen. It is not yet possible to make any general statement as to the likelihood that a given proportion of bones from a given site will contain useful amounts of DNA, simply because enough work has not yet been done.

Future Technical Work

The agenda for future technical work is immense, but it is worth outlining the obvious questions that are now being addressed.

1. What conditions favour the survival of DNA in archaeological bone?
2. What proportion of archaeological bone is likely to contain sufficient DNA to give useful genetic information?
3. Can nuclear, and in particular single-copy, DNA be recovered from bone?
4. How serious and frequent a problem is field and handling contamination?
5. How much can the existing techniques be improved in efficiency, sensitivity and specificity?

Future Applications

Future applicaitons will firstly be limited by the quality of the surviving DNA, the effort required to recover it, and the extent to which populations can be adequately sampled. A second limit is imposed by lack of knowledge linking specific DNA sequences to significant information at the phenotypic level; for instance, we do not know which genes control features like height, hair colour, intelligence, etc. The potential of obtaining genetic information by direct analysis of surviving DNA is, however, immense, and it will take many years before it can be clearly appreciated. Some steps towards the development of this potential, which seem to us to be worthwhile in the nearer term, are outlined below.

1: *Sex Identification*
This depends on the ability to detect clearly sequences on the Y chromosome (for humans). Such work allows for objective evaluation of the accuracy of DNA analysis, since often the sex is known from independent measures. On the other hand, in many cases, such as fragmentary bones, other methods do not work. Sexing animal bones would provide information on the economics of animal hunting and stock management.

2: *Species Identification*
Again, this allows for objective testing of the DNA analysis. The immediately applicable sequences to test would be mitochondrial DNA, but single copy sequences could also be tested. One obvious application is in distinguishing sheep and goats, but the whole question of the different species of sheep and their former geographical ranges could be examined.

3: *Domestication Change*
For example, of sheep, cow, or dog. MtDNA sequences may well show changes in time which could then be followed through the archaeological record. Certain phenotypic changes, e.g. for wool production in sheep, are quite definite, but not enough is known at present about the relevant changes in DNA sequence.

4: *Genetic Disease (Humans)*
Several human genetic diseases have now been characterised to the extent that the mutant gene involved has been identified and the sequences of mutants worked out. While most are rare, some populations contain

surprisingly high frequencies of mutant alleles. Some forms of the blood disorder thalassaemia are extremely common in the Mediterranean and the mutant haemoglobin sequences which cause them are well known. It would be very interesting to evaluate their frequency in ancient populations (especially since thalassaemia is associated with a survival advantage in regions where malaria is endemic). Other diseases are associated with skeletal changes which can and have been described in archaeological remains, so allowing the material to be selected. Examples are osteogenesis imperfecta, ankylosing spondylitis, and some forms of dwarfism such as achondroplasia. Detecting these mutations does, however, depend on the ability to detect single-copy nuclear sequences.

5: *Kinship Studies*
Differences in mtDNA sequences (which are quite common in modern humans) would demonstrate that two individuals did not have the same mother, or very likely the same maternal lineage. The use of other highly variable DNA sequences in kinship studies on ancient DNA is not clear - forensic methods for genetic fingerprinting may not apply directly to ancient DNA as they depend upon better preserved material. Other methods, such as HLA genotyping, may be more useful if and when ancient nuclear DNA analysis has become more developed.

6: *Archaeological Correlations*
An early and promising application would be to look for correlations between DNA sequence variation (using mtDNA polymorphisms) and spatial or cultural information in archaeological burial practices. For example, in Anglo-Saxon cemeteries, do the already established correlations between cultural and skeletal epigenetic data also apply to the mtDNA data? Or again, in Neolithic chambered tombs, is there more similarity between the mtDNA sequence data within one group than between groups? Can one arrive at a measure of the degree of kinship within and between these groups? These questions constitute a major population genetic study on the polymorphic structure of ancient populations, and such a study would inevitably develop once the relevant information from ancient DNA becomes available sufficiently easily.

7: *Population 'Markers'*
The analysis of a given population occasionally discloses one or a few polymorphic variants which serve to distinguish the population. Preliminary analysis is therefore necessary. MtDNA may well furnish useful examples and is now being used in extensive analysis on modern human populations (Stoneking 1990). Similar work on ancient populations could then enable population movements to be tracked, enabling some of the most basic questions of archaeology to be tackled

References

Ajie, H O, Kaplan, I R, Slota, P J and Taylor, R E, 1990. 'AMS radiocarbon dating of bone osteocalcin', *Nuclear Instr. & Methods* B52, 433-438.

Ambrose, S H, 1990. 'Preparation and characterisation of bone and tooth collagen for isotope analysis', *J. Archaeol. Sci.* 17 (4), 431-451.

Cattaneo, C, Gelsthorpe, K, Phillips, P and Sokol, R J, 1990. 'Blood in ancient human bone', *Nature* 347, 339.

Hagelberg, E, Sykes, B and Hedges, R E M, 1989. 'Ancient bone DNA amplified', *Nature* 342, 485.

Hanni, C, Laudet, V, Sakka, M, Bègue and Stéhelin, D, 1990. 'Amplification de fragments d'ADN mitochondrial à partir de dents et d'os humains anciens', *C R Acad. Sci. Paris*, 310 (III), 365-370.

Hedges, R E M, in press. 'The detection of DNA in ancient bone', in *Proceedings of 2nd Seminar of Nordic Physical Anthropology, Lund 1990* (eds. Boldsen, J, Iregren, E and Sellevold, B J), Report Series of the Inst. of Archaeol. Univ. Lund.

Hedges, R E M and Law, I A, 1989. 'The radiocarbon dating of bone', *Appl. Geochem.* 4, 249-253.

Hedges, R E M and Wallace, C J A, 1978. 'The survival of biochemical information in buried bone', *J. Archaeol. Sci.* 276, 255-7.

Herrman (pers. comm.)

Horai, S, Hayasaka, K, Murayama, K, Wate, N, Koike, H and Nakai, N, 1989. 'DNA amplification from ancient human skeletal remains and their sequence analysis', *Proc. Japan Acad.* 65, (B10), 229-233.

Lengyel, I A, 1975. *Palaeoserology: blood typing with the fluorescent antibody method*, Académiai Kaidó.

Päabo, S, 1989. 'Ancient DNA: Extraction, characterisation, molecular cloning, and enzymatic amplification', *Proc. Natl. Acad. Sci.* 86, 1939-43.

Schwarcz, H P, Hedges, R E M and Ivanovich, M (eds.), 1989. 'Proceedings of First International Workshop on Fossil Bone, July 1988', *Appl. Geochem.* 4, (3).

Smith, P R and Wilson, M T, 1990. 'Detection of haemoglobin in human skeletal remains by ELISA', *J. Archaeol. Sci.* 17, 255-268.

Stoneking, M, Jorde, L B, Bharia, K and Wilson, A C, 1990. 'Geographic variation in human mitochondrial DNA from Papua New Guinea', *Genetics* 124, 717-733,

Tuross, N, 1989. 'Albumin preservation in the Taima-taima mastodon skeleton', *Appl. Geochem.* 4, 255-9.

Ulrich, M M W and Perizonius, W R K, Spoor, C F, Sandberg, P and Vermeer, C, 1987. 'Extraction of osteoclacin from fossil bones and teeth', *Biochem, Biophys. Res. Commun.* 149, 712-9.

Weiner, S and Bar-Yosef, O, 1990. 'States of preservation of bones from prehistoric sites in the Near East: a survey', *J. Archaeol. Sci.* 17, 187-196.

KING DJEDKARE ISESI AND HIS DAUGHTERS

Eugen Strouhal and Mohammad Fawzi Gaballah
with contributions by
Přemysl Klír and Alena Němečková
and the collaboration of Shelley R. Saunders and Willy Woelfli

1. Introduction

In mastabas in the southern part of the Fifth Dynasty royal cemetery at Abusir recently examined by the Czechoslovak Institute of Egyptology, Charles University Prague, physical remains of some members of the family of King Djedkare Isesi were found and submitted to an anthropological analysis. In Mastaba B, Princess Khekeretnebty was buried with her probable daughter, Tisethor (Strouhal 1984); in Mastaba K, Princess Hedjetnebu, a sister of Khekeretnebty, both the King's daughters 'of his own body' as testified by inscriptions and proved by archaeological and anthropological identification (Strouhal 1992). Mastaba L, situated in the next row of mastabas, belonged to an anonymous lady whose morphological distance from the aforementioned persons was clearly greater. She did not belong to the King's nuclear family, like the others, but could have been related to it more distantly.

In connection with these findings we started a search for the physical remains of the father of the two princesses, King Djedkare Isesi himself. We were aware that they had been discovered by an archaeological team led by the architect Abd El Salam M Hussein and the Egyptologist A Varille in the pyramid called 'Haram esh-Shawaf' ('Pyramid of the Sentinel') at the south-east angle of the plateau overlooking the village of Saqqara during the season 1945-46. The excavations proved that the pyramid had been built for King Djedkare Isesi and that the original name of it was 'Nefer' ('Beautiful').

An archaeological report was never published (Moursi 1987). We have at our disposal only some basic data contained in the anatomical report by the Egyptian anatomist and anthropologist M A Batrawi (1947). Fortunately, in the former anthropological collection of Professor Batrawi, which is housed in the Department of Anatomy, Kasr El Ainy Medical Faculty, we succeeded in finding a small box labelled in Batrawi's handwriting 'Remains of Mummy (Royal) Djed-Ka-Rê, Vth Dynasty, described by A Batrawi, ADSAE, no. 803-A', which contained the fragments published by the author (Batrawi 1947). Because Batrawi did not attempt an anthropological study but limited his paper to a listing of the preserved fragments and a description of their embalming method, we decided to re-examine the remains from the demographical and morphological points of view, in the case of the latter with a view to establishing a possible similarity between them and the remains of his daughters. The authenticity of the King's remains was tested by radiocarbon dating. Supportive evidence was sought in blood group determination and histological analysis. The material was studied in the Department of Anatomy in Cairo and radiographically examined at the Radiological Clinic of Professor Dr Hoda A El-Deeb in Cairo in December 1988.

2. Authenticity of the Remains

In spite of the fact that the exact findspot of the King's remains was never described or drawn on a plan, Batrawi (1947) gave three reasons for his belief 'that that body must have been the original and the only body which was buried inside the pyramid', viz.:

1. The original entrance into the pyramid was found to be closed by a huge granite block. There was no other passage except for a narrow tortuous channel made by ancient plunderers which would not allow insertion of an intrusive burial.

2. Several 'shapeless masses' composed of wrapping-cloth soaked in resin were recovered from a small pit, dug in the floor of the burial chamber, in which a piece of a Canopic jar inscribed with the King's name was also found. It seems probable that some of these wrappings contained viscera, originally preserved in Canopic jars.

3. The preserved human remains showed no duplication and were all in the same state of preservation, betraying their origin from one and the same body, buried originally in the pyramid.

In order to test this hypothesis and to exclude definitively the possibility of an intrusive burial, a few samples were removed from the remains and together with samples from both the princesses they were submitted for radiocarbon dating at the Eidgenössische Technische Hochschule Zürich (thanks to the generosity of Professor Dr Willy Woelfli, the Director of the Institut für Mittelenergiephysik) using the AMS-technique (Woelfli 1987). The results, kindly submitted to us in a personal communication of November 30, 1990, were summarised by Professor Woelfli as in Table 1.

A few comments were added by Professor Woelfli to his tabulated results. With the weighted means two errors are listed. One is the weighted statistical error derived from errors of the single measurements, the other reflects the variance of the single results. The applied χ^2 test betrays in the first weighted mean a too-high variance resulting in only 0.13% probability of the normal distribution of the single values around the mean.

The result from ETH-5334 stands so markedly apart from the statistical grouping that it was eliminated from consideration. It was a sample of body tissue that was so soluble in water that it was not possible to purify it chemically. Moreover, we have to suspect that the King's body was saturated by embalming solutions that could not be removed. Therefore, the resultant date has to be considered a mixed one, as suggested also by the abnormally high $\delta 13_C$ value. Excluding this sample from the grouping, we get the second weighted mean with a higher (3.44%) probability of normal distribution. Even this is not very convincing, but we must take into consideration that the daughters were, minimally, twenty years younger than Djedkare, thus causing the greater value of the variance compared with that of the statistical error.

We may conclude - with all caution - that King Djedkare Isesi and both princesses are proven to be contemporaries. This result is a further strong argument for accepting the authenticity of the King's remains.

3. Material and Methods

The fragments listed by Batrawi (1947) were carefully prepared in order to obtain tissue for histological examination and to expose bones which could have been, in some cases, joined together. In some fragments, where the soft tissue layers were too thick, the integrity of the piece was preserved and radiography was applied as a method of investigation.

The newly arranged fragments comprise the following regions or items:

1. The anterior half of the calva consisting of the incomplete squama frontalis, the left parietal bone and the anterior lateral quarter of the right parietal bone (Plate 9,1).

Table 1. Radiocarbon dating of samples related to Djedkare Isesi and his daughters

Person	Laboratory No.	Sample	Conventional C^{14}-age (BP)[1]	$\delta 13_c$	Calibrated age range (BC)[2]
Djedkare	ETH-4340	DI-O mummy wrapping (linen)	4025 ± 55	-25.4	2864-2460
	ETH-5334	DI-1 body soft tissue	4385 ± 80	-14.4	3340-2787
	ETH-5335	DI-2 charcoal	4200 ± 75	-21.2	3014-2580
	ETH-5336	DI-3 mummy wrapping (linen)	4235 ± 75	-22.8	3031-2612
Hedjetnebu	ETH-5337	12/K/87 mummy wrapping (linen)	4205 ± 65	-24.8	2920-2600
Khekeret-nebty	ETH-6949	170/B/76 mummy wrapping (linen)	4020 ± 65	-22.3	2869-2403
First weighted mean (of 6 examples)			4152 ± 28/55 (χ^2=4.0, P=0.13%)	-	2910-2580
Second weighted mean (without ETH-5334)			4121 ± 30/48 (χ^2=2.6, P=3.44%)	-	2886-2507

Explanations

1 = all error values are ± 1σ

2 = ranges are ± 2σ (probability 95%)

2. The posterior medial quarter of the right parietale which has contact points with the previous fragment but cannot be joined to it (Plate 9,2, right).

3. The incomplete squama occipitalis, mostly from the right side (Plate 9,2, left).

4. Soft tissues of the left posterior inferior part of the face with the temporal region containing the shrunken auricle and a part of the left neck wall (Plate 9,3). Inside the fragment parts of the left zygomatic, ethmoid, sphenoid and temporal bone (pyramid) can be found.

5. The incomplete alveolar process of the right maxilla with the anterior lateral quarter of the processus palatinus and the teeth row from right upper C to M_3 in situ (Plate 9,4).

6. The left half of the mandible with the second and third left lower molars in situ (Plate 10,1).

7. The sternum with attached left clavicle (without its lateral third), left first rib (posterior end missing) and medial ends of six left and four right ribs. The fragment is covered externally by muscles, subcutaneous tissue, skin, textile and grey stucco (Plate 10,2), internally by the periosteum and possibly pleura (Plate 10,3).

8. The vertebral column from C_1 (left half) to T_5 held together by ligaments, muscles and the remains of skin on the rear. A fragment of the distal third of body T_6 connected with the whole body T_7, preserved separately (Plate 11,1). An isolated perforated intervertebral disc from the lumbar section of the spine (Plate 12,3, right) as well as six rib-fragments.

9. Postmortally broken fragment consisting of the distal end of the left humerus with proximal ends of the antebrachial bones (Plate 11,2).

10. Metacarpals and proximal phalanges of both first fingers, the metacarpal and all phalanges of the left second finger, the metacarpal of the left third finger and the left trapezium and trapezoideum.

11. A part of the right os coxae containing the acetabulum, incisura ischiadica and the edge of the tuber ischiadicum (Plate 11,3).

12. The proximal sixth of the left femur with the originally adhering incomplete rim of the left acetabulum (Plate 11,4).

13. The lateral (fibular) half of the left foot from the os naviculare to the tips of the third-fifth toes (Plate 12,1).

14. Several fragments of compact bone originating from the long bones of the extremities.

15. Several fragments of soft tissue, some with skin covered by adhering fine linen wrapping, e.g. an outside convex piece from one of the shoulders with remnants of muscle and sinew on its concave inside (Plate 12,2).

16. Several fragments of pads and bags composed of many layers of textile soaked in a black substance (most probably resin). One of the pads has been smeared on its flat, somewhat concave, side with a layer of lime in which a green faience tubular bead has been fixed (Plate 12,3, left). A spheric oblong

object (100 x 78 x 42 mm), possibly a small bag, with one smoothly convex side, the other side being flat and covered by a 25 mm thick layer of plaited strings into which a grey material (possibly lime) has been pressed and over which a black matter (probably resin) has been smeared (Plate 12,4). In addition, many loose pieces of finely woven ochre or brown linen wrapping have been preserved.

It was possible to use standard anthropological methods of investigation (Martin and Saller 1957, 1959, Brothwell 1963, Strouhal and Jungwirth 1984) only on a limited number of preserved features. Therefore, some unusual measurements and descriptive observations had to be undertaken, aimed at comparison with the purported daughters of the King.

4. Demographic Data

Sex Determination. The general robusticity of the cranial remains was medium to gracile. Both tubera frontalia and parietalia were only slightly developed; the right linea temporalis was present as an indistinct sign on the frontal scale only; the supramastoid crest was developed slightly. The profile of the forehead was oblique, arched fluently without any angular break, and the forehead was low. In addition the occipital scale was not angulated but curved fluently. No distinctive protuberantia occipitalis externa was developed, but there was a slight transverse eminence continuing laterally into an indistinct narrow torus occipitalis. The mastoid process was medium-long, anteroposteriorly broad, and mediolaterally rather thick, the incisura mastoidea being medium-large but deep. The nuchal muscular relief showed medium to strong development.

The mandible was mediumly robust with a mediumly developed muscular relief. The left mandibular angle, partially broken off, preserved evidence for an original eversion whose extent could not be ascertained. In spite of a missing lower border the shape of the chin was clearly square with outstanding, slightly elongated, tubera mentalia and a well differentiated pointed protuberance. The mylohyoid line was only slightly marked. The left condyle was rather big.

The postcranial bones were also mostly medium-robust to gracile with a medium to slightly developed muscular relief. The right tuber ischiadicum, whose edge has been preserved, was probably medium-sized, the right incisura ischiadica maior narrow and deep. On the lower side of the medial end of the left clavicle a small costoclavicular facet can be observed. The right humerus was perforated (6 x 5 mm).

The majority of diagnostically significant features point to the sex of the individual being male. At the same time, his body-build tended to gracility and somewhat underdeveloped muscular relief.

Age Determination. With regard to cranial sutures, the coronal one was completely fused inside, but outside only in C_1 and medial C_2 sections. Similarly the sagittal suture was completely gone inside, while outside remnants of suture can be found in S_3-S_4 sections. The right medial quarter of the lambdoid suture was still open, the right lateral quarter being closed inside and in a state of incipient closing outside. A slight sulcus sagittalis and shallow grooves (breadth 4-5 mm, depth 0.5-1 mm) along the anterior edges of both parietal bones (which were thinned up to 2-3 mm) can be observed.

Abrasion of the upper right canine and first premolar was very progressed with removal of more than half of the crown height but without any opening of the pulpar chamber (grade 7 of Strouhal and Jungwirth 1984). At the same time, the upper right second and third molars showed only enamel abrasion (grade 2), pointing to a premature loss of their antagonists. Similarly, the lower left second molar had only a facet of incipient enamel abrasion (grade 1) and the lower third molar a facet of incipient dentine abrasion (grade 3), most probably for the same reason. This finding points indirectly to carious dentition in the individual when at a younger age. Alveolar resorption was progressed (6 mm in the upper, 5 mm in the lower jaw, grades 3 and 2 of Brothwell 1963). Upper right M_1 was lost infra vitam and his alveolus closed with an incipient atrophy of the alveolar bone. On the upper right second and third molars, dental calculus was deposited (grade 2 of Brothwell 1963) in the supragingival level.

Of the preserved postcranial bones, the cervical and upper thoracic parts of the spine could not be inspected directly, because they were still covered by ligaments. From the radiogrammes, it was possible to detect the practical absence of osteophytes (Plate 13,1-2), which were, however, present at the top of the dens epistrophei (3 mm long). At the same time, cartilaginous parts of the ribs were already ossified and the corpus sterni fused with the processus ensiformis (Plate 13,3), while the manubrium sterni remained separate. The left femoral head showed a partially filled-up foveola and an area of arthritic erosion. No age or pathological changes were found in the few other preserved joints (Plates 13,4 and 14,1).

From the histological report (see further), we may quote that brown pigment was still present in the hair, but atherosclerotic changes could be demonstrated in some minute blood vessels.

Summing up the features, an advanced age of between 50 and 60 years is suggested. This agrees with the original determination of about 50 years by Batrawi (1947). The single feature that does not agree with this determination is the absence of the osteophytosis of the spine, which seems generally to have begun in ancient Egyptians around thirty years of age. Together with the aforementioned limited development of the muscular system, this points to a physically inactive way of life, with no heavy or excessively burdensome work on the individual's part.

We may add that the histomorphometric analysis at the midshaft compacta of the left humerus of Djedkare Isesi performed by Professor Shelley Saunders yielded an age estimate of 52.8 $\pm$ 8.5 years (written communication of October 29, 1991) which fits well the determination mentioned above. The King's age at death lay thus in the range of 45-60 years.

5. Other Descriptive Features

No metopism was observed. The profile curve of the cranial vault was slightly depressed postbregmatically, as in the skull of the Princess Khekeretnebty, but the skull was not drawn up and back as in the remains of Hedjetnebu. The obelic region was flattened to a fairly strong extent. Tiny wormian bones were present in the lateral parts of the coronal suture and probably also in the lambdoid suture. An Inca bone was not developed. The cranial outline in the vertical norm was possibly ovoid.

The left zygomatic bone bulged prominently to the side. The lower edge of the piriform aperture was anthropine, in the form of a low crest. The subnasal region showed a wavy relief. Both the upper and lower dental rows were most probably elliptic and rather small. The depth of the palate was medium and concave anteriorly.

From the upper right teeth, I_2 was lost postmortally, C and P_1 were present and healthy, from P_2 a secondarily broken-off torso survived, M_1 was lost infra vitam and both M_2 and M_3 were present without pathological changes. Of the lower left teeth, the incisors and the canine were lost postmortally, the alveolus of P_1 was damaged and the tooth lost postmortally, as were P_2 and M_1, while M_2 and M_3 remained in situ and healthy. The upper right M_2 was fusiform with a strongly reduced hypoconus. Its neighbouring M_3 had only three cusps. The lower left M_2 showed the usual cruciform pattern of furrows, dividing the four cusps. On the adjoining M_3 the number of cusps and pattern of furrows were probably the same. No Carabelli cusps were present on the upper right M_2 and M_3. No crowding appeared in the lower frontal dentition.

The preserved left auricle had a rolled up posterior half. Its original height was 57 mm, the breadth being reduced by rolling to 17 mm only. The external auditory meatus was also narrowed by it (height 5 mm, breadth only 2 mm).

6. Comparison of Metric Features

A few of the currently used craniometric features, complemented by some atypical ones, could be measured in the remains of King Djedkare Isesi, as well as in the skulls of his daughters Hedjetnebu and Khekeretnebty, of his granddaughter Tisethor and of the possibly more distantly related woman L (Table 2). One would have expected the King's dimensions, being those of a male, to be greater than those of the four females. In reality, however, this was the case only with the maximum frontal breadth, the height and breadth of the ascending ramus, and to a lesser extent with the minimum frontal breadth and the height of the mandibular body between M_2 and M_3. In the other features, the King's values were close to those of the compared females (differing maximally 1 mm). This appears in such significant features as the development of the mastoid process, both the height and thickness of the mandibular body and even the thickness of the cranial vault. The state of these features is not just the result of the underdevelopment of the muscular system, but also expresses the striking gracility of the King's cranial structure.

Measurements of the four preserved molars of the King showed mostly similar values in comparison with homoloug molars of the four females (Table 3). Both the King's upper and lower third molars were mesiodistally longer, while the diameters of his lower second molar and the buccolingual breadth of his third lower molar were paradoxically both slightly shorter than those of the females. This finding suggests that in the King's gracility and shorter dimensions genetic background also played a role.

A few of the currently used osteometric features, complemented by some atypical ones, were able to be determined in the King's and the four females' remains (Table 4). In the case of these features a still greater sexual differentiation might have been expected than that in the craniometric features. This was actually the case with seventeen of the measurements, viz. the ventral vertical diameters of the vertebrae C_2, C_3, C_6, T_5 and T_7, the length and breadth of the manubrium sterni, the thickness of the anterior end of the first rib, the circumference of the mid-diaphysis of the clavicle, the lower epiphyseal breadth of the humerus, the breadth of the upper edge of the tuber ischiadicum, the transverse breadth of the trochanter minor and the three dimensions of the femoral head. These results fitted well the male sex of the King. In the eleven remaining measurements, however, the King's values were close to those of the compared females (differing maximally 1 mm) with the exception of the still smaller ventral vertical diameter of the vertebra T_4. Some other osteometrics of the King (Table 5), which had no counterparts in the skeletal parts preserved with the four females, were also consistent with the sex being male, while at the same time showing a tendency to gracility as well as to shorter stature.

We should note that gracility and short stature were the most outstanding features of the four compared females (Strouhal 1990).

7. Blood Groups (by P. Klír)

In addition to the previously tested samples of tissue from the bodies of Princess Khekeretnebty and her probable daughter Tisethor (Tesař and Klír 1984), as well as from those of the Princess Hedjetnebu and Lady L (Klír 1992), the remains of King Djedkare Isesi were also recently examined.

Table 2. Craniometric comparison between Djedkare Isesi, his family members and Lady L

No.	Measurement	Djed-kare	Hedjet-nebu	Khekeret-nebty	Tisethor	Lady L
9	Minimum frontal breadth	95	93	91	91?	90
10	Maximum frontal breadth	114	108	109	110	108
16	Breadth of the foramen magnum	27?	28	27	-	26
60	Maxilloalveolar length	52?	47	51	47	53
ML	Length of the mastoid process	30?	26	29 R	28	26
MT	Thickness of the mastoid process	14?	10	13 R	10	11
69.1	Height of the mandibular body	29?	29	29	23	27
69.3	Thickness of the mandibular body	11?	8	9	10	11
70	Height of the ascending ramus	65?	58	57?	58	58
71	Minimum breadth of the ascending ramus	33	27	28	30	28
79	Gonial (mandibular) angle	118?	118	122	121	124
-	Cranial thickness at left tuber frontale	5	-	5	-	7?
-	Cranial thickness at right tuber frontale	5	-	5	5	7
-	Cranial thickness at left tuber parietale	5	6	5	6	7
-	Height of the mandibular body between M_2 and M_3	27	25	24	24	22
-	Thickness of the mandibular body between M_2 and M_3	14	12	13	14	14

Table 3. Odontometric comparison between Djedkare Isesi, his family members and Lady L

Tooth	Measurement	Djedkare	Hedjetnebu	Khekeretnebty	Tisethor	Lady L
M2 (lower)	Mesiodistal diameter	10	9	-	9.5	-
	Buccolingual diameter	9.5	10	-	9.5	-
M3 (lower)	Mesiodistal diameter	9.5	-	8	-	-
	Buccolingual diameter	10	-	10.5	-	-
M2 (upper)	Mesiodistal diameter	8.5	9	10	10	-
	Buccolingual diameter	8	8.5	9.5	9.5	-
M3 (upper)	Mesiodistal diameter	11	-	9.5	9.5	-
	Buccolingual diameter	8.5	-	9	9	-

Table 4. Osteometric comparison between Djedkare Isesi, his family members and Lady L

Bone/ number	Measurement	Djed-kare	Hedjet-nebu	Khekeret-nebty	Tisethor	Lady L
Vertebrae						
C_2	Ventral vertical diameter	39	34	34	35	33
C_3	Ventral vertical diameter	12?	11	-	-	-
C_4	Ventral vertical diameter	13	12	9	-	11
C_5	Ventral vertical diameter	14?	10	-	-	-
C_6	Ventral vertical diameter	14?	12	11	10	-
C_7	Ventral vertical diameter	15	14	11	-	-
T_1	Ventral vertical diameter	15	14	13	-	-
T_2	Ventral vertical diameter	15	16	14	12	-
T_3	Ventral vertical diameter	15	16	-	-	-
T_4	Ventral vertical diameter	13	16	-	-	-
T_5	Ventral vertical diameter	17	15	-	-	-
T_6	Ventral vertical diameter	20	16	-	15	16
Sternum						
2	Length of the manubrium	61	-	-	49	-
4	Breadth of the manubrium	61	-	-	47	-
1st rib						
-	Thickness of the anterior end	9?	6 R	-	-	7 R
-	Thickness in the middle	4	4 R	-	-	4 R

Table 4 (cont.) Osteometric comparison between Djedkare Isesi, his family members and Lady L

Bone/ number	Measurement	Djed- kare	Hedjet- nebu	Khekeret- nebty	Tisethor	Lady L
Clavicle						
6	Circumference of the mid-diaphysis	40	30 R	28	27	31
Humerus						
4	Lower epihyseal breadth	62? R	51 R	-	-	53 R
Coxae						
22	Maximum diameter of the acetabulum	48	-	-	45 R	47
-	Breadth of the upper edge of the tuber ischiadicum	30	-	-	-	25
-	Thickness between the centre of the acetabulum and the lower edge of the incisura ischiadica maior	26	21	-	18	22
-	Thickness of the iliac wing on the linea arcuata above the apex of the inc. ischiad. maior	23	17	-	17	17
Femur						
-	Proximo-distal length of the trochanter minor	25	24 R	-	20	14
-	Transversal breadth of the tranchanter minor	16	14 R	-	13	13
-	Prominence of the trochanter minor (viewed from medial side)	11	10 R	-	-	11
18	Vertical diameter of the caput	45	37	-	-	39
19	Transversal diameter of the caput	45	38	-	-	39
20	Circumference of the caput	146	123?	-	-	125

Table 5. Other osteometrics of Djedkare Isesi

Bone/ number	Measurement	Value	
Sternum			
-	Length of the sternum	156	
3	Length of the corpus sterni	85 ?	
-	Length of the processus ensiformis	10	
5	Breadth of the corpus sterni	42 ?	
-	Breadth of the processus ensiformis	17	
Clavicle			
-	Height of the medial end	29	
-	Antero-posterior thickness of the medial end	23	
Hand			
2	Length of the 1st metacarpal	50 L	50 R
2	Length of the 3rd metacarpal	65 L	- R
3	Length of the 1st proximal phalanx	34 L	35 R
3	Length of the 2nd proximal phalanx	37 L	- R
3	Length of the 2nd medium phalanx	23 L	- R
Femur			
16	Minimum antero-posterior diameter of the neck	25	

Explanations for Tables 2 - 5 :

Pair measurements were done usually on the left side or on both sides (L = left, R = right).

Numbered measurements are those of Martin and Saller (1957, 1959); others are introduced by the present authors.

Two methods were used: the absorption method (Mueller 1975:117, Lengyel 1975:17-18) with anti-A and anti-B sera (titre 1:64) and the absorption-elution method with anti-A and anti-B sera of a higher titre (1:256). The absorption phase lasted 24 hours in a temperature of 4°C. For testing the decrease of titre in the absorption method after the absorption phase, 1% solution of erythrocytes of groups A and B was used. In the case of the absorption-elution method this solution of erythrocytes was mixed with SAGH (serum antiglobulinum humanum).

Samples were first cleaned and, before the actual examination, crushed. All samples were tested by both methods except hair which - due to the small amount of material - was analysed by the absorption-elution method only. All tests were repeated several times.

From the remains of King Djedkare Isesi fragments of soft tissue from the left side of the face and from the left os lunatum were used for blood typing. Both samples consistently showed blood group A, on repeated analyses.

The identical blood group A also characterised Princesses Khekeretnebty and Hedjetnebu, the probable daughter of Khekeretnebty, Tisethor and Lady L (Tesař and Klír 1984, Klír 1992).

8. Histological Analysis (by A. Němečková)

Samples of soft tissue from the princesses have been identified in our previous contributions. From Khekeretnebty, muscular and brain tissue (Němečková 1984), from Hedjetnebu, brain tissue and hair (Němečková 1990) have been identified. No abnormality or pathological change were evidenced.

Three fragments of tissue from King Djedkare Isesi are dealt with in the present report. Macroscopically they were dry, of light brown colour, originating from the scalp.

At first we processed the specimens by classic histological methods. Small pieces of tissue were fixed in 10% formaldehyde and transferred over metacrylate into paraffin. For staining, hematoxylin and eosin, orcein, and the method after Verhoeff's hematoxylin were applied.

For observation in an electron microscope we carried out a special processing. Having softened the tissue according to Sandison (1955), we transferred the specimens in glutaraldehyde solution in an 0.5 M phosphate buffer. Minute particles of the tissue were postfixed in 1% osmium dioxide in an 0.1 M phosphate buffer and in 0.1 M saccharose prior to being dehydrated and embedded in EPON 812. Ultra thin sections were made on an ultramicrotome and were contrasted with uranylacetate and lead citrate.

By classic histology we succeeded in processing specimens of the skin. In the corium, collagen and elastic fibres were well represented. Desiccation of the tissue had resulted in a considerable deformity of blood vessels, whose preserved walls revealed remains of the media and adventitia (Plate 14,2,A). In some minute blood vessels a pathological substrate in the form of an increased fibrous centre revealed degenerative atherosclerotic changes (Plate 14,2, arrow, and Plate 14,3). A cross-section of the hairs showed tiny granulae of brown pigment.

Remnants of epidermis were also investigated by electron microscopy. Inside the epithelium forming epidermis we found cells with projections that were covered with intercellular substance (Plate 14,4,A). The cells were interconnected by broken desmosomes into which indistinct tenofilaments in the cytoplasm were fixed (Plate 14,4,C). In most cells we found keratohyalin granulae differing in shape and forming clusters. These granulae were coated with a distinct membrane (Plate 14,4,B).

The mummified tissue which has been studied showed structures quite comparable with those of recent times, notwithstanding the great chronological distance. Connective tissue, fibrous protein collagen and elastin proved to be the main solid structures of our specimens. These fibres appear to resist well both chemical and climatic influences. As to the adipose connective tissue, we have not found as many fatty cells within it as were noted in any of our previous investigations (Němečková 1977, 1984).

9. Discussion

The radiocarbon dating which has proved the contemporaneity of King Djedkare Isesi and his daughters may strike Egyptologists as producing too-high a result. The second calibrated age range (2886-2507 BC), compared with the accepted ranges for the Fifth Dynasty, whose penultimate sovereign was Djedkare Isesi (2505-2345 BC in its higher or 2460-2310 BC in its lower version) (von Beckerath 1975: 967-971), appears to be 381-162 or 426-197 years too old. On the other hand, this result agrees well with the radiocarbon dating of sixty-four samples of organic remains taken from the major Saqqara, Abusir and Giza monuments of the Third to Sixth Dynasties, whose dates were older by at least three centuries than those established by traditional historical reconstruction (Haas et al. 1987: 597).

The advanced individual age of the King's remains, as now determined (45-60 years), can be compared with the available data on his regnal span. The Turin Canon indicates 28 years (Gardiner 1961: 435). An alabaster vase made for his Sed-festival (which took place usually in a king's 30th regnal year) is known from the Louvre Museum (Maragiglio and Rinaldi 1975: 64). A fragment of the Abusir papyri (Louvre E 25416a) yields the date of the sixteenth census of cattle carried out under Djedkare Isesi. As the census took place every two years, he should have reigned at least 32 years (Maragiglio and Rinaldi 1975: 64). In Manetho's list he is called "Tancheres" and is assigned as many as 44 regnal years (Gardiner 1961: 435). Neither of these dates contradicts the King's individual age.

We have not yet mentioned one very important member of the family which we have been investigating: the King's wife and mother of the two princesses. Most probably she was not the Great Queen (the first or main wife) for whom a pyramid - the largest funerary monument for a queen of this period - was built northeast of the King's pyramid (Maragiglio and Rinaldi 1975: 98). In the few published reliefs she appears as the royal wife of Djedkare Isesi but her name has not survived. A fragment of a probable decree mentions a daughter probably connected with this queen (Moursi 1987, Abb. 13).

Another king's wife, Meresankh IV, was buried in mastaba D 5 at Saqqara North, not far from the mastaba, D 3, of her son Raemka (Callender 1989: 4). This queen was also most probably not related to the princesses buried in the remote traditional cemetery of the previous Fifth Dynasty kings at Abusir.

Yet another, hitherto unknown wife of King Djedkare Isesi, very probably buried not far from the burial-place of her daughters Hedjetnebu and Khekeretnebty and of her granddaughter Tisethor, has to be postulated theoretically at Abusir. One of the candidates, from the same group of mastabas, could be the anonymous Lady L. This hypothesis can only be tested when excavations in the neighbourhood of this burial group have been completed. They might reveal a mastaba of another king's wife, the mother of the two princesses.

10. Summary

In considering the hypothesis of Batrawi (1947) on the authenticity of King Djedkare Isesi's remains, we may conclude, on the basis of the circumstances of their discovery and from the results of the radiocarbon dating, that their contemporaneity with the burials of the two princesses has been confirmed. In spite of several feminine features, medium to gracile body-build and mediumly developed muscular relief, the sex of the King's remains is male and the majority of ageing features - with the exception of the spine which is almost completely devoid of osteophytosis - point to an age at death within the range of 45-60 years. The King's way of life was physically relatively inactive and free of hard or heavy work. Despite the difference in sex, a striking similarity between the King's and his daughters' values was found in cranial and dental measurements, emphasising his gracility. In the post-cranial measurements the sexual difference was more marked, but again the tendency to gracility and shorter stature was obvious. The same features were characteristic of the princesses. Of the descriptive features, the postbregmatic depression of the King's skull was similar to that of Khekeretnebty. The King shared the same blood group, A, with his daughters, granddaughter and Lady L. The histological finding of atherosclerosis confirms his advanced age and reflects perhaps stresses arising from royal concerns. In the final section of the paper, there was discussion of the high radiocarbon dates, the relation of the King's age at death to his known regnal span and the question of the identity of his secondary wife - the mother of the said daughters.

References

Batrawi, A, 1947. 'The Pyramid Studies. Anatomical Reports', *ASAE* 47, 97-111.

Beckerath, J von, 1975. 'Chronologie', *Lexikon der Ägyptologie*, I, 967-971.

Brothwell, D R, 1963. *Digging up Bones*, British Museum (Natural History), London.

Callender, G, 1989. *A Prosopography of Mr.s-ᶜnh IV*, Private print, Thornleigh, Australia.

Gardiner, Sir Alan, 1961. *Egypt of the Pharaohs*, Clarendon Press, Oxford.

Haas, H, Devine, J, Wenke, R, Lehner, M, Woelfli, W, Bonani, G, 1987. 'Radiocarbon Chronology and the Historical Calendar in Egypt', in *Chronologies in the Near East*, Aurenche, O, Evin, J and Hours, F (eds.), BAR International Series 379, Oxford.

Klír, P, 1992. 'Blood Groups', in Strouhal, E, 1992.

Lengyel, I A, 1975. *Palaeoserology*, Akademiai Kiadó, Budapest.

Maragioglio, V, Rinaldi, C, 1975. *L'Architettura delle Piramidi Menfite*, Parte VIII, Officine Grafiche Canessa, Rapallo.

Martin, R, Saller, K, 1957, 1959. *Lehrbuch der Anthropologie*, 3. Auflage, Band I-II, G Fischer, Stuttgart.

Moursi, M, 1987. 'Die Ausgrabungen in der Gegend der Pyramide des Ḏd-K3-R' "Issj" bei Sakkara', *ASAE* 71, 185-193.

Mueller, B, 1975. *Gerichtliche Medizin*, Teil I, Springer, Berlin/Heidelberg/New York.

Němečková, A, 1977. 'Histology of Egyptian Mummified Tissues from Czechoslovak Collections', *ZÄS* 104, 142-144.

Němečková, A, 1984. 'Histological Analysis', in Strouhal, E, 1984, 181-182.

Němečková, A, 1992. 'Histological Comment', in Strouhal, E, 1992.

Sandison, A T, 1955. 'The Histological Examination of Mummified Material', *Staining Technology* 30, 277.

Strouhal, E, 1984. 'Princess Khekeretnebty and Tisethor: Anthropological Analysis', *Anthropologie* 22, 171-183.

Strouhal, E, 1992. 'Anthropological and Archaeological Identification of an Ancient Egyptian Royal Family (5th Dynasty)', *International Journal of Anthropology* 7, 43-63.

Strouhal, E, Jungwirth, J, 1984. *Die anthropologische Untersuchung der C-Gruppen- und Pan-Gräber-Skelette aus Sayala, Ägyptisch-Nubien*. Denkschriften, 176 Band, Österreichische Akademie d. Wissenschaften, phil.-hist. Kl, Verlag d. Österr. Akad. d. Wissensch., Wien.

Tesař, J, Klír, P, 'Blood Groups', in Strouhal, E, 1984, 182.

Woelfli, W, 1987. 'Advances in accelerator mass spectroscopy', *Nuclear Instruments and Methods in Physics Research*, B29, 1-13.

THE CORRELATION OF SKELETAL REMAINS AND BURIAL GOODS AN EXAMPLE FROM NAGA-ED-DÊR N 7000

Patricia V Podzorski

Background

The correlation of data derived from the analysis of human remains with that derived from artefacts recovered from archaeological sites yields more information about ancient peoples than if these two areas of study remain separate. For example, a recurrent theme in Predynastic archaeology is that women were accorded differential burial treatment from men, although quantitative substantiation is seldom given (Baumgartel 1960, 1975; Castillos 1982; Hassan 1988). One site for which we possess both osteologic and artefact data is cemetery N 7000 at Naga-ed-Dêr. The site of Naga-ed-Dêr, which is located on the east bank of the Nile just north of Abydos, consisted of a series of cemeteries ranging in date from Predynastic to Christian times and was excavated by the Hearst Egyptian Expedition of the University of California (Reisner 1908: vi-vii). Albert Lythgoe conducted the work on the early Predynastic cemetery between 1902 and 1904 (Lythgoe 1965: xii-xiii). Cemetery N 7000 was located on a gravel terrace at the base of limestone cliffs above a small wadi. 635 graves were excavated in this cemetery, most ranging in date from the Naqada I (Amratian) (c. 3900-3650 BC) to the end of the Naqada II (Gerzean) (c. 3650-3300 BC), with a few later graves as well (Lythgoe 1965; Hassan 1988: 138, 140).

The analysis presented below incorporates the work of several individuals. Albert Lythgoe was responsible for the field notes, photographs and some of the osteologic observations made during the excavations. Sir Grafton Elliot Smith, who was on leave from his post as lecturer in anatomy at the Khedivial Medical School in Cairo, gave Lythgoe expert assistance and contributed most of the demographic and pathological information recorded (Lythgoe 1965). Renée Friedman, from the Department of Near Eastern Studies at the University of California at Berkeley, studied the ceramics from the early Predynastic cemetery at Naga-ed-Dêr and the relative chronology used for the cemetery is taken from her work (Friedman 1981). My own contribution to the analysis consisted of collecting all the documentary evidence on the human skeletal material and combining that information with a study of the fragmentary remains of approximately 150 individuals which are now housed in the R H Lowie Museum of Anthropology on the Berkeley campus of the University of California (Podzorski 1990).

An important concern when dealing with skeletal remains from older excavations is assessing the accuracy of the original excavator when making determinations of age and sex (Mann 1989: 246). The Hearst Expedition at Naga-ed-Dêr was unusually fortunate in two respects. First, for most of the work at N 7000 the Expedition had the help of a trained anatomist (Elliot Smith) who was able to work with the skeletal remains. Second, preservation of soft tissues in some parts of the site was so good that it was possible to identify the sex of many individuals directly (e.g. N7379, N7036) (Lythgoe 1965: Fig. 103a, Fig. 8h). In addition to direct evidence of sexual characteristics, clues from materials not usually preserved in archaeological sites were available to the excavators. For example, in some instances facial hair (e.g. N7387) was used to determine the sex of an individual (ibid.: Fig. 105d). Sexing of skeletal remains by Smith and Lythgoe was done on the basis of the pelvis when possible, although their notes do not specifically mention which traits of the pelvis were used (e.g. N7011A) (ibid.: 6).

It was more difficult for the original excavators to determine the age at death of the individuals recovered. Ages for well preserved bodies appear to have been based upon the apparent size of the individual, dental wear, presence of grey hair, etc. Ageing of subadults was more precise than for adults. Most older individuals were simply identified as 'Adult', with occasional modifiers such as 'middle aged' or 'old'. From an examination of the human remains that are now preserved in the Lowie Museum of Anthropology (Podzorski 1990), it appears that Smith and Lythgoe's original comments were very often correct. Few emendations of their conclusions were necessary and we may have a fairly high degree of confidence in the accuracy of the documentation on the sample.

Analysis

Information derived from human remains can be related to the artefact assemblages found with those remains in an almost endless number of ways. Only three approaches will be discussed below. The first is an examination of the size of the tombs in relation to the age and sex distribution of the occupants. The second approach attempts to estimate overall tomb richness and diversity using two statistics: a simple count of the artefacts in the tomb and a ratio based on the overall assemblage of grave goods from this cemetery. Lastly, the relationship of specific types of artefacts to individuals is examined.

Tomb Size

To facilitate comparison of tombs by size the area of the tomb floor was calculated in square metres by multiplying length by width. Using this method the figure calculated for the floor area of round or oval tombs is slightly larger than the actual area, while the floor area recorded for rectangular and square tombs is more accurate. Measurements were calculated to four decimal places and then rounded to the nearest tenth of a metre for convenience of comparison in the text.

Fig. 1. FREQUENCY OF TOMBS BY FLOOR AREA IN SQUARE METRES, SINGLE INHUMATIONS ONLY, NAQADA I-III (n=485).

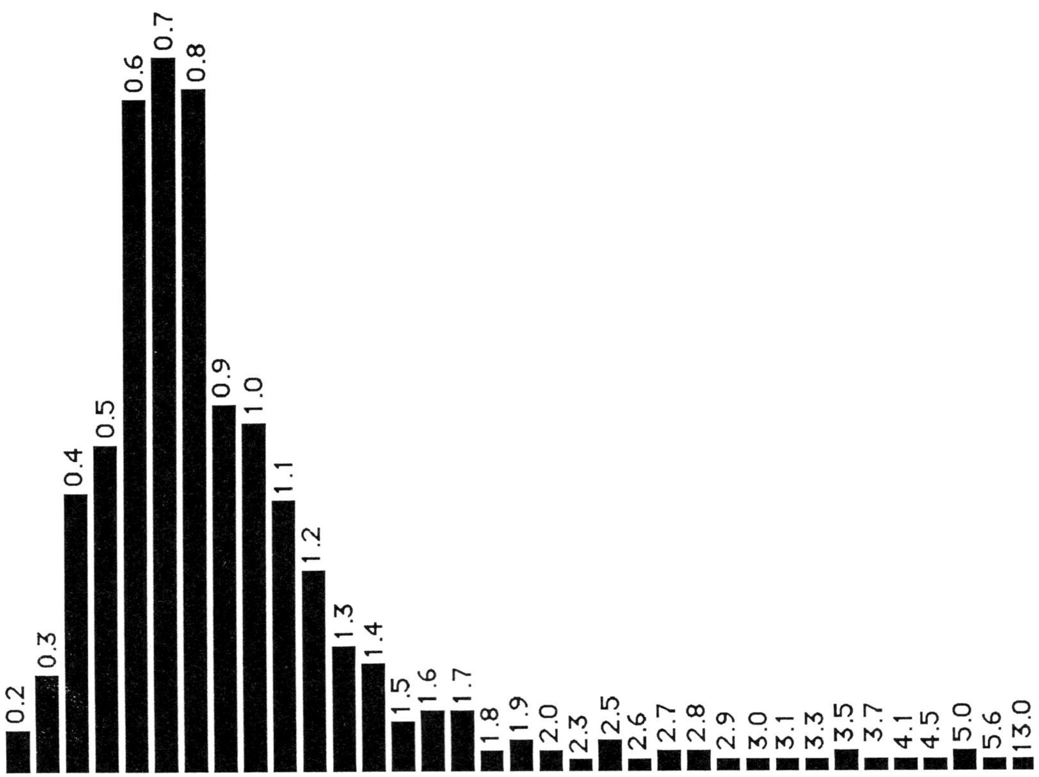

The range in tomb size for the entire Predynastic period at Naga-ed-Dêr N 7000 was 0.2 to 13.0 sq. m. (see Fig. 1). Of 485 single inhumations 215 (44%) fall between 0.6 and 0.8 metres. This was the most common size of tomb. 81% of the tombs fall between 0.4 and 1.2 sq. m. About three-quarters of the tombs which date to the Naqada I and early Naqada II range in size from 0.2 to 1.2 sq. m. Approximately half of the tombs of 1.3 to 1.6 sq. m. can be attributed to the earlier periods (Naqada I-IIb). However, almost all of the tombs measuring 1.7 sq. m. and larger (n=36) date to the later Naqada II and the Naqada III. Of the 485 single inhumations from all phases of the Predynastic, 146 were identified as female, 130 as male and 209 were unsexed.

Tomb size appears to have been functionally determined. It was a basic matter of accommodating the body and the objects which were to be buried along with it. The smallest tombs (0.2 and 0.3 sq. m.) belonged primarily to children.

The age of the individual shows a strong correlation with tomb size: the younger the individual, the smaller the tomb. Date also seems to have been a factor, however, as the smallest tomb for a single infant burial in the later half of the Naqada II was 0.6 sq. m. as compared to 0.2 sq. m. earlier in the Predynastic. Most children (67 of 83 or 81%) found in single burials were in tombs between 0.2 and 0.8 sq. m. One notable exception to this generalisation was found in tomb N7438 (Naqada IId) (Lythgoe 1965: Fig. 119b). The floor area of this tomb was 2.5 sq. m. and it contained the remains of a child of approximately five years of age. This was a rich tomb, but the fact that the pottery was found on two different levels within the grave may indicate reuse, although if this were the case, absolutely nothing of the earlier burial remained. All other tombs larger than 1.4 sq. m. held the remains of adults. About 40% of all children under the age of fifteen were buried in solitary inhumations. Single burial increased to 65% among those individuals between fifteen and twenty years of age. This figure is similar to that for adults (402 of 661 or 61%) and may indicate that these older juveniles were considered adults in their own culture.

Fig. 2. TABLE COMPARING TOMB SIZE BY DATE FOR FEMALES AND MALES

Date	No.	MALES Tomb Size*		No.	FEMALES Tomb Size*	
I-IIa	41	0.35-1.275	(3.225)	38	0.325-1.2	(2.52)
I-IIb	6	0.54-1.035	(1.68)	10	0.51-1.32	(1.68)
I-IIc	4	0.665-1.26	(2.275)	4	0.44-0.7	(2.275)
Ia/b	3	1.05-1.575	(3.68)	1	0.85-0.85	(3.68)
Ic	4	0.54-1.3	(2.835)	7	0.6175-1.75	(2.835)
IIa	10	0.45-1.44	(2.925)	13	0.51-1.43	(1.69)
IIa-IIb	9	0.44-1.1	(0.825)	17	0.5525-1.235	
IIb	27	0.56-1.87	(2.563)	30	0.39-1.5	(1.86)
IIb-IIc	3	0.69-1.08	(2.24)	6	0.325-1.755	(2.3)
IIb-IId	3	0.68-1.21	(1.7)	3	0.6825-0.96	(1.7)
IIc	9	0.825-1.733	(3.19)	12	0.51-1.875	(3.5)
IId	3	0.6175-5.035	(2.73)	3	1.08-3.72	(1.5)

* Numbers in parentheses indicate maximum size of multiple burial tombs containing at least one individual of the gender indicated. During the Naqada IId multiple burials are in fact larger than single burials, but they are so thoroughly plundered that the gender of the occupants could not be determined.

Fig. 2 shows the range in tomb size according to gender. When the categories above are grouped together by date we see that during the Naqada

I the seven single inhumation burials of men ranged from 0.5 to 1.6 sq. m. The eight identified tombs of women from that period ranged in size from 0.6 to 1.75 sq. m. The difference in size is only about one-tenth of a square metre in favour of the women's tombs. Tomb sizes for single burials of males ranged from 0.4 to 1.9 sq. m. (n=46) in the early half of the Naqada II (IIa-IIb). Single burials of women from the same period ranged from 0.4 to 1.5 sq. m. (n=60). In this instance the maximum size for the tombs of women was slightly less than the maximum size for men's tombs. In the latter half of the Naqada II (IIc-IId) the size of single male burials ranged from 0.6 to 5.0 sq. m. (n=12), while those of women during the same period ranged from 0.5 to 3.7 sq. m. (n=15).

It is apparent that there was an increase in the maximum size of single inhumation tombs for both men and women over time. While the numbers available seem to indicate an increase in the size of men's over women's tombs during the second half of the Predynastic period, this correlation is not secure. When the maximum tomb sizes for all identified segments of the Naqada II are examined, it can be seen that the highest values alternate between the male and female categories. Men appear to have the largest tombs in the Naqada IIa, IIb, IIb-IId and IId segments. The sizes of the women's tombs are larger during the Naqada IIa-IIb, IIb-IIc and IIc. Thus it seems prudent to infer that the apparent larger maximum size of men's tombs as compared to those of women is an artefact of the available sample rather than a reflection of actual cultural practice.

Tomb Richness and Diversity

Two statistics were used to indicate the diversity and richness of the artefact assemblages in the tombs. The data for this study were compiled from the published account of the excavation (Lythgoe 1965) and it should be noted that there are some discrepancies between that source and the objects now housed in the Lowie Museum of Anthropology. The first measure of burial richness presented is a straight count of each object in the tomb and the second measure is a ratio based upon the total artefact assemblage from the site. In the simple count the only objects not counted individually were beads. A few graves contained jewellery made up of over 700 beads. For this survey of burial richness each type of bead (type being identified by material) was counted as one (thus a necklace of seven beads, three ostrich eggshell and four carnelian would be counted as two). In a very few cases the number of types of beads was low while the actual count of beads was high. In an attempt to compensate for this disparity the artifact count was based on the number of jewellery items represented if that number was greater than the number of types of beads. For example, one adolescent woman from grave N7474 wore a bead bracelet on each wrist and a bead necklace. Some 600 beads were recovered, but in only two materials, blue glaze and carnelian (ibid.: 299). Instead of counting two items (glaze and carnelian), a count of three (two bracelets and one necklace) was taken. Jewellery items were not usually counted because most burials were disturbed and loose beads were all that remained in the graves. The simple count of objects is designated as the 'C Range' below (see Figs. 3 and 4).

To estimate the diversity of grave goods a ratio based on the total number of types of artefacts recovered from the cemetery was used. This is designated as the 'R Range' (see Figs. 3 and 4). In all, fifty-eight categories were used (see Appendix 1). These fifty-eight categories do not comprise the total Predynastic Egyptian artefact repertoire known. In order to keep the number of categories below a thousand individual pottery types and some artefact types were not distinguished; instead general classes were counted. For this reason there are only eight pottery types (B, P, R, D, W, S, C and N) (Petrie 1921; Needler 1984: 69) and six bead classes (local stone, foreign stone, whole shell beads, glazed materials [including glazed steatite and faience], metal and organic materials [bone, horn, ostrich eggshell and teeth]). The categories cover a wide range of artefacts from raw materials to fabricated objects. The more varied the types of objects from a grave, the higher the ratio (expressed as n/58). The ratio was intended to emphasise diversity when compared to the simple count of objects. Tombs with several similar pots or beads have a lower diversity index and a higher total artefact score.

Fig. 3. TABLE OF ARTEFACT DIVERSITY AND RICHNESS FOR ADULTS OF UNKNOWN SEX

Date	No.	R Range	Mode	Med	Mean	C Range	Mode	Med	Mean
Ia/b	8	0-3/58	1	1	1.25	0-5	1	1	1.5
Ic	14	0-4/58	1,2	2	1.8	0-4	1	2	2.0
I-IIa	52	0-2/58	0	0	0.3	0-3	0	0	0.3
IIa	14	0-19/58	1	1	2.8	0-21	1	1.5	3.4
IIb	25	0-3/58	1	1	1.36	0.8	1	1	2.16
IIc	32	0-11/58	1	2	2.75	0.27	1	3	5.0
IId	33	0-26/58	1	2	4.1	0-49	1	2	8.6

As expected, there was a general increase in both the diversity and quantity of grave goods over time (see Fig. 3). For adults of unidentified sex the diversity index ranged from a maximum of 4/58 in the Naqada I to a high of 26/58 at the end of the Naqada II and the total number of artifacts in these graves went from a maximum of five in the Naqada I to a high of forty-nine at the end of the Naqada II. The mean of the total artifact count was two at the end of the early Predynastic and over eight at its end. The median number of objects per tomb was one in the early Naqada I and two at the end of the Naqada II.

Fig. 4. TABLE OF ARTEFACT DIVERSITY AND RICHNESS FOR FEMALES AND MALES

Date	Sex/ No.	R Range	Mode	Med	Mean	C Range	Mode	Med	Mean
Ia/b	M (9)	0-8/58	0	1	1.55	0-8	0	3	1.7
	F (3)	1-3/58	3	3	2.3	1-4	1,3,4	3	2.7
Ic	M (17)	0-5/58	1	1	1.2	0-5	1	1	1.2
	F (25)	0-7/58	0	1	2.0	0-10	0	2	2.3
I-IIa	M (64)	0-4/58	0	0	0.3	0.4	0	0	0.3
	F (56)	0-3/58	0	0	0.3	0.3	0	0	0.3
IIa	M (18)	0-5/58	1	0.5	1.55	0-5	1,2	1.5	1.5
	F (20)	0-6/58	1	1	1.75	0-6	1	1	1.9
IIb	M (40)	0-9/58	2	2	2.3	0-13	4	3	3.4
	F (43)	0-8/58	2	2	2.65	0-11	2	2	2.8
IIc	M (17)	0-11/58	0,3	3	3.1	0-20	2,3,6	3	5.0
	F (30)	0-8/58	2	1	2.0	0-11	2	2	2.8
IId	M (5)	0-6/58	All	1	3.0	0-6	4	4	3.4
	F (5)	1-9/58	4	4	4.4	1-12	4,12	4	6.6

When broken down by gender the basic trend of increasing diversity and quantity of grave goods is seen for both sexes (see Fig. 4). The values for male and female artefact diversity and quantity during different phases of the Predynastic are erratic and it may be that the sample of securely dated tombs of persons of known gender is too small to give reliable results. At the end of the Naqada I the maximum value of the diversity index was 5/58 for males and 7/58 for females with means of 1.2 and 2.0 respectively. In the early Naqada II the diversity index reached a high of 9/58 for males and 8/58 for females, but with means of 2.3 and 2.65 respectively (index for males higher but mean number of types larger for females). In the later Naqada II this trend continued with the maximum value for the male index being 11/58 while the female index peaked at 9/58. Again, the mean of the artefact types was higher for women (4.4) than for men (3.0). This indicates that while a wider variety of objects was found in men's tombs, more objects were placed in women's tombs. The total count of artefacts presents a slightly different picture. During the Naqada II the median number of objects was higher in men's tombs than in those of women except in the Naqada IId. The mean number of artefacts was higher for men in the Naqada IIb and IIc, but in the Naqada IIa and IId it was higher for women. For the largest sample available, that broadly dated to the Naqada I to IIa, the number of grave goods was slightly larger for men than women (4 vs. 3) and the means, medians and modes for the C Range and R Range were the same for both sexes.

Although some children were buried with more grave goods than some adults, generally adults possessed more grave goods than children. It should also be remembered that in all periods there were individuals who were buried with few or no grave goods. A study of tomb contents for cemetery N 7000 revealed that almost half of the single inhumation burials (230 of 485) contained no grave goods. While some of these empty graves may be attributed to the work of tomb robbers, the magnitude of these figures is still significant. This class of grave is under-represented in the study because it is difficult to give an exact date to a grave which has no artefacts. One can sometimes make reasonable assumptions concerning the date of such a grave based on its location within the cemetery; however this is not always possible.

Specific Objects (See Appendix 2 for a listing of the burial number for each object mentioned in the text)

Only a few gender or age specific items were noted among the remains from N 7000. One of two types of objects exclusive to male burials throughout the period of use of this cemetery was the 'fishtail' flint. Of six fishtail flints reported, four came from male burials and two were buried with individuals of unknown sex. The other object apparently exclusive to male burials was the disk mace head. Only two of these were found in the cemetery and both had been buried with men. The tentative association of objects with males occurs in only one case. Of twelve instances of worked flint objects, two were associated with male burials and the other ten examples belonged to individuals of unknown sex.

The exclusive correlation of females with burial goods was found in only one case. Three of four graves with hairpins belonged to women. The sex of the owner of the fourth grave was undeterminable.

There are a few types of objects which appear to demonstrate a preference for burial with women, although they are not exclusively limited to individuals of that gender. There was a strong correlation between women and the tusks and 'tusk' amulets from this cemetery. These objects were often found associated with leather bags and were not apparently worn. Of fifteen burials with tusks (Naqada I to early II?) or 'tusk' amulets (Naqada II) ranging in date from the Naqada Ia/b to IIc, seven were associated with the bodies of women, two with burials of men and six were found with bodies of unknown sex. In three instances these objects were found with the bodies of children under five or six years of age. The number of tusks or 'tusk' amulets with a body varied from one to three. All three juvenile burials had only one tusk or 'tusk' amulet. Two tusks or 'tusk' amulets were the most common number found (one male, five female, two unknown) and a set of three was rare. Only one female and one male burial contained three 'tusk' amulets or tusks.

Bead bracelets, necklaces and anklets were found throughout the cemetery. Among adults and older juveniles (14-21 years) there was a strong preference for association with female burials (17 of 25). Only four male burials and four adult burials of unknown sex were associated with bead jewellery. Although there are so many unidentified individuals that definitive conclusions are dangerous, another possible case of association with women involved the blue stone beads which may be lapis lazuli, azurite or sodalite (Naqada IIc-IId). Of ten cases identified, four were associated with female burials and the other six cases belonged to individuals of unknown sex. Baskets are the last category of object examined which appear to have a higher incidence of occurrence with female burials. Of seventeen examples, seven were found with women, two with men and eight were from the burials of unsexed individuals. Two baskets were buried with children under the age of five.

Two classes of objects that were often associated with children and older juveniles were solid bangles and bead jewellery. Ring bangle bracelets or anklets of ivory, bone, horn, leather, wood and shell (Naqada I-II) showed a strong correlation with children, especially those under the age of five. Of fifteen bracelets or anklets, twelve were found on subadults; in fact, eleven were on children under the age of eight. One teenage girl wore a shell bracelet on each wrist, two older women had bangle bracelets and one woman wore a wooden anklet. Beads also appeared commonly on children. Twenty-eight of forty-four beaded bracelets, necklaces or anklets found *in situ* were buried with persons under the age of twenty-one.

All other classes of objects studied revealed a fairly even distribution between male and female burials. Only a few will be specifically mentioned. Flint knives of various forms (Naqada II) were found in six graves, four with persons of unidentified sex, one male and one female. The remains of eight bone or ivory combs were identified (Naqada I-II), two with male burials, two with female burials and four with persons of unidentified sex. Objects of unfired clay or mud in the shapes of humans, animals and possibly foodstuffs were identified in nine tombs (Naqada I-II). Examples were divided among three males, four females and two persons of unknown sex. In two cases children under the age of five had clay objects in their tombs.

Stone palettes used for grinding eye pigment were recovered from thirty-five burials. They were found in a rather limited variety of forms at this site, only five being identified. Twice as many female as male burials were identified with palettes (14 vs. 6); however, fifteen burials of unidentified sex also contained palettes and so this apparent preference for burial with women may not be real. When broken down by form the sample becomes quite small and therefore is probably not reliable. The most common form was the diamond or rhomboid palette with fourteen examples. One-third of these (5) were found with female burials, one came from a male burial and eight were found in burials of persons of unidentified sex. Five fish palettes were identified, four buried with women and the other with an individual of unknown sex. No male burials with fish palettes were noted. Double-bird-head palettes were found with two male and two female burials. One male burial contained a 'pelta' palette (Petrie 1920: 37). Irregular or fragmentary palettes were found with twelve individuals (four women, two men and six individuals of unknown sex). There were three examples of one individual being buried with more than one palette: one male and two female. Children were buried with examples of all but one ('pelta') of the types of palettes found at this cemetery.

For the pottery only the C, D and W wares will be considered. Of twenty White Cross-lined (C ware) vessels from the Naqada I, six were found with males, seven with females and seven with burials of unknown sex. In one case an eight- to ten-year-old male was buried with a C ware piece. It may be significant, however, that the three bowls decorated with animal motifs (hippopotami and crocodiles, hippopotamus only, and horned animals) were all found with male burials. The more common geometric designs were found with both male and female burials. The red line decorated wares (D wares) were found in thirty graves (Naqada IIb-III). Five of these were identified as male, six female and nineteen were with persons of unknown sex. In one instance a child under the age of five was buried with a D ware piece. An analysis of the principal design motifs of spiral circles, wavy lines, boats and birds indicated no apparent preference for gender. The last ceramic category to be considered, the Wavy-handled jars, were found in twenty-two graves (Naqada IIb-III). The graves which contain these vessels were often heavily plundered and the sex of only three individuals (two male and one female) could be identified. It is interesting to note that no subadults were buried with W ware jars.

In the latter half of the Naqada II period it became common to line graves with wooden panels. Some of the larger graves contained multiple inhumations; however, in thirteen instances single burials were contained in wood-lined graves. Nine of these graves contained persons of unknown sex. Tombs with wood linings for two males and two females were also identified. In multiple burials from seven graves lined with wooden panels, three females, one male and seventeen unidentified individuals were recorded.

Discussion

Over the years there has been some discussion in the literature concerning the theory that differences in the size and richness of Predynastic graves were related to the gender of the tomb occupant. Guy Brunton observed that during the early Badarian period the tombs of women were generally larger and richer than those of men, but that later in the Badarian period men's tombs became larger and that men were more frequently buried in rectangular graves later in the Predynastic (Brunton 1948: 9, 17). Elise Baumgartel theorized that Egyptian society of the Naqada I period was matriarchal (Baumgartel 1975: 30) and that women occasionally ruled during the Naqada II as well (Baumgartel 1960: 142). In his survey of Predynastic and Early Dynastic Egyptian cemeteries, Juan José Castillos reported that in most cemeteries, including that of Naga-ed-Dêr, women's tombs were generally wealthier than those of men (Castillos 1982: 68), although he also states that most rectangular tombs belonged to men and that rectangular tombs tended to be more richly endowed than other forms (*ibid.*: 67). Most recently Fekri Hassan has continued the theory of matrilineality, in part based on the reported preponderance of female graves from Merimde (Hassan 1988: 169).

From the information obtained through the study of skeletal biology and grave goods at Naga-ed-Dêr it appears that there is no compelling evidence in support of the theory that women were given preferential or 'better' treatment in burial than men. In fact, the slightly larger maximum size of some men's tombs and the greater diversity of grave goods from male burials may indicate just the opposite. For single inhumations, grave size was more strongly affected by date than by gender. The summary of the range of tomb size over time indicates that graves in all sizes were available to both sexes. An examination of the richness and quantity of grave goods indicates that, overall, the average woman's grave probably contained about the same number of objects as the average man's. Men's graves did hold a wider variety of objects. However, the graves of women were not remarkably poorer than those of their male counterparts. The burials of children exhibited economy of grave size, fewer objects in general and were more often in multiple burials with other children or adults, than were adult burials. Children were, however, more likely to be buried with jewellery than adults.

Only a limited number of specific types of grave goods appears to have been the exclusive property of either men or women. Both the disk mace heads (males only) and the hairpins (females only) probably carried a functional association: men were involved in warfare and, as we know from the physical remains at the site, women generally wore their hair longer (Podzorski 1990:

85). The function of the fishtail flints (male only) is uncertain. It may have been a ritual object or it may have functioned in some more mundane capacity (Harvey 1988: 73). Neither wealth in general nor ritual objects of any particular class were found exclusively in the graves of women.

A few objects sometimes thought to be associated with male or female roles were also examined. Flint knives, clay figurines, combs and a number of other artefacts were found in approximately equal numbers in the graves of both men and women.

During all phases of the Predynastic period many individuals were buried without grave goods. In the Naqada II, among those graves containing objects, a significant increase in the quantities of goods occurred, and towards the end of the period there was an increase in the size of tombs of both men and women. This seems to indicate a period of increasing cultural complexity and unparalleled local prosperity, a prosperity in which both sexes appear to have shared.

References

Baumgartel, Elise, 1960. *The Cultures of Prehistoric Egypt* II, Oxford University Press, London.

Baumgartel, Elise, 1975. 'Some Remarks on the Origins of the Titles of the Archaic Egyptian Kings', *Journal of Egyptian Archaeology* 61, 28-32.

Brunton, Guy, 1948. *Matmar. British Museum Expedition to Middle Egypt, 1929-1931*, Bernard Quaritch, Ltd, London.

Brunton, Guy and Caton-Thompson, Gertrude, 1928. *The Badarian Civilization and Predynastic Remains near Badari*, Bernard Quaritch, Ltd, London.

Castillos, Juan José, 1982. *A Reappraisal of the Published Evidence on Egyptian Predynastic and Early Dynastic Cemeteries*, Benben Publications, Toronto.

Cowgill, George, 1989. 'The Concept of Diversity in Archaeological Theory', in *Quantifying Diversity in Archaeology*, Robert Leonard and George Jones (eds.), Cambridge University Press, London, 131-141.

Dunnell, Robert, 1989. 'Diversity in Archaeology: A Group of Measures in Search of Application?', in *Quantifying Diversity in Archaeology*, Robert Leonard and George Jones (eds.), Cambridge University Press, London, 142-49.

Friedman, Renée, 1981. *Spatial Distribution in a Predynastic Cemetery: Naga ed Dêr 7000*, MA Thesis, University of California, Berkeley.

Harvey, Stephen P, 1988. 'Fish-tail knife', in *Mummies and Magic. The Funerary Arts of Ancient Egypt*, Sue D'Auria, P Lacovara and C Roehrig (eds.), Museum of Fine Arts, Boston.

Hassan, Fekri, 1988. 'The Predynastic of Egypt', *Journal of World Prehistory* 2, 135-185.

Lythgoe, Albert M, 1965. *The Predynastic Cemetery N7000. Naga-ed-Dêr*, Part IV, Dows Dunham (ed.), University of California Press, Berkeley and Los Angeles.

Mann, George E, 1989. 'On the Accuracy of Sexing of Skeletons in Archaeological Reports', *Journal of Egyptian Archaeology* 75, 246-9.

Needler, Winifred, 1984. *Predynastic and Archaic Egypt in the Brooklyn Museum*, Brooklyn Museum, New York.

Petrie, W M Flinders, 1920. *Prehistoric Egypt*, Bernard Quaritch, London.

Petrie, W M Flinders, 1921. *Corpus of Prehistoric Pottery and Palettes*, Bernard Quaritch, London.

Podzorski, Patricia V, 1990. *Their Bones Shall Not Perish. An Examination of Predynastic Human Skeletal Remains from Naga-ed-Dêr in Egypt*, SIA Publishing, New Malden, England.

Reisner, George A, 1908. *The Early Dynastic Cemeteries of Naga-ed-Dêr*, Part I, J C Hinrichs, Leipzig.

APPENDIX 1: ARTEFACT CATEGORIES

Pottery:
- B (Black Topped Red)
- P (Polished Red)
- R (Rough Faced)
- D (Decorated)
- W (Wavy-handled)
- S (Smooth Faced, Hard Orange)
- C (White Cross-lined)
- N (Incised/Nubian)

Palettes, cosmetic
Rubbing/Grinding Pebble
Beads:
- Local Stone
- Foreign Stone
- Shell, whole
- Glazed
- Metal
- Organic

Comb, ivory or bone
Hairpin, ivory or bone
Garment Pin, ivory or bone
Bangle Bracelet, organic
Bangle Anklet
Hair Balls
Bag, leather or cloth
Sandal
Basketry
Tomb Lining, wood panels or twigs
Animal Bones
Grain/Seeds
Horn
Bivalve/Snail Shell
Stone Vessel
Ivory Vessel
Faience Vessel
Seal, cyliner or stamp
Mace head
Mortar
Pestle
'Marbles'
Tool, wood, ivory or bone
Tool, metal
Mud or Clay 'Loaf'
Flint Chips
Flint worked
Tusk, worked
'Tusk' Amulet
Amulet, shell, horn or bone
Amulet, stone
Charcoal
Horn, Ivory or Bone Object
Reed Object
Clay Object
Stone or Mineral Object
Metal Object
Leather or Cloth Object
Ochre
Malachite
Galena
Resin

APPENDIX 2: LOCATIONS OF SPECIFIC TYPES OF ARTEFACTS

Fishtail flint - N7014, N7016B, N7120, N7271A, N7454A, N7625B

Disk macehead - N7373A, N7489A

Worked flint - N7121A, N7185B, N7189, N7284, N7304, N7418A, N7442, N7519, N7538, N7539, N7568, N7595A

Hairpin - N7130C, N7201B, N7304, N7532

Tusks and 'Tusk' amulets - N7014, N7072, N7083, N7087, N7130D, N7136, N7150, N7188, N7202B, N7235, N7293, N7444, N7469B, N7598, N7623B

Beads - N7003C, N7035B, N7036A, N7038B, N7061A, N7068A, N7068E, N7110C, N7115A, N7116C, N7116F, N7124, N7130B, N7130C, N7130D, N7132A, N7135, N7164, N7197E, N7214A, N7229, N7265, N7327B, N7333B, N7357, N7375C, N7386B, N7426, N7429C, N7446, N7458B, N7474B, N7484, N7485B, N7501, N7558, N7560A, N7565, N7566, N7604B, N7610B, N7623B, N7624B, N7626B

Blue stone beads - N7235A, N7290, N7304, N7338, N7461A, N7525A, N7534, N7538, N7540, N7546

Baskets - N7130D, N7202A, N7203A, N7269B, N7293, N7331, N7368A, N7394D, N7439, N7454C, N7459, N7464, N7469B, N7526, N7595A, N7595B, N7626B

Bangles - N7006B, N7011B, N7063, N7115A, N7115D, N7124, N7130D, N7130E, N7132A, N7375C, N7379, N7394E, N7558, N7566, N7626B

Flint knife - N7151, N7304, N7454C, N7491, N7583, N7595A

Comb - N7129C, N7150, N7266B, N7327B, N7328, N7394D, N7417A, N7626A

Mud or clay figures - N7008A, N7016A, N7024A, N7035A, N7037C, N7124, N7245, N7394A, N7462

Palettes - N7003B, N7008B, N7008C, N7021, N7102, N7114, N7115A, N7121B, N7129C, N7202B, N7234B, N7247A, N7290, N7299, N7375C, N7392A, N7402, N7434, N7438, N7447, N7484, N7485A, N7485B, N7487A, N7509, N7519A, N7522A, N7534, N7539, N7548, N7590, N7623A, N7626B, N7627A, N7630

C ware - N7014, N7016A, N7016C, N7036C, N7052, N7128A, N7128B, N7128C, N7129C, N7130A, N7130B, N7134, N7155, N7179E, N7202A, N7203A, N7375A, N7392B, N7625B, N7626B

D ware - N7086, N7113, N7117, N7151, N7248, N7254A, N7265, N7266A/B, N7268A, N7272, N7290, N7301, N7338, N7402, N7413A, N7415A, N7418A/B, N7461A, N7482, N7485B, N7502A, N7510, N7521A, N7522A/B, N7523, N7539, N7543A, N7544, N7546A, N7547A

W ware - N7113, N7151, N7271A, N7273, N7290, N7303, N7304, N7309, N7402, N7418A, N7442, N7510, N7512, N7517, N7519A, N7520A, N7522A, N7526, N7531, N7534, N7538, N7539

Tomb lining - N7138, N7185, N7190, N7235, N7265, N7271, N7292, N7338, N7418, N7442, 7454C, N7519, N7521, N7522, N7526, N7531, N7533, N7534, N7539, N7548

CRANIAL INJURIES IN THE GUANCHE POPULATION OF TENERIFE (CANARY ISLANDS): A BIOCULTURAL INTERPRETATION

Conrado Rodriguez Martin, Rafael González Antón,
Fernando Estévez González

Introduction

According to Alexandersen (1967), traumatic lesions of bones have been observed in skeletal remains of man from all periods, injuries to the skull being more frequent than injuries to any of the other individual bones in the skeleton. Ortner and Putschar (1985) state that injuries to the skull are rather common in archaeological specimens and that most fractures of the skull appear to be the result of intentional violence rather than accidents. Manchester (1983) says that the incidence of intentional trauma varies from skeletal sample to skeletal sample. He adds that, although injury in general is more common in the male, the sex difference in skull injury is perhaps more striking. However, Alciati, Fedeli and Pesce (1987) have found that, although there is a higher incidence of skull injury among males, it is also well attested in females. In their opinion, this means that women in antiquity actively took part in warfare.

Cranial Injuries in the Guanche Population of Tenerife

The Canaries consist of seven islands, mountain peaks of volcanic origin that rise out of the Atlantic Ocean close to the coast of Northwest Africa. The origins of the first people to colonize the archipelago lay somewhere in the latter area. The date of their immigration is not certainly known but archaeologists are of the view that it might have taken place at the beginning of the Christian Period. In general, cattle raising, fishing and agriculture formed the basis of their ecomony.

Tenerife is a very high island with many mountains, ravines and valleys. The climate varies in the different parts of the island, being humid in the north and dry in the south. The main economic activities among the Guanches (the pre-hispanic inhabitants of Tenerife) were shepherding (the most important in the opinion of archaeologists and prehistorians), agriculture, hunting and gathering.

The Spanish conquest of the Canaries began on the island of Lanzarote in 1402, progressing to the other islands during the next hundred years and finishing with the conquest of Tenerife in 1496.

Material and Methods

The number of skulls examined amounted to 492. These skulls belong to the collection of the Museo Arqueológico de Tenerife. We selected only Tenerife crania with known provenance and in good condition for examination. It is important to realize that all the skulls and skeletons of the museum's collection are pre-hispanic. The dates of these specimens fluctuate between AD 200 and AD 1450. Details of their sexual distribution are as follows: 259 males; 149 females; 21 of unknown sex, with a male-female ratio of 1.74. The ages of the specimens, excluding the subadults, range from 14/17 to 55/60.

All the specimens were macroscopically examined for the presence/ absence of injuries. We also measured the lesions (including the depth) and described them. Following Mensforth, Surovec and Cunkle (1987), all the fractures that were identified in our initial survey were then re-examined. This procedure ensured that other pathologies would not be mistakenly diagnosed as traumatic injuries.

Results

Incidence and Sexual Distribution of Cranial Injuries

We have observed thirty fractures in this serie. This represents 6.99% of the total, which is relatively high. The sexual distribution of the injuries is as follows: males, twenty-two cases in 259 individuals = 8.55%; females, eight cases in 149 individuals = 5.37%.

Geographic Distribution

It is interesting to note that the majority of fractures occur in the south of the island (twenty-six cases in 265 individuals or 9.18%, with 11.63% for males and 7.55% for females, while in the north there appear only four fractures; this is 2.44%, 2.3% for males and 2.9% for females). The incidence-ratio south-north is 4.02, markedly skewed in favour of the south. The highest incidence in the south was seen in San Miguel de Abona, in a place called Barranco de Orchilla, a deep ravine where the incidence of cranial injuries was 17.65% and the lowest in El Rosario (in a place called El Chorrillo), with an incidence of 7.35%, while in the north the locality with the highest incidence was La Laguna with 3.22%. It is interesting to note that in the northern part of Tenerife the incidence was superior in females.

Age and Skull Fractures

We have not observed a single case among sub-adults and only three cases among people over 35. The age-group/skull-fracture relationship is as follows:

- 14-17: one case (female)
- 18-23: eight cases (7 males and 1 female)
- 20-25: four cases (all males)
- 25-30: eight cases (6 males and 2 females)
- 30-35: six cases (4 males and 2 females)
- 35-40: one case (male)
- 40-45: two cases (females).

Twenty-six cases are attested (86.66%) in people between the ages of 18 and 35. This range of age corresponds with the period of greatest activity in the life of the individuals and also with the ideal age for hand-to-hand fighting. We have divided the skull injuries into five different types:

1. Depressed fractures or small crushing fractures: fourteen cases representing 46.66%. We have subdivided this group into two categories:
   - circular in shape, possibly due to 'bolas' (a missile stone used by the Guanches): three cases (10%).
   - irregular, possibly due to missile-stone: eleven cases (36.66%).
2. Simple linear fractures: six cases (20%).
3. Penetrating wounds: four cases (13.3%).
4. Gross crushing injuries: two cases (6.66%).
5. Incised wounds: one case (3.33%).

There are also three cases with more than one contemporary lesion. It is interesting to describe them:

1. Male, 20-25, with three different lesions:
   - depressed fracture, irregular in shape, in the left side of the frontal bone.
   - depressed fracture, circular, in the right parietal.
   - a penetrating wound in the left parietal.
2. Female, 30-35, with a simple linear fracture in both parietals and penetrating wound in the occipital bone.
3. Male, 20-25, with a simple linear fracture in both parietals and penetrating wound in the left side of the frontal bone.

We can see that the most important group is that of depressed fractures (nearly 50%), which are clearly caused by deliberate violence (face-to-face or hand-to-hand fighting and shooting of missile stones). Most cases occur in individuals between 18 and 30 and males. The second group is that of simple linear fractures (20%) and here the age is not so clearly related to the ideal age of fighting, fluctuating between 18 and 45. Some fractures could, therefore, have been caused by accidents. Two other groups clearly arising from deliberate violence comprise penetrating wounds (four cases, all between 20 and 30) and incised wounds (one male between 30 and 35).

In sum, most of the fractures appear in individuals under the age of 35, the majority being males, and in our opinion are of a non-accidental nature.

Topographical Distribution

The frontal bone is the most affected with sixteen cases (seven cases on the left side and nine on the right). The parietals are affected in fourteen cases: eight on the left side, two on the right and four on both. The fronto-parietal region is affected in two cases (right side). Additional parts affected, far from these two regions, are the malar and occipital bones and the glabelar region.

Of the areas affected, we can observe a clear predominance of the anterior and lateral regions of the skull, which in the opinion of almost all authors is suggestive of lesions caused by violence.

Pathology

We have observed the pathology of skull injuries on the basis of the following data:

1. Effect on the inner table: in all cases the outer table was affected and also the diploe but not always the inner table: in six cases the lesion was limited to the outer table and diploe, four cases were simple linear fractures, one was an incised wound and one a depressed circular fracture.
2. Healing: in only five cases have the lesions of perimortem been diagnosed (three depressed fractures, one penetrating and one linear). We have followed the criteria of Maples (1986) in the differential diagnosis of postmortem fractures: concentric circular fracture lines; radiating lines and stellate fractures that are not commonly produced on a dead and non-elastic skull.
3. Radiating fracture lines from the central point only appeared in three cases.
4. The size of the lesions changes depending on type:
   - in depressed fractures the mean was 3 cm which clearly corresponds with the size of a missile-stone or 'bolas'.
   - simple linear fractures have a mean of 13.35 cm.
   - penetrating wounds have a diameter of 1.8 cm (which corresponds with the diameter of a Guanche javelin).
   - in crushing injuries the mean of the area was 45 cm^2.
   - incised wounds: the single case is 6.8 by 2.3 cm in an L Shape.
5. Signs of infection: we have observed clear signs of infection in 20% of cases.
6. Defect in the bone: in 20% there are indications of very clear defect in the bone and these defects persisted for the remainder of the individual's life.

Survival and Complications

Only five cases were diagnosed of perimortem; the remainder show clear signs of healing. This does not mean, of course, that the individuals did not suffer postraumatic complications and that after some time could not have died owing to these complications. The complications that may be suspected in this sample are:

- loss of the eyes: two cases with lesions of the bone eye orbit, almost certainly affecting the eyeball.

- epilepsy: due to the depth of the lesions and the consequent effect on and disposition of the inner table, it is possible that epilepsy could have occurred in seven cases (23%) which is very similar to the incidence given by Rogers (1984) for this kind of lesion (30%).

Cultural Interpretation

It is difficult to reconstruct the details of the daily life of the Guanches, since the ethnohistorical sources, which were written by the Spaniards after the Conquest, are poor in data regarding Tenerife.

With regard to the origin of fractures, we can divide them into two main groups:

1. *Accidental*: these fractures are most commonly found in the post-cranial skeleton. It is clear that there were accidents due to the difficult orography of the ground (high mountains, big valleys and deep ravines) and the Guanche way of life, which was mainly shepherding. These data are confirmed by Abreu Galindo (1602). We have in our sample some examples of simple linear fractures that could be classified as 'accidental', and these fractures appear mostly in individuals over the age of 35, who according to I. Schwidetzky (1963) are to be classed as 'old people'.
2. *Fractures due to violence*: most skull fractures may be ascribed to this category. In general, fractures may be explained by the fact that the Guanches engaged in a great deal of fighting. Note Abreu Galindo's statement (op. cit.) that 'War among the Guanches is due to conflicts over territory and pastures for the cattle'. Tenerife is a small island, which was divided in Guanche times into nine kingdoms with variable limits depending on their respective power at any one time. Thus, the geopolitical background was an influential factor in war. Fractures appear with higher incidence in males. In Guanche society the division of work was clearly based on sex and age. War seems to have been the exclusive preserve of men, although Abreu Galindo (op. cit.) says 'Women went to war to bring food for the warriors and if these warriors were killed the women had to bury them in the burial caves'. So, if the women did not take part actively and directly in the battles, these fractures in females may have been caused by accidents suffered during the course of war (and perhaps in some cases participation in war) or by physical punishment suffered as the result of social custom; for example, since the men were not allowed to speak to women under penalty of death, it is possible that in some cases women and men were punished because they had not respected this custom.

As for the weapons used, we can affirm that they were made of stone and wood, because metals do not (and did not) exist in the Canaries. The historians of the sixteenth and seventeenth centuries (Torriani, Espinosa and Abreu Galindo) confirm these data.

Discussion

Verneau (1887) states that 25% of his male crania from Tenerife showed some kind of lesion. Luschan (1896) also found 25% of cranial injuries among the Guanches but did not identify the location and kind of lesion. We can not compare our results with those of Verneau and Luschan, as they only give the incidence of lesion, which in both cases seems very high, and omit to provide information on topography, type, duration of survival, etc.

We very largely agree with Hooton (1925), who affirms that the pathological features of the various series of Guanche skulls are almost entirely of traumatic origin. The most common lesions are healed depressed areas of all sizes, usually found on the frontal bone, often on the parietal and most commonly on the left side due to right-handed blows. The sexual incidence was 9.5% for males and 8.6% for females (our female incidence is 3% less than that of Hooton). For the American anthropologist, healed depressed fractures of the skull were observable in six males and in two females; three of the former were fractures of the left parietal, one of the right parietal and one of the frontal bone. Healed linear fractures were noted in six males. Bosch Millares (1944 and 1961-62) divides the cranial injuries in the pre-hispanic population of the Canaries into three types:

1. Incised wounds caused by wooden swords.
2. Crushing injuries caused by maces or clubs.
3. Penetrating wounds caused by wooden knives, obsidian knives ('tabonas'), arrows or javelins.

We agree with this classification but would add two more types: depressed fractures and simple linear fractures. We also agree with the latter author when he affirms that most skull fractures among the Guanches occur on the cranial vault, especially on the frontal and parietal bones, being rare in the facial bones. Perez (1980-81) agrees with Bosch on the types of lesion but the problem is that she too fails to give statistics. Our statistics for North Tenerife are consistent with those of Garcia Sánchez (1979) in Pino Leris (la Orotava), where he found 2.56% of cranial injury, with the main location on the left side of the cranial vault. Gutiérrez González (1986) found an incidence of 4.92% of cranial injury in Tenerife, which is lower than our statistic (6.99%). His sexual distribution was similar to our incidence. He could not find skull fractures in subadults.

Conclusions

1. The incidence of cranial injury among the Guanches is high, especially in the south of Tenerife.
2. Skull fractures are much more frequent than post-cranial fractures.
3. Fractures appear only in adults with the male-female ratio showing that they are nearly twice as common in males.
4. The main type is the depressed fracture, irregular in shape, which was clearly caused by violence deliberately inflicted.
5. The anterior and lateral parts of the skull are much more frequently affected than the posterior part and the facial skeleton. This supports the hypothesis that the fractures were caused by deliberate violence.
6. Healing occurs in most cases and only 17% of perimortem was diagnosed.
7. Warfare played an important part in the life of the population of Tenerife during the prehistoric period.

References

Abreu Galindo, Fr J de, 1977 (1602). *Historia de la Conquista de las Siete Islas de Canaria*. Sta Cruz de Tenerife, Goya.

Alciati, G, M Fedeli, V Pesce Delfino, 1987. *La Mallatia dalla Preistoria all'etá antica*. Bari, Laterza.

Alexandersen, V, 1967. 'The evidence for injuries to the jaws', in Brothwell, D R, and Sandison, A T (eds.), *Diseases in Antiquity*. Springfield, Charles Thomas, 623-34.

Bosch Millares, J, 1944. 'Las armas y fracturas de cráneo de los Guanches', *El Museo Canario*, V, 6-29.

Bosch Millares, J, 1961-62. 'La medicina canaria en la época prehispánica', *Anuario de Estudios Atlánticos*, 7, 539-620; 8, 11-63.

Brothwell, D R, and Sandison, A T, 1967. *Diseases in Antiquity*. Springfield, Charles Thomas.

Courville, C B, 1967. 'Cranial injuries in prehistoric man', in Brothwell D R, and Sandison, A T, *Diseases in Antiquity*, Springfield, Charles Thomas, 606-622.

Diego Cuscoy, L, 1968. *Los Guanches. Vida y Cultura del Primitivo Habitante de Tenerife*. Sta Cruz de Tenerife, Publ. Museo Arqueológico, 7.

Espinosa, Fr A de, 1980 (1596). *Historia de Nuestra Senora de Candelaria*. Sta Cruz de Tenerife, Goya.

Garcia Sánchez, M, 1979. 'Paleopatologia de la población aborigen de la cueva sepulcral de Pino Leris (La Orotava, Tenerife)', *Anuario de Estudios Atlánticos*, 25, 567-584.

González Antón, R, 1983. 'Conquista y aculturación de los aborigenes de Tenerife', *Gaceta de Cararias*, 3, 35-48.

González Antón, R and A Tejera Gaspar, 1981. *Los aborigenes canarios (Gran Canaria y Tenerife)*. La Laguna, Secretariado de Publicaciones. Universidad de La Laguna.

Gutiérrez González, F, 1986. *Variaciones anatómicas y patologia orbitaria de la población prehispánica de las Islas Canarias*. La Laguna, Tesis Doctoral (unpublished).

Hooton, E A, 1925. *The Ancient Inhabitants of the Canary Islands*. Cambridge, Mass., Harvard African Studies VII.

Luschan, F von, 1896. 'Uber eine schädelnsammlung von den Canarischen Inseln', in Meyer, H, *Die Insel Tenerife*. Leipzig, Hirzel, 285-319.

Manchester, K, 1983. *The Archaeology of Disease*. Bradford, University of Bradford.

Maples, W R, 1986. 'Trauma analysis by the forensic anthropologist', in Reichs, K J, *Forensic Osteology*, Springfield, Charles Thomas, 218-228.

Mensforth, R P, S A Surovec, and J R Cunkle, 1987. 'Distal radius and proximal femur fracture patterns in the Hamann-Todd skeletal collection', *Kirtlandia* 42, 3-24.

Meyer, H, 1896. *Die Insel Tenerife*. Leipzig, Hirzel.

Ortner, D J, and W G J Putschar, 1985. *Identification of Pathological Conditions in Human Skeletal Remains*. Washington, Smithsonian.

Perez, P J, 1980-81. 'Nueva aportación paleopatológica acerca de la población prehispánica canaria', *El Museo Canario*, XLI, 29-45.

Reichs, K J, 1986. *Forensic Osteology*. Springfield, Charles Thomas.

Rodriguez Martin, C, 1985. 'Fracturas por agresión producidas en cráneos prehispánicos de Tenerife', *CCPC Parlamento de Canarias*.

Rodriguez Martin, C, 1989. 'Los traumatismos craneales en la prehistoria de Canarias', in *Historia y Sociedad* (in press).

Rogers, S L, 1984. *The Human Skull*. Springfield, Charles Thomas.

Schwidetzky, I, 1963. *La Población Prehispánica de las Islas Canarias*. Sta Cruz de Tenerife, Publ. Museo Arqueológico, 4.

Torriani, L, 1978 (1592). *Descripción e Historia del Reino de las Islas Canarias*. Sta Cruz de Tenerife, Goya.

Verneau, R, 1887. 'Rapport sur une mission scientifique dans l'Archipel Canarien', *Arch. des Mis. Scient. et Litt.*, 3, 13, 592-812.

THE NUBIAN PATHOLOGICAL COLLECTION IN THE NATURAL HISTORY MUSEUM, LONDON

Theya I Molleson

The Nubian Pathological Collection

The Nubian pathological collection of bone specimens at the Natural History Museum, London, is part of a larger collection of bones collected for Sir Grafton Elliot Smith during the Archaeological Survey of Nubia.

The Archaeological Survey of Lower Nubia was undertaken by the Egyptian Government in order to discover and record the historical material which would otherwise have been lost when the district was submerged by the filling of the New Aswan Reservoir to the level of 113 metres above the sea. More than 6,000 bodies were examined and only some of those with external evidence of pathology were retained for museums. The majority became part of the collection of the Hunterian Museum at the Royal College of Surgeons.

In May 1941 the Royal College of Surgeons suffered severe bomb damage and large parts of both the collections and the records were destroyed. Most of the human skeletal material that survived was transferred over a period of years (1948-1968) to the Natural History Museum. It included the Nubian Pathological Collection, together with its card catalogue, although a few specimens were retained by the College. In 1968 the Nubian specimens from University College London were added to the collection.

The collection now comprises 124 specimens, which have been catalogued (retaining where possible the original RCS number), photographed and radiographed. Provenance and date have been checked against the cemetery and grave numbers listed in the reports of the Archaeological Survey of Nubia (Reisner 1910); so far over 100 subjects have been identified. New assessment of the pathologies represented based on the radiographic examination has been carried out and will be presented in detail elsewhere.

The collection is well known to anthropologists and palaeopathologists and has been used many times to illustrate examples of bone disease (see, for example, Brothwell and Chiarelli 1973; Brothwell 1981; Brothwell and Sandison 1968; Ortner and Putschar 1981; and also Buikstra et al. in this volume).

Formation of the Collection

The excavations for the Archaeological Survey of Nubia were directed by George Reisner in the first season, 1907-1908, and by C M Firth in the next three seasons, 1908-1909, 1909-1910 and 1910-1911 (Reisner 1910; Firth 1912, 1915, 1927; Adams 1977, 71-4). Thanks to the climate and geographical conditions, the remains had been preserved in remarkable condition - stone, wood, metal, woven fabrics and even delicate animal organisms. The chronology, moreover, of this remarkable civilization is, comparatively speaking, reasonably well worked out (Reisner 1910; Adams 1977, 3-8 and 16, fig. 2).

The problem of first consideration to the Archaeological Survey of Nubia was the collection of the material. The work fell naturally into three divisions:

1. The discovery of cemeteries, buildings and other remains left by man.
2. The excavation of these remains when located.
3. The recording of the material found.

Reisner's rate of progress was impressive as can be seen from the summary of his itinerary for the first season:

1907	Sep	2-8	-	Survey from Shellal to Dakka
		9-18	-	Cairo for stores and equipment
		19	-	Shellal with 20 men
		20	-	Camp set
		21-24	-	Reconnoitered 9 cemeteries,
			-	photographed 5, took on 61 men
		25 - 3 Oct	-	Excavation of cemetery 2,
			-	workforce increased to 190 men
	Oct	4-15	-	Cemetery 3 and vaults of 2 excavated
		7-12	-	Reconnoitered cemeteries 10-16
		18	-	Excavating cemeteries 2, 3, 5, 7, 8
	Oct-Dec		-	Work gangs moved on as they finished
			-	a cemetery
	Dec	11	-	Packing at Shellal
		17	-	Moved to Khor Ambakol to excavate
			-	cemeteries 12-18 (last done on Jan 1)
	Jan	2	-	Dabod, cemetery 22
		2-8	-	Gangs shifted
		11-21	-	Camp moved to Dabod for cemeteries 22-24
		21-26	-	Cemetery 26
		30-31	-	All on cemetery 25
	Feb	13	-	Cleared 8 cemeteries
	Feb		-	Moved to Birein, Khartum and Wadi Qamar;
			-	recorded cemeteries 30-36 and 37-42
		24	-	All on cemetery 43
		25	-	Cemeteries 44 and 45 excavated
		26-29	-	All on cemetery 45
		26	-	Moved to Dehmit; cemeteries 43-49
	Mar	7	-	Moved to Seati; cemeteries 50-53
		10	-	Ginari
		13	-	Metardul; cemetery 50
		17-29	-	Ginari; recorded cemeteries 53-56
		24	-	87 men laid off
		29	-	Returned to Shellal
		30 - 2 Apr	-	2 men packed up antiquities.

The sites were excavated as a whole, except that to avoid duplication of material certain late cemeteries were excavated 'in skeleton'. This was done in particular for Roman and Christian graves after excavation of several cemeteries (2, 3, 5, 8, 15, 24, 25), which contained nearly 3,000 graves of the Ptolemaic-Roman and Christian Periods.

The anatomical work was undertaken by Dr Grafton Elliot Smith, then on the staff of the School of Medicine in Cairo. He made a special study of the racial characteristics of the people being excavated. He was able to show that at least in the earlier part of the period the people of Lower Nubia, in physique as in culture, were indistinguishable from the people of contemporary periods in Egypt. Later there were immigrations of Africans from the south who mingled with the indigenous population to an extent that Smith came to recognise a Nubian people.

Elliot Smith had had eight years experience of Egyptian material derived from various sources and especially from the large series of skeletons of every period from both Upper and Lower Egypt obtained by the Hearst Egyptian Expeditions of 1905 to 1906. This gave him confidence in the recognition of physical types. He had also learned to recognise clustering of various non-metric traits - genetic anomalies of the cranial sutures, teeth, vertebrae and scapula - as evidence for family groupings among the burials.

He was assisted in the anatomical study of the skeletons and mummies, as they were unearthed, first by Dr F Wood Jones then by Dr D E Derry, both of whom paid attention to the evidence for pathology in addition to taking field measurements of the crania and of the limb bones for estimation of stature. It was possible, however, to take measurements in a small percentage of cases for so many of the graves had been disturbed by local peasants digging for mud to put on their allotments. Nor were they adverse to robbing the graves of the treasures buried with the body.

The fieldwork of the second season was resumed on 1 October 1908 and completed on 24 March 1909. Dr D E Derry assisted by Mr H W Beckett collected and measured the anatomical material (Firth 1912). In the third season, 1909-1910, anatomical work was deferred and arrangements were made for all the bones which were in sufficiently good condition to travel to be packed and shipped to England (Firth 1915). In the fourth and last season, 1910-1911, no anatomist was available for the field work but material was sent to Cairo and studied there by Elliot Smith (Firth 1927).

The recording of the finds included a topographical map of each site and a detailed map of each cemetery. The cemeteries were numbered consecutively from 1 to 55 in the first season and to 123 in subsequent seasons. Records were kept methodically. To save time in the field and to facilitate classification of their notes afterwards, millimetre squared cards were prepared with all the items of information to be sought printed alongside blank spaces. These data cards - perhaps the first ever printed - were of a blue colour 'to protect the eyesight from the glare of the sun'. (Smith and Jones 1910). A record was made of each grave with drawings and notes. The tomb-card was marked with the number of the cemetery and, below, the number of the grave. When there were several burials, they were lettered A B C etc. The notes on the age, sex, and race of the skeleton were supplied by Elliot Smith and Wood Jones. Practically every grave which contained anything at all was photographed at least once. Except for those which were published, it is doubtful if any of the records have survived.

The arrival of the collections in London must have created great interest (overshadowed only by the discoveries at Piltdown) and various pathologists, experts in their own field, were, it seems, allowed to remove especially delightful cases to their own offices. This before they were fully catalogued. Many specimens doubtless were never returned to the Royal College and are now lost to the collection. Two skulls were re-donated having been given by Elliot Smith to Dr T S P Strangeways. A third group of specimens has no reference in the reports of the Archaeological Survey of Nubia nor any numbering that fits the system used by the excavators. These are usually particularly impressive specimens and may have been added to the collection from sources other than the Nubian excavations. We do not know what modern comparative specimens Elliot Smith brought with him from Cairo. Two, at least, were found by Englebach at Shurafa on a Flinders Petrie Expedition in 1911-1912 (Derry 1913a, 1913b).

The Nubian Pathological Collection in the Royal College of Surgeons

A descriptive catalogue of the first consignment of pathological material presented to the Royal College of Surgeons by the Survey Department, Egypt, 1908, was prepared in February 1909 by Professor S G Shattock, then Pathological Curator. The catalogue relating to the second consignment of Nubian bones has not been found. If, in fact, one was prepared, it was probably destroyed during the bombing of the College in May 1941. In order to conform with the classification of pathology followed for the first consignment, the specimens in the second were inserted at the appropriate place by the addition of a suffix letter; eg. 5A, 5B, 5C are skulls, like 5, exhibiting cranial trauma.

The catalogue of the Nubian Pathological Collection of the Royal College of Surgeons was arranged in two parts, of which the first, and by far the largest, listed the fractures in anatomical categories. The second part included growth disorders, infective conditions, tumours, spondylitis deformans and congenital anomalies. Date or precise provenance were not of particular significance and were not always recorded. The terminology used to describe the pathologies would not necessarily be that used by present pathologists.

Some cases were catalogued out of anatomical order or lack cemetery and grave number. It is not certain that they form part of the original Nubian Pathological Collection; none of these cases was published in the reports of the Nubian Archaeological Survey. Certainly, without provenance, no inferences can be made concerning the histories of the pathologies shown.

The Nubian Pathological Collection in the Natural History Museum, London

From 1948 onwards much of the skeletal collection of the Royal College of Surgeons was transferred, together with the card catalogues, to the Natural History Museum, London. The Nubian Pathological Collection was accessioned in 1963. In 1968 the Anthropology Department, University College London (where Grafton Elliot Smith had been Professor of Anatomy and Derry a lecturer), donated to the Museum a number of skulls, some of which are believed to have come from Nubia. Included is a 'tam o'shanta' skull typical of osteogenesis imperfecta, without provenance, but it raises again the question of the diagnosis of the condition of a post cranial skeleton already in the collection (Brothwell 1973). (Plate 15, 1 and 2.)

Recently an attempt has been made to identify, from the published records of the excavations, the provenance and date of each case. This is to be the basis for a new interpretation of the pathologies represented including a radiographic study of the bones.

The origin of 115 specimens has been identified. Thirty one cemeteries are represented covering all periods from Predynastic to Byzantine times:

Cemetery		*Age*	*RCS Number*+
2	Hesa	Byzantine:	6, 13, 16, 55, 58, 66, 70, 87, 114, 117, 122, 125, 129, 130, 183, 183A, 1.2028, 1.2029
3	Hesa	Ptolemaic:	19
5	Biga	Christian:	46, 63, 71, 74, 107, 134, 137, 175, 190, 200, 209, 607, 1.2027, 1968.8.8.60
		Coptic:	(209B)
7	Shellal	Predynastic:	(178), 188C
		Early Dynastic:	95
		Old Kingdom:	121
		Roman:	(3), 5, 5A? 7, 8, (9), (12), 14, 16A, (18), (18B), 73
		New Kingdom-	18C, 61, 85, 139, 140,
		C-Group:	(171), 175A, A182, (183B)
		Byzantine E-Group:	18C
		Coptic:	4
14	K Ambukol	Ptolemaic-Roman:	72
15	K Sherifal	New Kingdom:	53, 56, 169
17	K Bahan	Early Predynastic:	(41), 45, (118)

Cemetery		*Age*	*RCS Number+*
22	Dabod	Old-Mid Kingdom:	123
23	Dabod, Naziria	Early Dynastic:	30, 113
		New Kingdom:	(2), 21, 182, (211)
24	Dabod	Ptolemaic:	1, 15, (17), 138
		Christian:	62
29	Badari (Tas)	Old-Mid Kingdom:	209D
30	Ras um Salim (Risqalla)	XXII Dynasty:	5B
35	Wadi Qamar	Christian:	99, 176
36	Khartum (Dimri)	New Kingdom:	93
39	Meris	Christian:	126
40	Siah	Ptolemaic:	(211A)
		Christian:	127
41	Meris-Markos	Early Dynastic:	174
		Old Kingdom:	159
		Late Ptolemaic:	111, 119
45	Shem Nishei (Demhid)	Early Dynastic:	(147), 155, 162, 163, 167, (159c)
58	Ginari	C-Group/New Kingdom:	20A, 182B, 182C, 205
72	Fagirdib (G Husein)	X-Group:	188B, 210
89	Koshtamna (Awam Umbukole)	Early Dynastic:	64A, 64B, 187A
92	Aman Daud	X-Group:	13A, 15A
101	Dakka	C-Group:	182E
110	Kubban	Re-used in New Kingdom:	(2B), 18A
112	Alagi	Byzantine X-Group:	(19B)
122	Qurta-Maharraga	Byzantine X-Group:	140B
123	Maharraga	New Kingdom:	119A
	Matmar	XXVI Dynasty:	5C
	Naga-ed-Dêr	Predynastic:	182A
	Shurafa	Roman:	171B, 1968.8.8.61

+ Wherever possible the number given in the Catalogue of the Royal College of Surgeons has been retained. Numbers in brackets refer to specimens no longer in the collection.

The provenance of 15 cases has not been identified: (41.1), 75A, 95A, 121, 123, 145A, 178A, 199, 204, 206, 214.2, 1968.8.8.17, and three cases without number.

Herewith the chronology (based on Reisner 1910) of the pathologies collected by the Nubian Archaeological Survey:

Period	*Cem.*	*Pathologies*
Prehistoric		#ulna
Predynastic	17	#clavicle, ulna
Before 3100 BC	7	osteitis deformans; synostosis radius-ulna
	89	#humerus; mastoiditis

Period	*Cem.*	*Pathologies*
Early Dynastic	7	#ulna
3100-2686 BC	23	#radius (Colles)
A-Group	41	#tibia
	45	#femur shaft, tibia-fibula (oblique); vertebrae (cut)
	89	chondroblastoma humerus
Old Kingdom	7	#metacarpal
3000-2500 BC	22	#metacarpal
B-Group	41	#fibula
Middle Kingdom	7	#clavicle, radius-ulna, femur shaft;
2500-1600 BC		rodent ulcer; parietal atrophy
C-Group		parietal cribra; metastasis?
XVIII Dynasty	29	achondroplasia; fusion tibia-fibula
	101	collapsed vertebrae (TB)
New Kingdom	15	#clavicle
1600-1350 BC	23	#skull, sternum; head ulcer; odontoma
	36	#ulna
	58	collapsed vertebrae; TB vertebrae fused vertebrae (DISH)
	110	#skull
Later New Kingdom		
1350-1100 BC		
XXII Dynasty	30	#skull/trephine?
XXVI	Matmar	#skull
	123	#radius-ulna
Ptolemaic	3	#nose
332 BC-AD 30	14	#humerus
	15	#metatarsal
	24	#skull, femur shaft
	40	ankylosed atlas-occipital
	41	#radius (Colles), radius-ulna
Roman	7	#skull (hanged), humerus
After AD 30	Shurafa	hydrocephalus; parietal perforation
Pagan	72	cancer sphenoid, cleft palate
	92	#skull
	112	#face
	122	#femur shaft
Christian	2	#skull, clavicle, humerus, radius
AD 500-1100		radius (Colles), ulnae, metacarpal,
Byzantine		pubis, ischium, femur neck
	5	#clavicle, humerus, elbow, ulna, femur neck; orbital cribra mucocele-eye, akylosed atlas; gout
	24	#scapula
	35	#ulna; microcephalic skull
	39	#pubis
	40	#ilium
Coptic	5	leprosy/psoriasis
	7	#skull
No date		tropical ulcer; actinomycosis of foot; osteogenesis imperfecta

The symbol # denotes a fracture.

Elliot Smith and Wood Jones reviewed the pathologies encountered during the first season's excavations in their report on the human remains for the Archaeological Survey of Nubia (Smith and Jones 1910). Over 240 cases were noted, 103 figured and 72 have been identified with numbered specimens in the collection; 60 of these are illustrated in the report. Some specimens were also described in the Bulletins (1907-1911). Additional specimens from later excavations were described by Firth in the Survey volumes published in 1912, 1915 and 1927.

Acknowledgements
The librarian at the Wellcome Institute provided invaluable assistance in locating the publications of the Archaeological Survey of Nubia. The photographs were taken by Philip Crabb, Photographic Studio, The Natural History Museum. I am especially grateful to the many visitors to the collection who by their interest and enthusiasm have encouraged me to persevere in trying to identify the specimens.

References
Adams, W Y, 1977. *Nubia, Corridor to Africa.* Allen Lane, London.

Bourke, J B, 1971. 'The palaeopathology of the vertebral column in ancient Egypt and Nubia', *Medical History* 15, 363-375.

Brothwell, D R, 1973. 'The evidence for osteogenesis imperfecta in early Egypt', in A K Ghosh, S K Biswar and R Ghosh (eds.), *Physical Anthropology and its Extending Horizons*, Longman, Calcutta.

Brothwell, D R and Chiarelli, B, 1973. *Population Biology of the Ancient Egyptians.* New York, Academic Press.

Brothwell, D R and Sandison, A T (eds.), 1967. *Diseases in Antiquity.* Charles Thomas.

Dawson, W R, 1927. 'Pygmies, dwarfs and hunchbacks in ancient Egypt', *Annals of Medical History*, 9, 315-326.

Derry, D E, 1909. 'Field Notes', *The Archaeological Survey of Nubia, Bulletin* 4, 22-28.

Derry, D E, 1913a. 'Parietal perforation accompanied with flattening of the skull in an ancient Egyptian', *Journal of Anatomy* 9, 417-429.

Derry, D E, 1913b. 'A case of hydrocephalus in an Egyptian of the Roman Period', *Journal of Anatomy* 47, 436-458.

Firth, C M, 1909. 'Description of Cemeteries 18-84 and 90-92', *The Archaeological Survey of Nubia, Bulletin* 4, 17-18.

Firth, C M, 1911. *The Archaeological Survey of Nubia, Bulletin* 7.

Firth, C M, 1912. *The Archaeological Survey of Nubia, Report for 1908-1909,* Vol. 2, Plates. Cairo.

Firth, C M, 1915. *The Archaeological Survey of Nubia, Report for 1909-1910.* Cairo.

Firth, C M, 1927. *The Archaeological Survey of Nubia, Report for 1910-1911.* Cairo.

Keith, A, 1913. 'Abnormal crania achondroplastic and acrocephalic', *Journal of Anatomy* 47, 189-206.

Jones, F W, 1908a. 'Post mortem staining of bone produced by the ante-mortem shedding of blood', *British Medical Journal*, 1908, 734-736, 769.

Jones, F W, 1908b. 'Examination of the bodies of 100 men executed in Nubia in Roman times', *British Medical Journal*, 1908, 736-737, 769.

Jones, F W, 1908c. 'The Pathological Report', *The Archaeological Survey of Nubia, Bulletin* 2, 55-69.

Morse, D, Brothwell, D R and Ucko, P S, 1964. 'Tuberculosis in Ancient Egypt', *American Review of Respiratory Diseases* 90, 524-541.

Ortner, D J and Putschar W G J, 1981. *Identification of Pathological Conditions in Human Skeletal Remains.* Smithsonian Institution Press.

Reisner, G A, 1909. 'The Archaeological Survey of Nubia', *The Archaeological Survey of Nubia, Bulletin* 4, 7-16.
Reisner, G A, 1910. *The Archaeological Survey of Nubia. Report for 1907-1908,* Vol. 1, *Archaeological Report,* Cairo.
Reisner, G A, Smith, G E and Derry, D E, 1909. 'The Archaeological Survey of Nubia', *The Archaeological Survey of Nubia, Bulletin* 3, 1-53.
Smith, G E 1909. 'Anatomical Report', *The Archaeological Survey of Nubia, Bulletin* 4, 19-21.
Smith, G E and Dawson, W R, 1924. *Egyptian Mummies*, London.
Smith, G E and Derry D E, 1910. 'Anatomical Report', *The Archaeological Survey of Nubia, Bulletin* 6, 9-30.
Smith, G E and Jones, F W, 1908. 'The Anatomical Report', *The Archaeological Survey of Nubia, Bulletin* 2, 29-54. Cairo.
Smith, G E and Jones, F W, 1910. *The Archaeological Survey of Nubia. Report for 1907-1908,* Vol. 2, *Report on the Human Remains*, Cairo.

A HISTORICAL FLORA OF EGYPT
A PRELIMINARY SURVEY

M N El Hadidi

Plants in Ancient Egypt have attracted the attention of numerous scholars. Among the notable classical contributions are those of Schweinfurth (1883) and Loret (1892), which dealt in some detail with 'Pharaonic Flora', based partly on substantial finds of plants in tombs and partly on old paintings and documented information in ancient papyri. Keimer's *Die Garten-pflanzen im alten Ägypten* (1924) is another concise work, prepared with the advice of G Schweinfurth, which aims to enumerate the cultivated plants of Ancient Egypt. Täckholm's *Faraos Blomster* (1951) is a richly illustrated volume in Swedish, which reflects the experience of its renowned writer with ancient Egyptian plants. Among the contributions on ancient Egyptian plants during the last decade are those of Hepper (1981), El Hadidi (1980, 1982, 1985), Germer (1985, 1988), Barakat (1986), Moens and Wetterström (1988), Hillman et al. (1989) and Manniche (1989).

A historical flora of Egypt may be considered to have its roots in the monumental work, *Flora of Egypt*. In the preface of Volume 1 (Täckholm, Täckholm and Drar, 1941), V Täckholm wrote (p 5): 'Egypt is a unique country, famous for its history, famous for its agriculture, famous for its pharmacology. Thus an Egyptian Flora should not follow any rules. It has to be written in the spirit of the country itself'. That was in fact the case with the plants dealt with in the four volumes published between 1951 and 1969. These volumes include all the available information about the species of vascular Cryptogams (Pteridophytes), Gymnosperms, monocotyledonous and a few of the dicotyledonous families of the Angiosperms (Flowering Plants). Whenever possible, the historical background and archaeological findings of the reported species are given.

Since the appearance of Volume 4 of the *Flora of Egypt,* the work has suffered from the loss of its principal author. Vivi Täckholm passed away in 1978, after which it was felt that the continuation of the *Flora* should involve some change in policy regarding its publication. The recently published issues have appeared as separate fascicules (*Taeckholmia* add. ser. 1-3, 1980-1988). Each fascicule includes a systematic treatment of the present-day species of a family of the dicotyledons.

It is hoped that a parallel work which includes species of historical interest will be prepared. This would be a documentary study of the plants and the vegetation of Egypt during the last 6-10 millennia. It would reflect the experience gained from intensive field studies on the present vegetation as well as that gained from materials recovered from archaeological sites. Among the tools for our research are the recent palaeoclimatic, phytogeographic and geobotanical information obtained from studies carried out in Egypt and adjacent countries.

The main features of this historical flora may be briefly outlined as follows:

1. An annotated check-list will be provided of the Vascular Plants, which are believed to have occurred in Egypt since the beginning of the Holocene.

2. Plant life in Ancient Egypt will be discussed with special reference to the history of human settlement during the different cultural periods in the different territories of the country.

I. The Check-list

At present, some 2000 species of vascular plants (mainly flowering plants) contribute to the natural vegetation of Egypt. These belong principally to Afro-Asiatic: Saharo-Sindian elements, African: Sudano-Zambesian elements; as well as Euro-Asiatic: Mediterranean elements. Some taxa with Western Asiatic affinities (eventually Irano-Turanian) are confined to mountainous Sinai.

During the early days of the Holocene (12000-10000 years BP), it is believed that the natural vegetation of Egypt included approximately 1000 species, which amount to less than half of the species known at present. With the adaptation of agriculture and the introduction of domesticates during the Neolithic-Predynastic period (6000-4500 years BP), an estimated 170 species were introduced and naturalised. These constitute with other native species the early weed assemblages associated with cultivated field plots.

Following the introduction of other crop plants during Pharaonic Egypt (4500-2000 years BP), the number of adventive species is believed to have reached between 225 and 255 species. During the last two millennia and up to the present, the number of recorded weeds increased to reach 470 species or about 25% of the total number of the wild species represented in the *Flora of Egypt*.

Statistical analyses of the present day weed assemblages show clearly that the number of species belonging to Mediterranean, Sudano-Zambesian or Saharo-Sindian elements has remained about the same since Graeco-Roman times. The actual increase in the number of species during the last two millennia is related to the introduction of more widely distributed taxa which found their way to Egypt as associates to other plants of economic value. The present-day weed assemblages include 69 Cosmopolitan, 87 Semi-cosmopolitan, 34 Pantropical and 22 Palaeotropical species.

It is obvious that the weed assemblages associated with an ancient crop such as wheat or barley, which are known to have been in cultivation some five millennia ago, must be different from those of maize, which was introduced into Egypt about 350 years ago. The weed assemblages of wheat (*Triticum* spp.) are rather complex, comprising between 87 and 121 species, belonging to several mono-, di-, or pluriregional elements. The weed assemblage of maize (*Zea mays*) does not exceed 40 species and includes Pan- and Palaeotropical species which were not previously known in Egypt.

Most of the taxa in the check-list will be systematically revised to specific or infraspecific levels, depending on the availability of archaeological or living material. The origin of each taxon will be considered and traced, especially those species which were introduced during the different cultural periods.

II. Plant Life in Ancient Egypt

Egypt may be divided both physiographically and bioclimatologically into four main sectors (Fig. 1):

1. A northern sector including the Mediterranean coastal plain and the Nile Delta.

2. The Nile Valley, which runs in a south-north direction and comprises the fertile lands on both sides of the main stream of the Nile.

3. The Eastern Desert and Sinai, which constitute a massif of moderate elevations with few peaks higher than 2000 m ASL. It is dissected by wadis, which form the major drainage systems to the Nile Valley and the Red Sea.

4. The Western Desert, which is a more-or-less flat expanse with depressions harbouring a series of oases of variable size and cultural occupations.

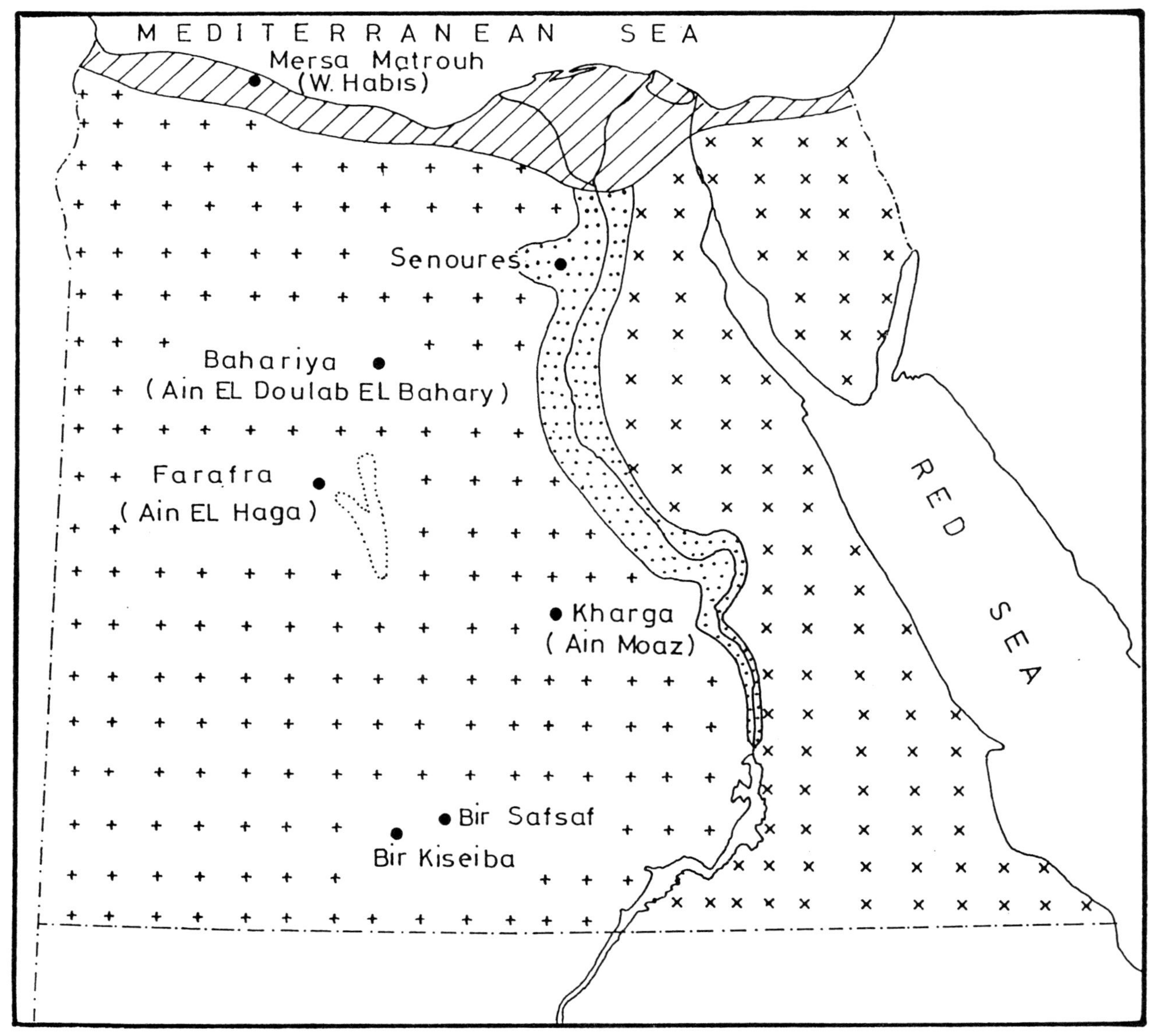

Fig. 1. Map of Egypt: four main physiographic and bioclimatological sectors.

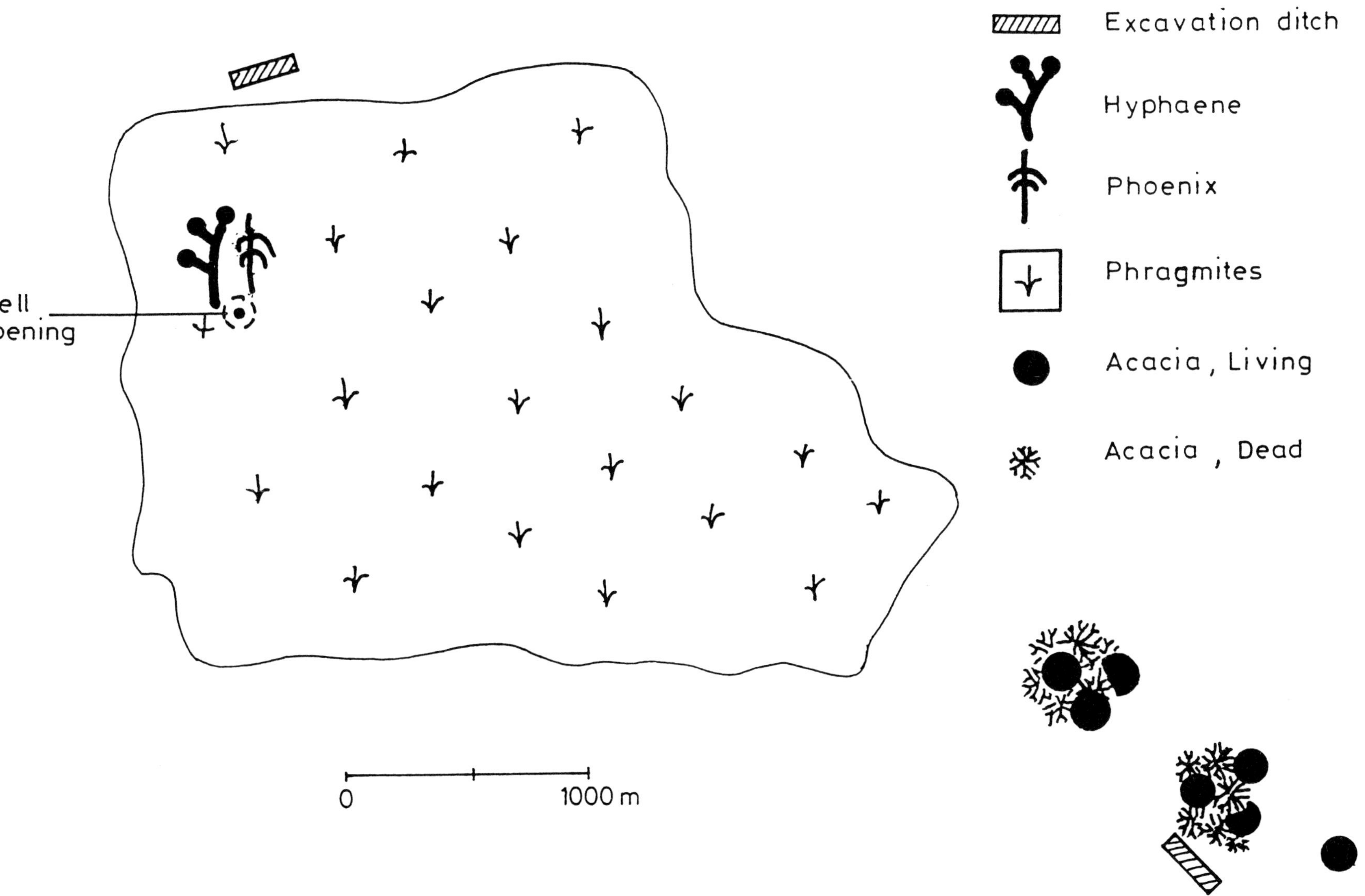

Fig. 2. Diagrammatic presentation of the vegetation of Bir Safsaf (Nubian Desert).

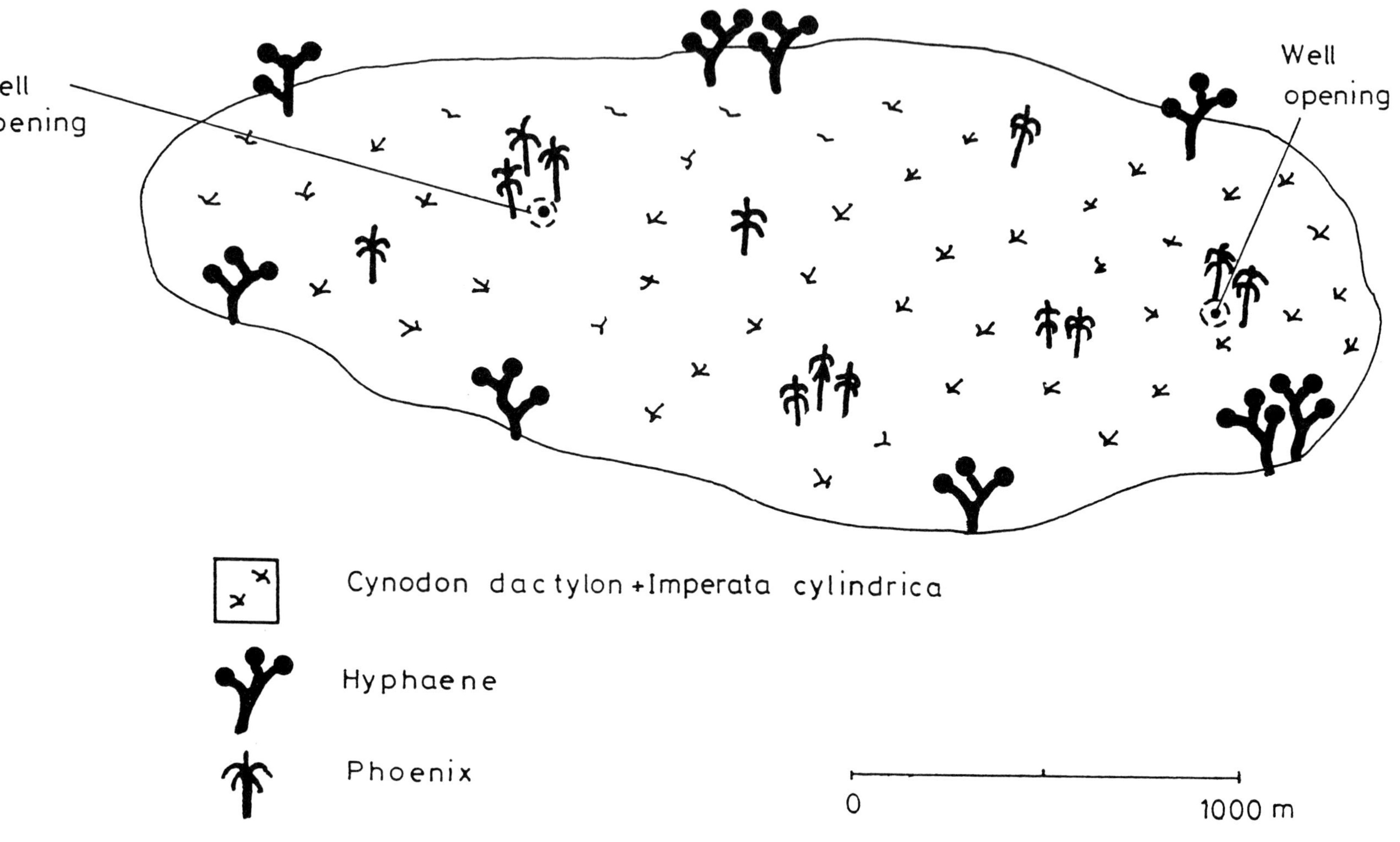

Fig. 3. Diagrammatic presentation of the vegetation of Bir Kiseiba (Nubian Desert).

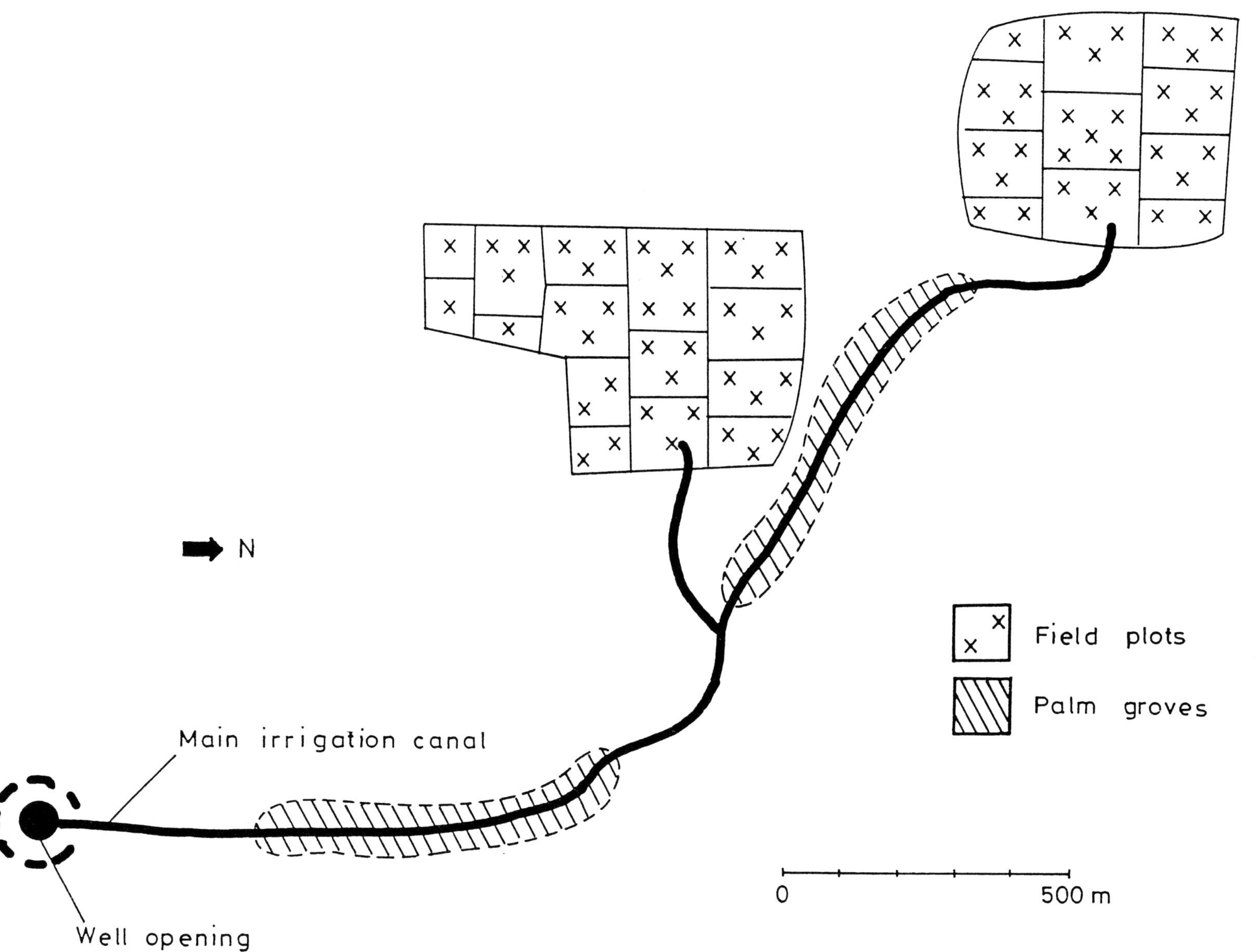

Fig. 4. Diagrammatic presentation of ancient Ain Moaz (Kharga Oasis).

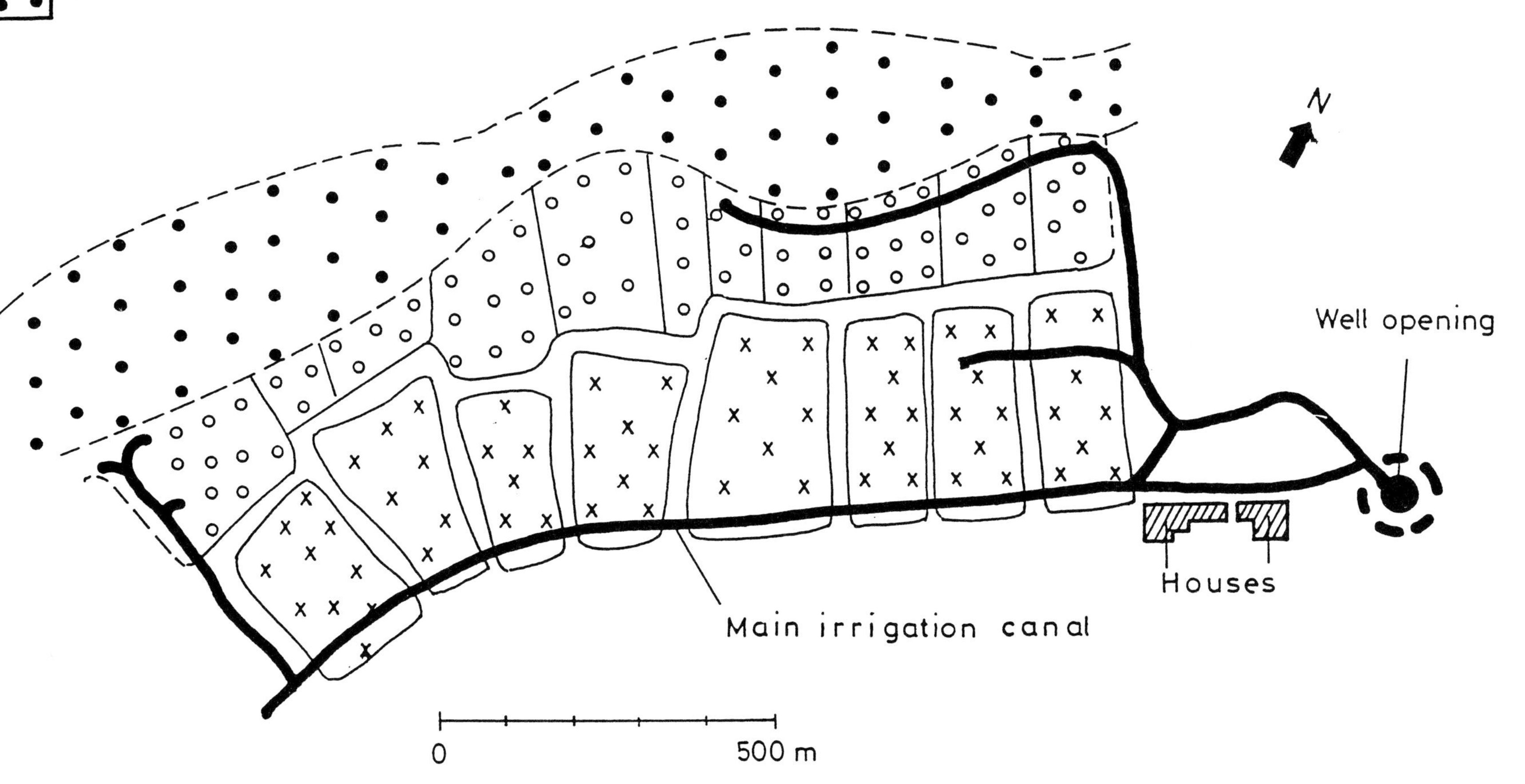

Fig. 5. Diagrammatic presentation of Ain El-Doulab El-Bahari (Bahariya Oasis).

TABLE 1

Species	Bir Safsaf (Nubian Desert)	Bir Kiseiba (Nubian Desert)	Ain Moaz (Kharga Oasis)	Ain El Haga (Farafra Oasis)	Ain El Doulab El Bahari (Bahariya Oasis)	Senouris (Faiyum, Nile Valley)	Floristic Category (Wickens, 1976)	Historical Background
Desmostachya bipinnata	+	-	-	+	+	+	SA-SI+S-Z	
Phragmites australis	+	+	+	+	+	+	PAL	
Alhagi graecorum	+	+	+	-	-	+	PAL	These taxa belong to the natural vegetation of Egypt and have been recovered from Late Palaeolithic and Neolithic sites.
Sorghum sudanese	-	+	-	-	-	-	PAL	
Cyperus rotundus			+	+	+	+	PAL	
Melilotus indicus			+	+	+	+	PAL	
Lolium temulentum			+	+	+	+	PAL	
Convolvulus arvensis			+	+	+	+	PAL	
Ammi majus			+	+	+	+	ME+IR-TR	
Calendula arvensis			+	+	+	+	ME+IR-TR	
Polygonum equisetiforme			+	+	+	+	ME+IR-TR	
Cichorium endivia			+	+	+	+	ME+IR-TR	
Malva parviflora			+	+	+	+	ME+IR-TR	
Plantago lagopus			+	-	+	+	ME+IR-TR	
Vicia monantha			+	-	+	+	ME+IR-TR	
Phalaris minor			+	-	+	+	ME+IR-TR	
Emex spinosus			+	-	+	+	ME+IR-TR+SA-SI	
Beta vulgaris			+	-	+	+	ME+IR-TR+ER-SI	These taxa are believed to have been introduced into Egypt during Predynastic times or later.
Lotus corniculatus			+	+	+	+	ME+IR-TR+ER-SI	
Dichanthium annulatum			+	+	+	+	ME+IR-TR+ER-SI	
Lolium rigidum			+	+	+	+	ME+IR-TR+ER-SI	
Polypogon monospeliensis			+	+	+	+	COSM	
Solanum nigrum			+	-	+	+	COSM	
Digitaria sanguinalis			+	+	+	+	COSM	
Anagallis arvensis			+	+	+	+	COSM	
Brassica nigra			+	+	+	+	COSM	
Chenopodium album			+	+	+	+	COSM	
Sonchus oleraceus			+	+	+	+	COSM	
Euphorbia peplus			+	+	+	+	COSM	
Euphorbia helioscopia					+	+	COSM	
Medicago polymorpha					+	+	COSM	
Lolium perenne					+	+	COSM	These taxa are of later introduction and have been identified from Graeco-Roman sites.
Avena fatua						+	COSM	
Gnaphalium luteo-album						+	COSM	
Capsella bursa-pastoris						+	COSM	
Veronica anagallis-aquatica						+	COSM	

ME: Mediterranean SA-SI: Saharo-Sindian IR-TR: Irano-Turanian
PAL: Palaeotropical S-Z: Sudano-Zambesian ER-SI: Euro-Siberian
COSM: Cosmopolitan

1. **The Mediterranean Coastal Plain and the Nile Delta**

These are floristically the least known parts of Egypt. Archaeobotanical information is rather fragmentary when compared with that available from the numerous excavated sites of the Nile Valley.

The wild barley *Hordeum spontaneum*, the progenitor of the cultivated forms of barley, is only known from the Wadi Habis in the west Mediterranean sector (El Hadidi et al. 1986: 292). This site is believed to be a relict of an earlier and wider occurrence in North Africa.

In answering the question why wild barley is found only in Wadi Habis, an archaeological-ethnographic viewpoint suggests that the sacred aura of the site (see El Hadidi et al. 1986: 293) possibly encouraged the preservation of its natural habitat. Today, the area is being subjected to rapid development due to increased population pressure. The natural vegetation at Wadi Habis is ploughed for the cultivation of cereals and other crops. One can therefore expect the destruction of the natural habitat of the wild barley in the near future, unless protective measures are taken now.

2. **The Nile Valley**

Early man penetrated the Nile Valley during the lower Palaeolithic period some 25000 years ago. The linear pattern of the Nile and the concentration of resources along its main course encouraged the establishment of numerous human settlements, which varied in size, character and density during the various periods or cultures. The artefacts attributed to the late Pleistocene period, between 22000-14000 years BP, show a remarkable decrease in faunal remains, with more reliance on fish and water fowl, accompanied by a gradual shift towards the utilisation of plants.

The adaptation of agriculture during the Neolithic-Predynastic periods (6000-4600 years BP) lead to the development of a unique economy in which cultivation and herding as well as the exploitation of Nile resources were inseparable. It is generally accepted that the Neolithic and Predynastic populations had shifted gradually to the cultivation of a wide range of food plants, including South-west Asian domesticates. One may expect evidence of the appearance of a considerable number of adventive species as a result of the free exchange of people and domesticates between South-west Asia and North Africa.

A simple weed assemblage including the following species has been recovered from an Old Kingdom site (Moens and Wetterström, 1988: 163) and seems to be related to cultivated plots of six-rowed hulled barley and emmer wheat:

Alhagi graecorum Boiss.
Lolium temulentum L.
Lathyrus sp.
Phragmites australis (Cav.) Trin ex Steud.
Cyperus sp.
Vicia sp.
Malva parviflora L.
Glinus lotoides L.
Polygonum sp.

At a later stage, cosmopolitan taxa would appear in the weed assemblages of winter crops. Among these are the following, which have been identified in Pharaonic sites:

Solanum nigrum L.
Brassica nigra (L.) Koeh.
Medicago polymorpha L.
Avena fatua L.
Panicum repens L.

The changes in the hydrology of the Nile during the last decades may lead to the gradual disappearance of some of the plants of antiquity. The sacred white lotus (*Nymphaea lotus*) was of common occurrence in the water bodies of the Nile Delta and the northern provinces of the Nile Valley. Hydrobiological changes following the construction of the Aswan High Dam in 1960 has had a drastic influence on the distribution of this water lily. The plant has not been recorded from the Faiyum and Beni Suef Provinces since 1970. At present, it is becoming very scarce in the northern parts of the Nile Delta. Unless protective measures are urgently taken, this remarkable water lily may soon become extinct.

Ceruana pratensis, 'Garawan', was used in ancient Egypt for making baskets and mummy coffins. Until the sixties of this century, the plant was of common occurrence on the silty banks of the Nile and the main irrigation canals. Since the construction of the Aswan High Dam, less silt has been deposited in its natural habitat and the species has become increasingly less common. Efforts to collect the plant during the last decade have not met with success, which suggests that the plant is not regenerating and is in the process of disappearing (Boulos, 1985: 131).

3. The Eastern Desert and Sinai

Floristically, the Eastern Desert and Sinai are perhaps the best known territories of Egypt. Among the notable works on the flora and vegetation of the Eastern Desert is that of Hassan (1985). An annotated list, including the whole flora of Sinai, was recently published by El Hadidi and others (1989). The historical role of the Eastern Desert and Sinai deserves intensive archeological study. The available archaeobotanical information is rather scanty when put in the context of its great significance as a migratory route between Egypt and South-west Asia.

4. The Western Desert and Oases

The northern and major part of the Western Desert is a limestone plateau which slopes gradually to the northwest towards the Siwa Oasis and the great Qattara depression. Its southern boundary is a high escarpment which leads to the depressions of the Kharga and Dakhla Oases. Within this limestone plateau, the greater hollows containing the Bahariya and Farafra Oases are situated. The southern part of the Western Desert is mainly sandstone and harbours a group of smaller oases.

Oases are depressions which receive very little or almost no rain. They owe their greeness to their perennial underground water supply which appears on the soil surface in the form of springs ('Ains') or is pumped from wells ('Birs'). Their size varies from a fraction of a square kilometre (non-inhabitated smaller oases with springs or 'Ains') to hundreds or thousands of square kilometres (inhabitated larger oases with several wells or 'Birs'.

A documentary study of the oases of the Western Desert (excluding the Siwa complex) was carried out between 1980 and 1986, in connection with a study of the history of agriculture since the Neolithic period. A group of oases believed to represent 'living antiquity sites' was selected. These oases are very isolated and are not subjected to modern human interference.

The smaller oases (e.g. Bir Safsaf and Bir Kiseiba, Fig. 1) represent the initial stage of oasis vegetation, which follows a general and similar pattern. At Bir Safsaf (El Hadidi 1980: 350), two palm groves (a date palm and a dom palm) grow near the water source. These are surrounded by a grass community of *Phragmites australis* and *Desmostachya bipinnata*. The Bir area is delimited by several *Acacia ehrenbergiana* mounds, 'Tarabeles'. Some of the acacias must be 500 to 700 years old; others have near their bases a considerable number of artefacts, with traces of human settlement dating back to the Neolithic period (Fig. 2).

The vegetation around Bir Kiseiba (Fig. 3), a stopping point on the Darb El Arbeen caravan route, shows some human influence. Tall palm groves cultivated by man and yielding edible dates grow together with dwarf spontaneous groves producing inedible dates. The several grasses present, including *Sorghum sudanese*, *Cynodon dactylon* and *Imperata cylindrica*, denote various human activities.

The ancient wells of Kharga, Dakhla and Farafra Oases date back to Graeco-Roman times or earlier. They provide examples of smaller oases with ancient patterns of agriculture. According to Shamloul (1986), some twenty of these wells are still operating around Kharga town (Kharga Oasis), each supporting the cultivation of a limited area ranging between 0.5-5 acres, and where irrigation follows a common pattern. At Ain Moaz (Fig. 4), the well-opening is located in the middle of an elevated mound. It is encircled by thickets of *Phragmites australis* and a few acacia trees which provide shade for the water source. The water flows through a principal canal, which divides into a number of side canals that distribute water to plots at different levels. The plots are cultivated for winter crops such as wheat or broad beans; during the summer, the plots are left fallow or cultivated for melons.

The weed assemblage associated with winter crops in the cultivated plots of Ain Moaz (Kharga Oasis) or Ain El Haga (Farafra Oasis) is perhaps the simplest known from the farmlands of Egypt (Table 1). It comprises about 30 species. The Palaeotropical species of this assemblage belong to the natural vegetation of the area and were recovered from several late Palaeolithic and Neolithic sites. The biregional (Mediterranean + Irano-Turanian) and pluriregional species of this assemblage (Mediterranean + Irano-Turanian + Euro-Siberian) are believed to have been introduced into Egypt during predynastic times or later. Several cosmopolitan species have been identified in archaeological sites of the Old Kingdom, c.2700-2250 BC (Moens and Wetterström, 1988: 163).

The weed assemblage recorded in winter cultivated plots of Ain El Doulab El Bahari of Bahariya Oasis (Fig. 1) includes other cosmopolitan species which have been identified from Graeco-Roman sites or later (Täckholm, 1951). The water source dates back to the first century AD and irrigates an area of about five acres. Orchards and field plots occupy higher levels, while excess irrigation water is drained to a lower-levelled Sabkha (Fig. 5).

At Senouris in Faiyum Province (Fig. 1), originally one of the oases of the Western Desert, the weed assemblage of winter crops is dominated by cosmopolitan species (Table 1). At present, the vegetation of the Faiyum area is very similar to that of the Nile Valley, denoting the intensive influence of man.

References

Barakat, H N, 1986. *Plant life in Doush area: Kharga Oasis. A comparative study of the present and Graeco-Roman periods*. MSc. Thesis, Faculty of Science, Cairo University.

Boulos, L, 1985. 'The arid eastern and south-eastern Mediterranean region', in Gomez-Campo, C (ed.), *Plant conservation in the Mediterranean area*, W Junk Publishers, Dordrecht, 123-139.

El Hadidi, M N, 1980. 'Vegetation of the Nubian Desert (Nabta region)', in Wendorf, F and Schild, R (eds.), *Prehistory of the Eastern Sahara*, app. 5, Academic Press, New York, 345-351.

El Hadidi, M N, 1982. 'The Predynastic Flora of the Hierakonpolis Region', in Hoffman, M A (ed.), *The Predynastic of Hierakonpolis - an Interim Report*, Egyptian Studies Association No. 1, Cairo University Herbarium, Egypt, 102-115.

El Hadidi, M N, 1985. 'Food Plants of Prehistoric and Predynastic Egypt', in Wickens, G E, Goodin, J R and Field, D V (eds.), *Plants of Arid Lands*, G Allen and Unwin, London, 87-92.

El Hadidi, M N and others, 1989. 'Annotated list of the Flora of Sinai (Egypt)', *Taeckholmia* 12, 1-100.

El Hadidi, M N, Abd El Ghani, M, Springuel, I and Hoffman, M A, 1986. 'Wild Barley, Hordeum spontaneum L in Egypt', *Biol. Conserv.* 37, 291-300.

Germer, R, 1985. *Flora des pharonischen Ägypten,* Mainz.
Germer, R, 1988. *Katalog der altägyptischen Pflanzenreste der Berliner Museen.* Otto Harrassowitz, Wiesbaden.
Hepper, F N, 1981. 'Plant material', Appendix G, in Martin, G T (ed.), *The Sacred Animal Necropolis at North Saqqâra,* Egypt Exploration Society, London, 146-51.
Hassan, L M, 1987. *Studies on the Flora of the Eastern Desert, Egypt.* PhD thesis, Faculty of Science, Cairo University.
Hillman, G, Madeyska, E and Hather, J, 1989. 'Wild plant foods and diet at Late Paleolithic Wadi Kubbaniya: the evidence from charred remains', in Close, A (ed.), *The Prehistory of Wadi Kubbaniya,* Vol. 2, Southern Methodist University Press, Dallas, 162-242.
Keimer, L, 1924. *Die Gartenpflanzen im alten Ägypten.* Berlin.
Loret, V, 1892. *La flore pharonique,* 2me. ed. Paris.
Manniche, L, 1989. *An Ancient Egyptian Herbal.* British Museum Publications, London.
Moens, M-F and Wetterström, W, 1988. 'The Agricultural Economy of an Old Kingdom Town in Egypt's West Delta: Insights from the Plant Remains', *JNES* 47, No. 3, 159-173.
Schweinfurth, G, 1883. *Neue Beiträge zur Flora der alten Ägypten.* Berlin, Deutsch. Bot. Ges. 1.
Shamloul, A M, 1986. *Plant life around the ancient wells in Kharga Oasis.* MSc. Thesis, Faculty of Science at Sohag, Assiut University, Egypt.
Täckholm, V, 1951. *Faraos blomster. Bokforlaget Natur och Kultur.* Stockholm.
Täckholm, V, Täkholm, G and Drar, M, 1941. *Flora of Egypt,* Vol. 1, Bull. Fac. Sci. Fouad 1 Univ., No. 17 (now Cairo University).
Wickens, G E, 1976. 'The Flora of Jebel Marra (Sudan Republic) and its geographical affinities', *Kew. Bull. Add. Ser.* 5, 40-49.

ANCIENT EGYPTIAN BREAD AND BEER
AN INTERDISCIPLINARY APPROACH

Delwen Samuel

Introduction

Ancient Egypt is famous for its pyramids, tombs and mighty pharaohs, but, as with any civilisation, the population was mainly composed of ordinary people going about their daily lives. A change in emphasis over the years has gradually shifted some attention from archaeological investigation of tombs, temples, inscriptions and art work, enthralling and valuable as this has been, towards domestic economy. One universally necessary part of daily routine is providing and consuming nourishment. The staples of diet in ancient Egypt for the common people, as well as the powerful rulers and nobility, were bread and beer. To understand how these foodstuffs were produced is to know about a significant part of culture and daily life in ancient Egypt.

Until recently, however, studies on baking and brewing have been sparse. Past investigators have tended to rely on the rich evidence provided by funerary decoration, and sometimes, on other archaeological finds. To advance our knowledge of ancient domestic activity, other approaches are needed.

A few settlement sites are now being excavated in Egypt, and one of these is Amarna, a New Kingdom (c 1350 BC) site on the east side of the middle Nile valley. Under the direction of Barry Kemp, University of Cambridge, a multi-disciplinary approach prevails and various specialist programmes are being pursued. Amongst them is a detailed study of two basic domestic activities, the production of bread and beer. This article is a summary of the project design, which aims to synthesize evidence from archaeology, archaeobotany, ethnography, experimentation and chemical analysis. The project is undertaken in collaboration with Scottish and Newcastle Breweries, who are providing funds and technical expertise and with the support of the British Academy. My thanks are extended to both.

Sources of Evidence

A. *Artistic Evidence*

All the Dynasties of ancient Egypt produced artistic records of regular and ritual events, including food production. Some periods were more prolific than others, so that our reconstructions tend to be heavily coloured by practice at a certain time. The Old Kingdom tombs contain more scenes of bread and beer production than any other. Best known perhaps is the Fifth Dynasty tomb of Ti at Saqqara (Épron et al. 1939, Plate 66). A few representations from the Middle Kingdom, like those from Beni Hasan (Newberry 1893, Plate 12, for example) and models from the tomb of Meketre (Winlock 1955, Plates 22, 23) and elsewhere (see Plate 16), and from the New Kingdom, for example the tomb of Ken-Amun (Davies 1930, Plate 58), show that methods did change over time. Archaeological evidence, in the form of bread moulds for example, shows this as well (see Section B).

Ancient art has been very much the predominant source of information on which Egyptologists have relied for discussions of bread and beer (Klebs 1915, 94; Montet 1925, 254; Wreszinski 1926; Wild 1975; among others). However, it leaves many questions unanswered. Tomb scenes tend to be iconographic, because the intention was to represent the activity, not provide an illustrated guide! The missing gaps cannot be filled intuitively by the modern observer unfamiliar with ancient Egyptian common practice. Accompanying inscriptions sometimes describe actions or ingredients, but there can be controversy about some interpretations (see Section F, chemical evidence), and even accepted names have rarely been independently confirmed.

Questions such as the proportions of ingredients cannot be answered at all using tomb art, nor do contemporary documents provide us with clear evidence. The artistic record is mute about the efficiency of the process, or the wastage, effort and grain supply involved in the production of a batch of bread or jar of beer. Artistic depictions are undoubtedly valuable guides, but clearly other sources of evidence are needed.

B. *Archaeological Evidence*

Perhaps the most fascinating artefacts connected with baking and brewing in ancient Egypt are preserved bread loaves and dregs of beer, left as offerings for the dead and still intact after the passage of millennia. We can unravel some of the processes involved in their production by studying them in detail. To start with, simple observation of loaves by eye and low power microscope can reveal diverse aspects of bakery, including how different loaves were formed; what some of the ingredients were, perhaps including flavourings; how finely the flour was ground and how well it was cleaned of chaff and contaminants. Depending on the technique, high power microscopy and chemical analysis may require destruction of a small sample, but these methods will provide far more detailed information on ingredients, contaminants and their alteration by baking or brewing (see also below, Section F). Since these loaves and beer residues were intended as funerary gifts, they were not necessarily made exactly as the products which sustained living people. Still, they are the closest we can come to daily food.

At present we know only that there were up to forty different kinds of bread (Wild 1975, 594) and perhaps seventeen types of beer (Darby et al. 1977, 543) with names such as black, clear and Nubian (Dykmans 1936, 175). By making a detailed inventory and analysis of them, we will get some precise ideas about what distinguished this multitude of breads and brews.

John McDonald (1982, 113) discusses a particular loaf held in the Museum of Fine Arts, Boston (No. 72.4757c), and hypothesises that its shape relates it to a bread labelled *herset* in a scene from the tomb of Rekh-mi-re (Davies 1943, Plate 38). This word relates to the colour red, and indeed the loaf has a reddish cast. He suggests the term is connected to bread made from a specific red ingredient. When this scene was checked, the word *herset* could not actually be distinguished (Kemp, pers comm).

On a recent visit to the museum, I was able to identify the main component of this loaf to be figs; the small reddish seeds mentioned by McDonald are fig seeds. The loaf has actually been broken, exposing the seedy interior. It has a smooth crust but is irregularly formed and looks very much as if it were shaped by hand.

Although it does not seem as if this particular loaf can, in fact, be linked to a term for red colouring, it may be possible to relate the actual composition of breads and beers to their shape, or colour, or other distinguishing features, in labelled depictions. If so, we should come much closer to enumerating the ancient Egyptian baking repertoire.

On a more usual archaeological note, we can look at the kitchens of the past. The site of Amarna is a good place for this, because it is rich in domestic artefacts. An entire kitchen was uncovered by Peet (1921, Plate 27.2) in the Workmen's Village. Many other areas associated with cereal processing have been uncovered, including a variety of mortar and quern emplacements and ovens (see, for example, Kemp 1987, 5, 10, 31-32, 36, 73-76). Similar contexts have been found at a few other sites. Bruyère (1953, 97) discovered a similar quern emplacement at Deir el-Medineh and Larsen (1936) found moulds, ovens and what he believes are oven lids, at Merimde Abu Ghâlib in the western Delta.

Examples of actual tools used to process cereals can be harder to find, either because they were made of ephemeral materials like wood, or because they were portable and too useful to be left behind after abandonment. Heavier tools of robust material have survived, however, and several limestone mortars were excavated at the Workmen's Village. Quern stones are fairly common on the site surface and have a variety of shapes and sizes. Sadly, recent increased disturbance is beginning to inflict severe damage on these quern stones, making a survey a matter of some urgency before they are destroyed.

One class of artefact associated with baking survives in abundance at Amarna, that of bread moulds. They appear in definite concentrations, at temple bakeries and the Military and Police post in the Main City (Rose 1987, 119), and around chapel ovens (Kemp 1984, 32-33) in the Workmen's Village. Although roughly made and thick walled compared to the Old Kingdom bread moulds, they are small and delicate! Their small size and pattern of distribution suggest that moulded bread was produced only for special occasions or particular purposes. Middle Kingdom bread moulds from the fort at Kuban (Emery and Kirwan 1935, 33-38) provide a useful comparison. The change in typology of bread moulds throughout the Dynastic period (Jacquet-Gordon 1981) is a further indication that methods of production slowly altered in ancient Egypt.

C. *Archaeobotanical Evidence*

Plant debris left behind in rooms, middens or other archaeological contexts was produced and distributed on the whole by specific human or animal actions. Analysing these remains, therefore, provides direct evidence for past activity. In ideal circumstances, these leavings remain on the site exactly where they fell during handling and the archaeologist can recover them and associate them directly with the surrounding artefacts and architecture.

Because of the extreme dryness of the Egyptian climate, plant remains are often in perfect condition and in certain contexts survive in overwhelming abundance. This is highly unusual, for plants are normally preserved on archaeological sites only by charring or waterlogging. Both processes destroy a portion of the original assemblage, whereas desiccation preserves nearly everything. Ancient Egyptian sites thus give us an unusually complete view of the botanical element in past human lives.

At Amarna, desiccated plant debris is associated with a wide variety of contexts, including cereal processing areas. Here, the ideal distribution does sometimes occur. One sample of soil was taken from the floor surrounding a mortar in the Workmen's Village, at West Street 2/3, a context labelled unit 2707 (see Kemp 1987, 8). The preliminary analysis of the plant remains found around this mortar is published in Samuel 1989, 280-286.

The first information gleaned from such a study is plant identification. Cereal of some sort might be expected around a mortar, although it is not the only material which could be pounded up. Around the mortar in West Street 2/3, by far most remains were various parts of emmer wheat spikelets. It is reasonable to infer the emmer was being processed shortly before the room fell out of use. Mixed in the assemblage were a few barley remains and some weed seeds. When the latter are fully identified, some idea of field conditions and subsequent grain purity will be established.

The condition of the plants can also be assessed. As noted in the discussion of this assemblage (Samuel 1989, 284), the spikelets had been broken up into a variety of pieces. Those which remained entire but had lost their grain tend to be broken in the same way: the thin glume phlanges were torn away, making a hole for the grain to fall out. This could have been caused by rodent gnawing, but comparison of modern emmer spikelets chewed by mice shows no trace of this breaching pattern. It seems more likely the pounding method was responsible. It remains to reconstruct that method, but whatever it was, it ought to produce the same breakage. Archaeobotanical analysis thus provides a further set of data on cereal processing.

D. *Ethnographic Evidence*

Ethnographic studies of traditional ways are sometimes used to interpret archaeological evidence. The best known example of ethnography applied to archaeobotany is the detailed work which Hillman (1984a, 1984b, 1985) did in Turkey, to work out the precise sequence of traditional Near Eastern cereal processing from harvesting the ripe crop to preparation of bulghur, and to describe all the resulting products and by-products. This work has become central to the interpretation of crop plant assemblages found within archaeological sites, especially those of the Near East. It has also demonstrated the value of ethnographic work.

Traditional techniques have been developed over generations of experience and are practised by skilled people. It is not possible or sensible that they be reinvented by present-day investigators of vastly differing culture. Instead, traditional methods can be observed and recorded, both to provide analogues for ancient procedure and as a record of human culture intrinsically valuable in its own right.

Present day parallels with ancient Egyptian methods of baking and brewing are to be found in Egypt and parts of Africa. The use of moulds seems to have died out, but other ways of baking still persist. For example, Henein (1988, 168-169) has a detailed description of *šamsi* bread, or bread risen and slightly fermented in the sun. Blackman (1927, 163) gives a more abbreviated description of the same thing. A similar procedure is suggested by Peet and Woolley (1923, 64), to describe bread platters found from houses in the Main City of Amarna. They based their explanation on analogy with local contemporary practice.

The basic method of ancient Egyptian beer brewing is clear from the artistic record: bread loaves are baked, soaked in water, fermented and strained. Lucas (1948, 20) points out that the same procedure is still done in parts of Africa and indeed in Egypt. Brewing can be a less accessible process to observe in Egypt, as it is frowned upon in Islamic society. In a description of contemporary Egyptian customs, Lane (1908, 96) mentions the brewing of 'bouzah', a fermented drink made from crumbled loaves, but gives few details. Richards (1939, 97-99) includes a detailed account of one particular brewing episode by Bemba people in northern Zimbabwe, beginning with the removing of unhulled cereal from the granary and ending with the finished fermented product. She recorded the time each step took to complete and noted that the process was affected by the ambient temperature and humidity. She also made the observation, common in ethnographic work, that it was difficult to get the Bemba brewers to calculate quantities and time. Traditional procedures tend not to be thought of in such terms and the process is acquired through experience without analysis.

Closer to Egypt, Burckhardt (1819, 218) was able to see brewing done by Berbers living at the confluence of the Blue and White Niles. These people brewed using strongly leavened sorghum or millet loaves, which they crumbled in water and fermented for two days. Burckhardt mentions three types of bouzah, classified according to their degree of fermentation. Brewing of bouzah seems to have been more extensive in his day than the present (Burckhardt 1819, 143).

Burckhardt's account is also useful because it describes the grinding of cereals using a saddle quern rather than a rotary mill (1819, 219). Pounding cereals in mortars together with grinding them on saddle querns is a practice which still survives in parts of Africa (Plate 17, 1) and is the closest analogue to the ancient Egyptian method of processing grain into flour.

There are one or two ancient ethnohistoric accounts which bridge the gap between Dynastic Egyptian practice and the present day. Lucas and Harris

(1962, 14) transcribe an account of brewing written in the late third or early fourth century AD by one Zosimos of Panopolis in Upper Egypt. Lutz (1922, 78) provides a slightly different translation. According to Lane (1908, 96), Herodotus observed the making of beer by the same methods.

All these accounts are of great help when considering the ancient Egyptian ways of processing cereals, baking and brewing. However, few were done with the express intention of direct comparison. The general method is clear enough, but questions of detail remain. For example, what was the traditional preparation of yeast? What are the proportions of different ingredients? As Richards points out (see above), those who prepare food do not think in terms of quantities and measurements, nor can they be aware of the precise nature of the product. All this requires an ethnographic study undertaken with such goals expressly in mind. It is hoped that this investigation into ancient Egyptian bread and beer will be able to incorporate relevant ethnographic observation.

E. *Experimental Evidence*

The previous sections have examined types of evidence for ancient Egyptian bread baking and beer brewing. These various approaches have different perspectives to offer, provide certain answers, and raise other questions. Taken together, a much more detailed picture of baking and brewing emerges, but without further work it cannot be tested. There must be some method of verification, otherwise a reconstruction can be only informed speculation.

Therefore, a key element in the Amarna bread and beer project is experimentation. The goal is to replicate ancient Egyptian techniques, using all available evidence to guide the experimental design. The archaeological record shows the position and installation of equipment, and sometimes authentic tools themselves can be used. The artistic record gives some idea of how these tools were used and which other equipment, not found from archaeological sites, was involved. Ethnography provides the opportunity to observe similar processes in action and benefit from the experience of skilled people. During experimentation, methods of processing are tested. Actually trying out the actions involved shows whether suggested techniques are workable, as well as demonstrating very clearly the amount of labour involved! The results can be checked and controlled by comparison with the archaeobotanical record: the by-products of experimental processing must match the ancient debris.

The idea of archaeological experimentation is not new. For example, some archaeobotanical questions have been explored in this way (Hersh 1981; Lüning and Meurers-Balke 1980; Moritz and Jones 1950; Beranovà 1986; among others). However, most of these (Lüning and Meurers-Balke 1980 excepted) tend to be small scale investigations, mainly using single trials to test the general feasibility of an idea or make a rough estimate of production.

The expertise and research interests of a group of people working at Amarna have made it possible to undertake a much broader and more extensive experimental programme than those quoted. Experimental work on ceramics has included production of bread moulds (Nicholson 1989, 243-246). These replicas are available for bread making experiments and a similar manufacture of beer jars is being planned. To test how grain was made into flour, replicas of a mortar and a quern emplacement have been constructed (Plate 17, 2), using stone equipment retrieved from excavation. The current approach has been to use them in a variety of theoretically possible ways and to look at the resulting product (Samuel 1989, 258-277). As ethnography becomes more integrated into the research programme, there should be less trial and error involved.

One desirable attribute of this type of experimentation is a defined and fully repeatable method. This delineates the exact experimental method, shows how each action relates to established evidence, and determines whether the results are compatible with what is already known. In addition, experimental work should be measurable and comparable to other types of data or similar experiments. The inclusion of detailed measurements means that aspects of cereal processing such as efficiency, wastage and yields can be explored. Such questions are not really accessible in any other way. Experimentation, as well as being a test of ideas, contributes information in its own right.

F. *Chemical Evidence*
Chemistry has several applications in this project. Some analysis and description of bread loaves (Leek 1972, 1973; Samuel 1989) and beer residue (Grüss 1929) composition has been started (see also above, Section B, archaeological evidence). Such work can be greatly expanded and add a corresponding mass of information. Chemical techniques can also be used to analyse residues encrusted on and embedded in ceramic artefacts. Whether moulds were tempered (Wilson 1989) may be determined by analysing bread mould sherd fabric for lipids. Deposits from oven linings can also be analysed for chemical composition and compared to modern tannour bread bakery, for example. DNA analysis may indicate the presence and type of yeast in beer dregs. Yeast is less likely to be found in ancient bread.

Egyptologists have pondered certain questions without being able to reach firm conclusions. Chemical analysis has the potential to provide definitive answers. The problem of translating the names for ingredients or substances is one example. According to Nims (1950, 261), the term *bnr* is generally agreed to refer to "dates". The word *bš3*, on the other hand, has been translated in several ways. Interpretations have included two-row barley, spelt, grain specially ear-marked for beer making, or malt (Nims 1958, 62-64). Spelt can be eliminated immediately, as it is a northern European wheat and has never been found from any period in Egypt (Dixon 1969, 131). Nims (1958, 64) presents a case for *bš3* as malt (sprouted grain used for brewing), using the Moscow Papyrus bread and beer problems. He extrapolates (1950, 262) from the established use of malting in Graeco-Roman times, and feels his case strengthened by evidence of malting in ancient Sumer (1958, 63), but there is no conclusive evidence for malt in Egypt during the Dynastic period. Chemical analysis is ideally suited to settle the question, for the enzymes and sugars produced during the malting process or their break-down products are distinctive, and if present in beer residues will be detectable.

Chemical monitoring will also be applied to the experimental production of bread and beer. This will allow control of experimental technique, show the degree and quality of beer fermentation and the nutritive value of the products. In other words, analysis of experimentally produced bread and beer will show exactly what has been made, and by which chemical processes.

Further Applications

This project aims to synthesise all available sources of evidence to answer specific questions and to reconstruct the general methods of ancient Egyptian baking and brewing. In addition, it has several wider applications. Firstly, it will result in the recognition of botanical residues connected with specific steps of cereal preparation. This will aid the identification of archaeological contexts whose function is not recognisable from architectural or artefactual evidence, either at Amarna or other ancient Egyptian sites. Secondly, a methodology for an experimental approach to aspects of ancient human diet will be developed. Thirdly, ancient health and nutrition can be explored. Evidence has been found for tetracycline in ancient Nubian populations (Basset et al. 1980, and see Rose et al. and Armelagos and Mills in this volume) and it has been hypothesised that the antibiotic was ingested in their beer. Current medical research shows a significant reduction in bacterial contamination of fermented food compared with non-fermented food (Mensah et al. 1990). Fourthly, a corpus of data will be compiled which can be compared to similar processes in other cultures. Speculation on similar ancient Mesopotamian processes and obvious parallels in production (see Millard 1988, for example) suggest this research could make a substantial contribution to understanding various aspects of ancient Mesopotamian bread and beer preparation.

The aim of the project is a comprehensive analysis of the ancient Egyptian staple diet, bread and beer. The study fascinates for both its academic and human interest and constantly generates new avenues to explore.

References

Bassett, Everett J, Keith, Margaret S, Armelagos, George J, Martin, Debra L and Villanueva, Antonio R, 1980. 'Tetracycline-labelled human bone from ancient Sudanese Nubia (AD 350)', *Science* 209: 1532-1534.

Beranovà, M, 1986. 'Origins of agricultural production in the light of co-ordinated experiments', *Archaeology in Bohemia 1981-1985*. Prague, 307-324.

Blackman, Winifred S, 1927. *The Fellahin of Upper Egypt*. London, George G Harrap.

Brùyere, Bernard, 1953. *Rapport sur les fouilles de Deir el-Médineh (années 1948 à 1951)*. Cairo, Institut Français d'Archéologie Orientale.

Burckhardt, John, 1819. *Travels in Nubia*. London, John Murray.

Darby, William J, Ghalioungui, Paul and Grivetti, Louis, 1977. *Food: The Gift of Osiris*. Vol 2. London, Academic Press.

Davies, Norman de Garis, 1930. *The Tomb of Ken-Amun at Thebes*. New York, Metropolitan Museum of Art Egyptian Expedition.

Davies, Norman de Garis, 1943. *The Tomb of Rekh-mi-re at Thebes*. New York, Metropolitan Museum of Art Egyptian Expedition.

Dixon, David M, 1969. 'A note on cereals in ancient Egypt' in Ucko, Peter J, and Dimbleby, Geoffrey W (eds.), *The domestication and exploitation of plants and animals*, London, Duckworth, 131-142.

Dykmans, G, 1936. *Histoire économique et sociale de l'ancienne Égypte*, Tome 2, *La vie*. Paris, Bibliothèque de l'école supérieur 15.

Emery, Walter B and Kirwan L P, 1935. *The Excavations and Survey between Wadi es-Sebua and Adindan (1929-1931)*, Vols. 1 and 2. Cairo, Government Press.

Épron, Lucienne, Daumas, François and Goyon, Georges, 1939. *Le tombeau de Ti*, Fascicule 1. Cairo, Institut Français d'Archéologie Orientale.

Grüss, Johannes, 1929. 'Saccaromyces Winlocki die Hefe aus den Pharaonengräbern', *Tageszeitung für Brauerei* 27 (59), 275-278.

Henein, Nessim Henry, 1988. *Mari Girgis, village de Haute-Egypte*. Cairo, Institut Français d'Archéologie Orientale. Bibliothèque d'Etude, Tome 94.

Hersh, Theresa Lillian, 1981. *Grinding stones and food processing techniques of the Neolithic societies of Turkey and Greece: statistical, experimental and ethnographic approaches to archaeological problem solving*. PhD thesis, Columbia University. Ann Arbor, Michigan, University Microfilms International.

Hillman, Gordon C, 1984a. 'Interpretation of archaeological plant remains: the application of ethnographic models from Turkey' in Van Zeist, Willem and Casparie, W A (eds.), *Plants and Ancient Man*, Rotterdam, A A Balkema, 1-41.

Hillman, Gordon C, 1984b. 'Traditional husbandry and processing of archaic cereals in recent times: the operations, products and equipment which might feature in Sumerian texts, Part I, The glume wheats', *Bulletin on Sumerian Agriculture* 1, 114-152.

Hillman, Gordon C, 1985. 'Traditional husbandry and processing of archaic cereals in recent times: the operations, products and equipment which might feature in Sumerian texts, Part II, The free threshing cereals', *Bulletin on Sumerian Agriculture* 2, 1-31.

Jacquet-Gordon, Helen, 1981. 'A tentative typology of Egyptian bread moulds' in Arnold, Dorothea (ed.), *Studien zur altägyptischen Keramik*, Mainz am Rhein, Phillip von Zabern, 11-24.

Kemp, Barry J, 1984. *Amarna Reports* I. London, Egypt Exploration Society.

Kemp, Barry J, 1987. *Amarna Reports* IV. London, Egypt Exploration Society.

Klebs, Luise, 1915. *Die Reliefs des Alten Reiches*. Heidelberg, Carl Winter.

Lane, E W, 1908. *The Manners and Customs of the Modern Egyptians*. Everyman's Library, Dent, London.

Larsen, Hjalmar, 1936. 'On baking in Egypt during the Middle Kingdom', *Acta Archaeologica* (Copenhagen) 7, 51-57.

Leek, Frank Filce, 1972. 'Teeth and bread in ancient Egypt', *Journal of Egyptian Archaeology* 58, 126-132.

Leek, Frank Filce, 1973. 'Further studies concerning ancient Egyptian bread', *Journal of Egyptian Archaeology* 59, 199-204.

Lucas, A, 1948. *Ancient Egyptian Materials and Industries*, 3rd edition. London, Edward Arnold.

Lucas, A and Harris, J R, 1962. *Ancient Egyptian Materials and Industries*, 4th edition. London, Edward Arnold.

Lüning, Jens and Meurers-Balke, Jutta, 1980. 'Experimenteller Getreideanbau im Hambacher Forst, Gemeinde Elsdorf, Kr. Bergheim/Rheinland', *Bonner Jahrbücher* 180, 305-344.

Lutz, H F, 1922. *Viticulture and Brewing in the Ancient Orient*. Leipzig, J C Hinrichs.

McDonald, John K, 1982. 'Conical loaf of bread' in *Egypt's Golden Age: The Art of Living in the New Kingdom 1558-1085 BC*, Boston, Museum of Fine Arts, 113, No. 97.

Mensah, Patience P A, Tomkins, Andrew M, Drasar, Bohumil S and Harrison, Tim J, 1990. 'Fermentation of cereals for reduction of bacterial contamination of weaning foods in Ghana', *Lancet* 336, 140-143.

Millard, Alan R, 1988. 'The bevelled-rim bowls: their purpose and significance', *Iraq* 50, 49-57.

Montet, Pierre, 1925. *Scènes de la vie privée dans les tombeaux égyptiens de l'ancien empire*. Strasbourg, Publication de la Faculté des Lettres de l'Université de Strasbourg, Fascicule 24.

Moritz, L A and Jones C R, 1950. 'Experiments in grinding wheat in a Romano-British quern', *Milling* 114(25), 594-596.

Newberry, Percy E, 1893. *Beni Hasan*, Part 1. Archaeological Survey of Egypt. London, Egypt Exploration Society.

Nicholson, Paul, 1989. 'Experimental determination of the purpose of a "box oven"' in Kemp, Barry J, *Amarna Reports* V, London, Egypt Exploration Society, 241-252.

Nims, Charles F, 1950. 'Egyptian Catalogues of Things', *Journal of Near Eastern Studies* 9, 253-262.

Nims, Charles F, 1958. 'The Bread and Beer Problems of the Moscow Mathematical Papyrus', *Journal of Egyptian Archaeology* 44, 56-65.

Peet, T Eric, 1921. 'Excavations at Tell el-Amarna: a preliminary report', *Journal of Egyptian Archaeology* 7, 169-185.

Peet, T Eric and Woolley, C Leonard, 1923. *The City of Akhenaten* Part I. *Excavations of 1921 and 1922 at el-Amarneh*. London, Egypt Exploration Society.

Richards, Audrey Isabel, 1939. *Land, Labour and Diet in Northern Rhodesia. An Economic Study of the Bemba Tribe*. London etc., Oxford University Press.

Rose, Pamela J, 1987. 'Report on the 1986 Amarna pottery survey' in Kemp, Barry J (ed.), *Amarna Reports* IV, London, Egypt Exploration Society, 115-131.

Samuel, Delwen, 1989. 'Their staff of life: Initial investigations on ancient Egyptian bread baking' in Kemp, Barry J, (ed.), *Amarna Reports* V, London, Egypt Exploration Society, 253-290.

Wild, Henri, 1975. 'Backen', *Lexikon der Ägyptologie* I, Wiesbaden, Otto Harrassowitz, 594-598.

Wilson, Hilary, 1989. 'Pot-baked bread in ancient Egypt', *Discussions in Egyptology* 13, 89-100.

Winlock, Herbert E, 1955. *Models of Daily Life in Ancient Egypt from the Tomb of Meket-re at Thebes.* Cambridge (Mass.), Harvard University Press.

Wreszinski, Walter, 1926. 'Bäckerei', *Zeitschrift für Ägyptische Sprache und Altertumskunde* 61, 1-15.

RECENT ARCHAEOBOTANICAL RESEARCH AT THE SITE OF MEMPHIS

Mary Anne Murray

During a study season in 1989, the archaeobotanical material from the excavations of the Egypt Exploration Society at the site of Kom Rab'ia at ancient Memphis was analysed for the first time. The samples had been recovered from a variety of contexts in the previous year by the expedition's archaeozoologist, Barbara Ghaleb. For the sampling strategy, see her contribution to this volume.

Kom Rab'ia is a very interesting site from an archaeobotanical point of view, since the preservation of the plant remains is generally excellent. It is believed to have been an artisans' settlement during both the New and Middle Kingdoms, ranging in date from the seventeenth to the twelfth centuries BC. With this long and continuous occupation, it should be possible to make interesting comparisons of the plant material through time and also, one hopes, with other settlement sites throughout Egypt.

The archaeobotanical material retrieved from the site consists of plant remains which have been preserved by charring. Soil samples recovered from many different context-types, such as pits, hearths, living floors, etc., are processed by means of machine flotation. The principle of this method of separation derives from the fact that the lighter charred plant material, including charcoal, floats when the soil samples are immersed in water. With some water pressure, the charred remains flow over a weir and into two sieves of 1mm and 300 micron mesh size. The plant material is thus divided into two size-components to facilitate analysis. Altogether ninety-nine samples have been floated, of which seventy-six are from Middle Kingdom and twenty-three from New Kingdom deposits.

Each floated sample is air dried, then sorted under a low-power binocular dissecting microscope. All items, such as the charred seeds of food plants, weed species, other plant parts and wood charcoal, are extracted and categorised. Specific identifications are made on the basis of morphological features, often with comparison of the ancient remains to modern comparative reference material and published literature on the subject. The positively identified species are then counted and recorded.

Much of the analysis has already been completed for about one quarter of the Memphis samples. Time constraints are often the greatest problem with archaeobotanical research. This particular analysis has been possible only for a two- month period during the 1989 study season, since we are unable to remove any material from Egypt. The project therefore is in the very earliest stages of analysis.

The initial sorting has so far revealed a wide range of species. The two main cereal crops found on this site and others in ancient Egypt are hulled barley (*Hordeum sativum* aka *H. vulgare* in the literature) and emmer wheat (*Triticum dicoccum*). Both barley and emmer were major crops in Egypt from the Neolithic (Wetterstrom 1982:363-4) and continued to be staples throughout Pharaonic times. Hulled barley appears to be more abundant than emmer in the samples sorted so far from Kom Rab'ia. This ratio, however, would have to be consistent for most of the samples before it could be considered a true reflection of the relative importance of these two cereals on this particular site.

Cultivated in Egypt since at least 5000-4500 BC, hulled barley is still used today as animal fodder and to brew beer. In dynastic times barley was grown for these purposes as well as for making bread (Wetterstrom 1982:6). The barley grains from this site have not yet been sorted into the categories of 2-rowed or 6-rowed barley, a determination which can be made from characteristic features of the grain. Likewise, the barley chaff has diagnostic features which often make discrimination between 2- and 6-rowed varieties possible.

Emmer is a husked wheat which was continually cultivated and was a staple crop in Egypt from the predynastic period to Roman times (Darby 1977:490). It was used primarily for making bread and beer. The straw from emmer was probably also important as a summer fodder, as it is today. By Graeco-Roman times emmer had been replaced by the more popular free-threshing wheats: bread-wheat (*T. aestivum*) and hard or macaroni-wheat (*T. durum*) (Wetterstrom 1984:53).

Lentils, *Lens culinaris*, are well represented in the samples. This is not surprising, as it is a winter crop that has been cultivated in Egypt since the predynastic period (Wetterstrom 1982:366). Several other members of the Leguminosae family are represented at Kom Rab'ia, possibly members of the genera *Vicia, Pisum* and *Lathyrus*, but because of an overlap of morphological characteristics these have been put temporarily into general categories until more time can be spent on establishing positive identifications.

The above are the most important staples identified in the samples. In addition, a wide range of other food-plants are attested at Memphis. A few whole olive pits (*Olea europaea*) and several fragments have been found. Since the olive is primarily a Mediterranean tree, it is very probable that these olives were imported and not cultivated locally. According to one source, the earliest archaeological finds of olives in Egypt date from the 18th Dynasty and during the New Kingdom olives and olive oil were imported from Syria, Greece (Manniche 1989:128) and Palestine, although apparently the tree was also cultivated in the Faiyum from at least late Pharaonic times (Wetterstrom 1982:368).

Grape seeds, *Vitis vinifera*, have also been recovered from the site. The species has been cultivated in Egypt since early in the Pharaonic period and like today was employed for making wine and used as table fruit and raisins. It is unclear whether the specimens from the site were grown locally in Memphis or came from the Delta or the oases, two areas where apparently the best vineyards were located at this time (ibid.).

Several specimens of date palm seeds, *Phoenix dactylifera*, have been identified. The date palm has grown wild in Egypt since the predynastic period (Manniche 1989:133) and has been cultivated in Egypt since the Pharaonic period. The date can be used fresh or dried or as a liqueur. Various other parts of the tree can be used for roofing, ropes, baskets, mats, bags, brushes and so on (Wetterstrom 1982:369).

Several seeds of what appears to be an as-yet undetermined species of pistachio (cf. *Pistacia sp.*) have been found. Apparently, hieroglyphic recipes found on the walls of the temples of Edfu and Philae call for a pistachio resin (Manniche 1989:58). Several species of Pistacia are highly resinous trees. However, more research must be done to determine firmly which species the Memphis seeds represent.

We also have several fruits of persea, *Mimusops schimperi*. During the Pharaonic period persea was considered a popular garden tree (Manniche 1989:122) and at one time grew all over Egypt but due to overcutting the species is now nearly extinct. The earliest finds of persea in Egypt are from a Third Dynasty tomb (Wetterstrom 1984: 59).

A single seed of *Ziziphus spina-Christi* or 'Christ's Thorn' was positively identified. This plant, which has a small apple-like fruit, has been grown since at least the predynastic period and at one time covered all of the Nile Valley. The earliest finds of Christ's Thorn are from a First Dynasty tomb and from inside the Third Dynasty Zoser complex at Saqqara (Darby 1?77:702). Today, it is cultivated as a shade tree (Wetterstrom 1982:369).

Some fairly well-preserved pods and seeds of *Acacia nilotica* are also present. During Pharaonic times various parts of this very versatile tree were used for tanning leather, for building materials, for fuel and fodder and for its medicinal properties (ibid.:373).

The edible tubers of both Wild Nutgrass (*Cyperus esculentus*) and Yellow Nutgrass (*Cyperus rotundus*) have been recovered. Wild Nutgrass is thought to be one of the most ancient foods in Egypt, having been in use since at least the fifth millennium BC (Darby 1977:649). Nutgrass is presently cultivated in Egypt for its edible tubers known as 'tiger nuts'.

Many weed seeds are also found in the Memphis samples, Canary-grass, or *Phalaris sp.*, being the most common. *Lolium sp.*, or Rye-grass, is another very common grass seed. Both genera are large seeded grasses that are almost exclusively segetals or weeds of crops. Several other species of the grass family, Graminae, are present but are as yet unidentified. Other weed seeds include species of the genera *Amaranthus* (Amaranth), *Papaver* (Poppy), *Scirpus* (Club-rush) and *Malva* (Mallow) and the families Caryophyllacae, Polygonacae, Boraginacae and Compositae.

So, what does this all mean? While a species list is not enough in itself any more, it is far too early to say too much about interpretation at this stage. However, a few general conclusions can be drawn.

Most of the plant material appears to be from domestic cooking, heating-fires and household trash. There are varying amounts of charcoal in each sample, suggesting that wood was indeed used as fuel. Much of what has been recovered appears to be waste fractions from various crop-cleaning residues, which were also commonly used as fuel. These include straw, chaff material, seeds from the weeds of crops, etc. From the presence of certain-sized weed seeds and chaff material, it appears that at least some of the remains are derived from sieving wastes. It cannot always be assumed that the evidence of cereal-cleaning residues means that the crop was necessarily cleaned on site. It is possible that the crops were processed elsewhere and that the waste fractions were later brought back to site for use as fodder, fuel, temper in pottery and mud brick, stable litter, and so on (Jones 1981:58). Eventually each sample will be assessed by what the ratios of its various components are able to tell us about the different crop-processing stages such as sieving, threshing, winnowing, etc. As yet, none of the samples represent exclusively any one specific stage of crop-processing or food-preparation.

It is probable that, along with trees and shrubs, animal dung was used as fuel in these domestic fires. It is perhaps likely that, over time, wood for use as fuel became increasingly scarce. Animal dung has often proved to be an obvious alternative. One or two of the samples appear to provide evidence of animal fodder rather than of human food. In one sample in particular, this is quite obviously the case. Some of the dung itself has survived along with barley and numerous leguminous forage plants. The need for fodder was no doubt great, and some of the fodder plants were probably cultivated in addition to those which grew wild, often in the form of weeds in cultivated fields.

Much of the food was probably grown locally, such as the cereals, lentils and other leguminous plants. Other foodstuffs, such as olive, grape and pistachio, may have been imported into Memphis from elsewhere in Egypt or from outside the country. It is also likely that some wild foods (like nutgrass tubers) continued to play a role in the diet at Memphis. All in all, the site of Kom Rab'ia appears to have been fairly well placed to satisfy its occupants' needs for food, building materials, fuel, fodder, etc.

A good deal more research must be done on the Memphis material before a detailed picture can emerge of the site-economy in general and crop-husbandry in particular. With time, a temporal and spatial analysis of the archaeobotanical material may be able to answer many questions; for example, with respect to the distinction between urban food 'consumers' and rural agriculturalists or 'producers', or perhaps it will be possible to detect how changes in human activity may have had an impact on agriculture and the natural environment, such as agricultural intensification, changes in resource management due to food stress and so on (Hillman pers. comm.). These questions are particularly intriguing here in Egypt, where so much of the archaeology of the past has been based on tomb-sites. It is to be hoped that, in addition to the archaeological evidence of how the wealthy died and were buried, bioarchaeological analysis will help to shed light on the day-to-day details of how most of the ordinary population lived.

References

Boulos, L and El Hadidi, M N, 1984. *The Weed Flora of Egypt.* American University in Cairo Press.

Darby, W, Ghalioungui, P and Grivetti, L, 1977. *Food: The Gift of Osiris.* Vol. 1. New York, Academic Press.

El Hadidi, M N and Boulos, L, 1979. *The Street Trees of Cairo.* American University in Cairo Press.

Jones, G E M, 1981. 'The carbonised plant remains from Meare West 1979:2', *Somerset Levels Papers*, No. 12, J M Coles (ed.), 57-60.

Manniche, L, 1989. *An Ancient Egyptian Herbal.* London, British Museum Publications.

Täckholm, V, 1974. *The Student Flora of Egypt*, Second Edition. Cairo, Cairo University Press.

Wetterstrom, W, 1982. 'Plant remains', in Donald S Whitcomb and Janet H Johnson, *Quseir Al-Qadim 1980, Preliminary Report*, Malibu, Undena Publications, 355-377.

Wetterstrom, W, 1984. 'The Plant Remains', in Robert J Wenke, *Archaeological Investigations at El-Hibeh 1980: Preliminary Report*, Malibu, Undena Publications, 50-77.

ANALYSIS OF ESSENTIAL OILS IN FUNERARY WREATHS FROM HAWARA

John Edmondson and Piotr Bienkowski

Abstract

A fragment of *Origanum* from Graeco-Roman funerary wreaths excavated from Hawara by Petrie in 1889-90 was subjected to gas chromatographic analysis and essential oils were detected. These were compared with reference samples, permitting the component oils to be characterised. This information, taken in conjunction with macroscopic features, enabled the identification of the material to be confirmed. The persistence of detectable amounts of essential oils in material probably dating from the first centuries BC/AD has not hitherto been reported; in this case a diagnostic chemotaxonomic character which separates two closely related species has been revealed.

Introduction

This study originated as part of a collaborative project between the Department of Botany and the Department of Archaeology and Ethnology at Liverpool Museum to establish or to confirm the identities of botanical material from the Egyptological collections.

The technique of gas chromatographic analysis is a non-destructive method of separating volatile chemical components from plant material and identifying them using a heated column through which they pass at different velocities, depending on their chemical composition. The extreme sensitivity of modern chromatographic methods allows extremely small amounts to be successfully detected.

The aim of this study was firstly to establish whether detectable amounts of essential oils remained in samples of *Origanum* (marjoram) from wreaths recovered during excavations almost one hundred years ago in Hawara, Egypt and secondly to try to draw taxonomic conclusions from these data. Recent research by Scheffer et al. (1986) has provided reliable data on the essential oils of various species of *Origanum* (Lamiaceae). Thanks to Professor Scheffer's kindness in agreeing to subject the wreath samples to chemical analysis at the Gorlaeus Laboratories, Leiden, it was possible to use these established techniques to determine whether fragrant oils were still present in the ancient plant remains in quantities sufficient to permit their chemical characterisation.

Recent History of the Funerary Wreaths from Hawara

The plant-material we investigated (see Plate 18) was bought by Liverpool Museum from the Norwich Castle Museum in 1956 together with much of the rest of their non-British collection of antiquities. Egyptian plant-remains included those from Middle-Kingdom Kahun as well as Graeco-Roman Hawara (cf. Bienkowski and Southworth 1986).

The botanical samples probably came to Norwich Castle Museum through the Kentish antiquary F C J Spurrell. Petrie gave many curios and objects he had excavated in Egypt to Spurrell. Some of these were presented to the Norwich Museum in 1904 by the latter himself and the rest by his sister after his death in 1915 (Caiger 1971, 6). Spurrell was linked with the work at Hawara in 'unpacking, arranging, and managing the collections' (Petrie 1889, 4).

The plant-remains were packed in small groups in small cardboard boxes with glass lids, sealed with self-adhesive tape; the contents could thus be seen without opening the box. They were lying either on cotton wool or directly on the base of the box, kept in place with paper mounting strips. The packaging and mounting were carried out by Percy Newberry in 1889 ('Mr Newberry has not only worked out the botanical collection, but has prepared and mounted the specimens, and formed series for different museums', Petrie 1889, 4).

When the Egyptian galleries at Liverpool Museum were re-opened to the public in 1976, the displays included a section illustrating the uses of grave-goods. A wreath similar to the specimen studied here has been on public display since this date.

Archaeological Context

The *Origanum* was found in the excavations at Hawara by Flinders Petrie (Petrie 1889, 1890, 1911, 1913). The Middle-Kingdom pyramid and surrounding area were used as a quarry and cemetery in the Graeco-Roman period (Petrie 1889, 3).

The circumstances of discovery of the plant-remains are not well documented. 'Over the mummies there were usually wreaths of flowers laid ...' (Petrie 1911, 15 and cf. Pl. XI:5,6). Newberry (in Petrie 1889, 47) reported that wreaths and a large quantity of fruits, seeds and leaves were found 'merely covered with dust and sand' but 'preserved with scarcely any change'. He implies that they were found in coffins. Plants were also found in the padding of crocodile mummies (Petrie 1911, 16). The custom of placing garlands of flowers on mummies goes back to the beginning of the New Kingdom.

Wreaths of sweet marjoram (*Origanum majorana*) were found in 1889 and 1890 (Petrie 1889, 51; 1890, 46). Newberry describes them as being 'in a wonderful state of preservation although they have lost all trace of their aromatic odour'. The absence of odour as sensed by the human nose is not universal among ancient specimens of aromatic plants; several twigs of 'myrtle' (*Myrtus communis*), with the leaves still attached, were found in the Hawara cemetery, still retaining their aromatic odour (Petrie 1889, 51).

The Hawara burials containing plant-remains probably came from the area north of the pyramid (Petrie 1889, 8 and Pl. XXV). The earliest wreaths were probably linked with plain box coffins, unpainted, with demotic scrawls. These contained elaborately bandaged mummies, with wreaths on heads, pectoral garlands and staves of flowers bound together (Petrie 1889, 14-15). They are certainly Graeco-Roman and are likely to date to the first centuries BC/AD.

The custom may have continued as late as the fourth century AD, but the evidence is inconclusive. Wreaths held in the hand and placed on the forehead moulded and painted as part of the mummy casing have come from Meir, dating to the first century AD (cf. also Grimm 1974, 120). It is unclear from Petrie's reports whether the placing of real wreaths continued on the 'Faiyumic' portrait mummies at Hawara, the majority of which date to the second and third centuries AD. If not, one might have expected Petrie to have noted a change. His journals from Hawara (now in the Griffith Institute, Oxford) do describe some portrait mummies as holding wreaths, but this is likely to refer to moulded rather than real wreaths. Where he does refer specifically to real wreaths, they are on non-portrait mummy cases.

Despite this lack of firm data (and Hawara remains our main source of evidence for Graeco-Roman burial customs), the sheer popularity of the wreath/garland motif might imply the continued use of real wreaths in burials throughout the Roman period in Egypt. The *Origanum* fragment described in this paper would therefore date broadly between the first century BC and the fourth century AD.

Botanical Context

The genus *Origanum* is a member of the family Lamiaceae, most of which contain rich assemblages of essential oils, categorised chemically as terpenes and related compounds. These are located in oil glands, particularly in the leaves and floral parts. Many species of Lamiaceae have consequently been widely employed as culinary herbs and as sources of fragrance.

Because so many different terpenes and sesquiterpenes are synthesised by the Lamiaceae, it is possible to identify samples by detecting the unique combinations of different substances which occur in the various genera. This technique, which is known as chemotaxonomy, uses highly sensitive analytical methods, one of which (gas-liquid chromatography employing head-space sampling) is particularly suited to being used on archaeological samples. The oils are extracted by gently heating a tiny sample of plant material which releases them as a gas.

The essential oil composition of *Origanum vulgare* subsp. *vulgare* has been investigated by Maarse (1971), who categorised the oils as monoterpene hydrocarbons, oxygenated monoterpenes and oxygenated sesquiterpenes; forty-six of these were identified in detail and a further six were detected but could not be characterised exactly. Comparative data involving different species of *Origanum* and different genera of Lamiaceae are scanty; a recent paper by Vokou et al. (1988) reports an analysis of the essential oil composition of *Origanum onites*.

In her compendium of the botany of ancient Egypt, Germer (1985) showed that plants played an important role in human activities. She considers their uses under several headings: ornaments; economic grains; fruits and vegetables; wine; oils; cosmetics; textiles; dyes; plaiting materials; papyrus writing materials; timbers; medicinal plants; smoking materials; and mummy garlands and related uses in religion and mythology. According to Germer, the literature is summarised under the scientific name *Majorana hortensis*; this is a taxonomic synonym of *Origanum majorana* var. *majorana*. Germer does not distinguish between the 'Rigani' of the Cypriots, *Origanum dubium*, a native of Cyprus and southern Turkey, and 'Amaracum', the species *O. majorana*, endemic to Cyprus, whose cultivated derivatives are known as var. *majorana* (Meikle 1985). This important distinction will be discussed below.

Chemical Analysis (by J J C Scheffer)

About 110 mg of the plant material was placed in a 10 ml vial that was tightly sealed with a rubber septum covered with teflon. Nitrogen was used as pressurisation gas. After an equilibration period of 20 minutes, during which the vial and its contents were kept at 150°C, a volume of 1 ml of supernatant gas was transferred from the vial into a Packard 428 gas chromatograph by means of a Dani HSS 3950 headspace sampler. The gas was analysed using a fused silica column (60 m long, 0.25 mm internal diameter) coated with Durabond-DB1 to a thickness of 0.25 micrometres, at an oven temperature programmed to rise from 45 to 220°C at 3°C per minute. Split ratio was 1:100. Retention times were compared with samples taken from named material of *Origanum majorana* from southern Turkey (Sarer et al. 1985). At least three peaks in the chromatogram were identified, as p-cymene, thymol and carvacrol respectively. These compounds are known to be components of the volatile oil content of various *Origanum* species, including *O. majorana*, though they also occur in varying proportions in other species.

A comparison of the relative sizes of the peaks on the chromatogram indicated that the proportions of these three substances were as follows: 21% p-cymene; 6% thymol; 5% carvacrol. Other volatile substances accounted for the remaining 68% of the extract. In fresh material of *Origanum onites*, by comparison, the volatile oils contained c. 65% of thymol and carvacrol (Ietswaart 1980). Seasonal variations in essential oil composition of *O. majorana* are well documented (Abou-Zeid 1973). The most significant chemotaxonomic feature is the small proportion of carvacrol.

Taxonomic Conclusions

The morphological features mentioned by Newberry were exclusively those of the leaves. He reported that 'a microscopical examination of the ovate greyish-green leaves, covered on both sides with thin down, shows that this species existed in exactly the same form two thousand years ago as it does now' (Newberry in Petrie 1889, 51). On re-examining some of the material, mature calyces were discovered by Dr J Ietswaart. The distal or abaxial lobe of the calyx is well developed in *Origanum majorana*, while the proximal or adaxial lobe is reduced to a small tooth or lobe. In the material from Hawara the adaxial calyx lobe was relatively conspicuous, whereas as illustrated by Ietswaart (1980, fig. 19c) the adaxial lobe is obsolete. This difference may be attributable to the period over which the *Origanum majorana* had been maintained in cultivation. We do not know the method by which the ancient Egyptians propagated this species, but as it is grown mainly as a pot-herb it could have been as easily propagated by division of the rootstock as by seed. Unconscious selection in cultivation can give rise to a narrowing of the genetic base; alternatively, the greater convenience which seeds offer for long-distance transport could have facilitated their repeated introduction from the area where the species grows in natural habitats: Cyprus and the adjacent parts of southern Turkey. It would be interesting to compare the morphology of populations of *O. majorana* cultivated in Egypt today with wild populations in Turkey and Cyprus.

The information yielded by the gas chromatographic analysis confirms that the Hawara wreath material contains several of the essential oils characteristic of *Origanum majorana* and closely related species of *Origanum* section *Majorana*. This section is based on the genus *Majorana* Miller, first validly published in his abridged 4th edition of the Gardener's Dictionary. Section *Majorana* includes three species according to Ietswaart (1980): *O. majorana*, *O. onites* and *O. syriacum*. A more recent treatment by Meikle (1986) differentiates between *O. majorana* sensu stricto and *O. dubium*. The latter was found by Holmes (1913a) to contain over 80% carvacrol; *O. majorana* sensu stricto contained only terpinene and thymol. Another member of section *Majorana*, *O. onites*, produces a high yield of essential oils and Vokou et al. (1988) have published comparative quantitative data for essential oils from ten sites ranging from Karpathos, Symi and Crete to Lesvos, Chios and Tilos.

Although there are significant variations in the percentage composition of the major components, carvacrol is consistently the largest single component ranging from 51% to 84.5%, with second place being disputed between g-terpinene and p-cymene falling within the ranges 2.3-13.6% and 5.1-12.2% respectively. Vokou et al. point out that this is not surprising, since these compounds are the immediate precursors of carvacrol in the biosynthesis of this phenolic compound. The other phenolic compound, thymol, is insignificant in *Origanum onites*; this provides additional evidence in support of the identification of the material from Hawara, which was shown to be relatively rich in thymol.

A sample of *O. syriacum* from southern Turkey was analysed at the same time as the material from Hawara. The quantities of material sampled were significantly different: 34 mg of *O. syriacum* compared with 117 mg of material from Hawara. The small contribution made by thymol in the analysis of *O. syriacum* was clearly visible, whereas in the Hawara material the thymol peak exceeds that for carvacrol. This may simply be a measure of the relative stability through time of the two compounds, however, and is unreliable as a measure of the relative abundance when the wreath was manufactured.

In reaching an identification of the Hawara wreath material, morphological evidence clearly points to a member of section *Majorana*. The leaf morphology and indumentum are diagnostic. To discriminate between *O.*

majorana sensu stricto and *O. dubium*, Meikle (1986) refers to the morphology of the inflorescence which is thyrsoid or sybcorymbose and compact in *O. dubium* and elongate, narrow, lax and extending far down the stem in *O. majorana*. The absence of obvious inflorescence branches in the Hawara material points to the former species, but such negative evidence is insubstantial. The crucial distinction is the aroma, which leads us to place much reliance on the chemotaxonomic data. Rigani (*O. dubium*) is a more coarsely pungent herb than Amaracum (*O. majorana*) and it can be stated on no less an authority than Pliny (Nat. Hist., quoted by Meikle 1985, 1270) that 'the most valued and most fragrant samsuchum or amaracum comes from Cyprus'.

The present-day native distribution of *O. majorana* is confined to the island of Cyprus. Its former area of occurrence may have been more widespread; palaeostratigraphic evidence from Turkey suggests that climatic changes have been significant, and it cannot be ruled out that Egypt once supported native populations of *O. majorana*, although the species is now certainly extinct in the wild in Africa. Three endemic species of *Origanum* persist in Cyrenaica, Libya, which is the home of many relict species which were very likely to have been more widespread prior to the well-documented episodes of deforestation and climatic change in the last two millennia. The key to further understanding of the wild origins of cultivated plants such as marjoram undoubtedly lies in a closer study of the wild populations.

Acknowledgements

We are greatly indebted to Professor J Scheffer (University of Leiden) for providing the facilities in his laboratory for the gas chromatographic analysis to be undertaken, and to Dr J Ietswaart (Free University of Amsterdam) for examining the gross morphology of the samples during the loan of the material to Leiden. We acknowledge the assistance of Dr Jaromir Malek for access to Petrie's Hawara Journals at the Griffith Institute, Oxford, and also of Mrs Barbara Adams, Dr Helen Whitehouse, Professor A F Shore and Dr Renate Germer. The Director's Research Fund, National Museums and Galleries on Merseyside, defrayed travel costs.

References

Abou-Zeid, E N, 1973. 'The seasonal variations of growth and volatile oil in the two introduced types of *Majorana hortensis* Mnch. grown in Egypt', *Pharmazie* 28, 55-56.

Bienkowski, P and Southworth, E, 1986. *Egyptian Antiquities in the Liverpool Museum* I. *A List of the Provenanced Objects.* Warminster.

Caiger, N D, 1971. *F C J Spurrell, Kentish Antiquary and Archaeologist.* Bexley Antiquarian Society.

Germer, R, 1985. *Flora des pharaonischen Ägypten.* Mainz am Rhein.

Grimm, G, 1974. *Die Romischen Mumienmasken aus Ägypten.* Wiesbaden.

Holmes, E M, 1913a, b. 'The Oils of Marjoram of Commerce', II. *Perf. Ess. Oil Res.* 4, 41; III. *ibid.* 4, 69-75.

Ietswaart, J H, 1980. *A Taxonomic Revision of the Genus* Origanum *(Labiatae).* Leiden.

Maarse, H, 1971. Samenstelling van der vluchtige olie van *Origanum vulgare* L. ssp. *vulgare* gedurende de ontwikkeling van der plant. Thesis, Groningen.

Meikle, R D, 1985. *Flora of Cyprus* 2, 1262-1270. Kew.

Petrie, W M F, 1889. *Hawara, Biahmu and Arsinoe.* London.

Petrie, W M F, 1890. *Kahun, Gurob and Hawara.* London.

Petrie, W M F, 1911. *Roman Portraits and Memphis* (IV). London.

Petrie, W M F, 1913. *The Hawara Portfolio: Paintings of the Roman Age.* London.

Scheffer, J J C, Looman, A and Svendsen, A G, 1986. 'The essential oils of three *Origanum* species grown in Turkey', in Brunke, E-J (ed.), *Progress in Essential Oil Research*, 151-156. Berlin.

Vokou, D, Kokkini, S and Bessière, J-M, 1988. '*Origanum onites* (Lamiaceae) in Greece: distribution, volatile oil yield, and composition', *Econ. Bot.* 42 (3), 407-412.

THE ARCHAEOZOOLOGY OF KERMA (SUDAN)

Louis Chaix

Introduction

The archaeological site of Kerma is situated in the north of the Sudan, approximately 600 km from Khartoum, close to the left bank of the Nile and about 12 km upstream from the Third Cataract. As early as 1820, it was well known by intrepid European travellers for its two impressive monuments of unfired brick rising out of the desert sand, though it was not until 1844 when the German Egyptologist Lepsius visited the site that a detailed description of the ruins was made (Lepsius 1849-1859).

It was the excavations conducted by G A Reisner between 1913 and 1916, sponsored by Harvard University and the Museum of Fine Arts, Boston, that brought renown to this remote region (Reisner 1923 a and b). More recently, since 1977, an archaeological mission from the University of Geneva, conducted by C Bonnet, has been excavating the site (Bonnet 1983, 1987, 1990).

Kerma is now recognized as having been a distinctly indigenous culture that developed during the third and second millennia BC. This culture belonged to a broad cultural horizon which spread over central and northern Sudan (Bonnet 1986).

Archaeozoological Data

The various sites of the Kerma culture, especially its capital, have revealed numerous vestiges which testify to the importance of the animal world in its economy as well as in its funerary and religious rituals. Similar observations have been made in Egypt, where the animal world also played a key role (Boessneck 1988; Houlihan 1986). After more than fifteen years of excavation and investigation on the Kerma site, we now have a clear picture of the relationships between the inhabitants of the kingdom and the various animal species, be they wild or tame (Chaix 1980, 1982, 1984, 1986a, 1986b, 1988, 1990). Our main sources are the bone remains found during the excavation of the ancient city and of the graveyard. In addition, some data have been provided by clay sculptures of various animals, as well as scanty representations (among which the friezes of the Eastern Deffufa deserve special mention) and a few graffiti discovered by the Geneva Mission (Reisner 1923a; Bonnet 1986).

The following description is based upon two areas. Firstly, the ancient city, where rich osteological material has been unearthed, the greatest part of which dates back to the middle phase of Kerma, between 2050 and 1750 BC. This material is unfortunately in poor condition, due to strong wind erosion and the destructive effect of variations in the water table. What is more, as the material unearthed in the living quarters consists of culinary remains, it is highly fragmented. Secondly, the eastern necropolis, situated a few kilometres east of the city. This site has provided us with an exceptional corpus of whole animals found in human burial sites and of various other documents of great interest (bucraniums, meat quarters, miscellaneous deposits). This desert area, uninfluenced by flooding, has allowed the natural mummification of organic remains, which in turn has made possible a detailed study of elements which have not been preserved in other contexts (fur, skins, stomach contents, coprolites, feathers), as well as plant-remains, which can be used as evidence to reconstruct the environment (seeds, leaves, fruits).

We shall first describe the economic aspects of the animal world and then the religious and ritual data.

1. Breeding and Hunting

All the research undertaken on this rich material (several thousand bone remains) has shown that the Kerma people were almost exclusively breeders and that hunting played only a very minor part.

In the city, several rubbish pits, which can be seen as closed sets indicative of the casting off of food products, have been examined. They show systematically the predominance of domesticated animals (more than 90%). An analysis of the many bones found in the various dwellings of the ancient city offers a similar picture. The livestock is represented principally by oxen (more than 50% of the total), then by caprines (sheep and goats, with 45%), and finally by a few dogs and donkeys (Fig. 1).

By comparison, wild animals are very rare. The presence at Kerma of representatives of the large African animals is attested to by giraffe and hippopotamus bones, to which may be added vestiges of elephant tusks and one canine from a lion, but these latter elements may well have been imported. Several bone remains show the presence of gazelles. The bony part of a horn found in the north-eastern chapel of the Western Deffufa belongs to the Nubian ibex. We can also mention the presence of a monkey closely related to the vervet. Systematic digging and sifting have yielded very few small-size remains. Bone remains of birds are extremely rare. While the use of ostrich feathers and eggs is often observed in Kerma, ostrich bones, on the other hand, have never been found. We can also mention some rare remains of a goose closely related to the Nile goose and the bones of as yet unidentified raptors.

Finally, crocodile dermal plates, Nile turtle plates and a fairly large number of fish vertebrae have also been found. The analysis of these vertebrae indicates the presence of Nile perch and catfish.

A. Domesticated Animals

a. The Ox (*Bos taurus L.*)

As already stated, the ox is the predominant species, representing more than 50% of the remains. We are fortunate in having at our disposal some excellent descriptive elements: the bucraniums and the skins preserved in the necropolis, on the one hand, and, on the other, elements of the post-cranial bones dug up in the city. These bones, and those of the other species too, are highly fragmented and often exhibit fractures and other marks specific to culinary vestiges. Unfragmented bones are very rare; only the most compact and the marrowless ones (carpal and tarsal bones, phalanges) have been preserved.

The oxen which have been discovered are of great size. Several measurements taken on various bones (distal humerus, centrotarsal, talus, and first phalanx) situate them within the range of variation of the Egyptian specimens from the site of Elephantine, dating from the Fifth Dynasty (Boessneck 1988). A full metatarsus, probably from a female specimen, suggests a height at the withers superior to 1.5 m. Other measurements, such as the length of the first phalanx, are within the lower limits of the range of variation of the Egyptian and European aurochs. The average is, however, significantly lower (Gautier 1968; Degerbøl and Fredskild 1970).

The many bucraniums found in the necropolis show that they were longhorn oxen, closely related to those found in dynastic Egypt (Epstein 1971). We can observe a certain morphological variability, but in general the skull has a flat, or slightly convex, post-frontal profile, whereas the ridge between the horns is low, with a single or double curve (Grigson 1976). Horned sheaths, which probably belonged to a single specimen, although they were found in two different burial grounds, present the typical posterior bending of the tip, which has often been observed among the Sanga breed reared by the Nuers of southern Sudan.

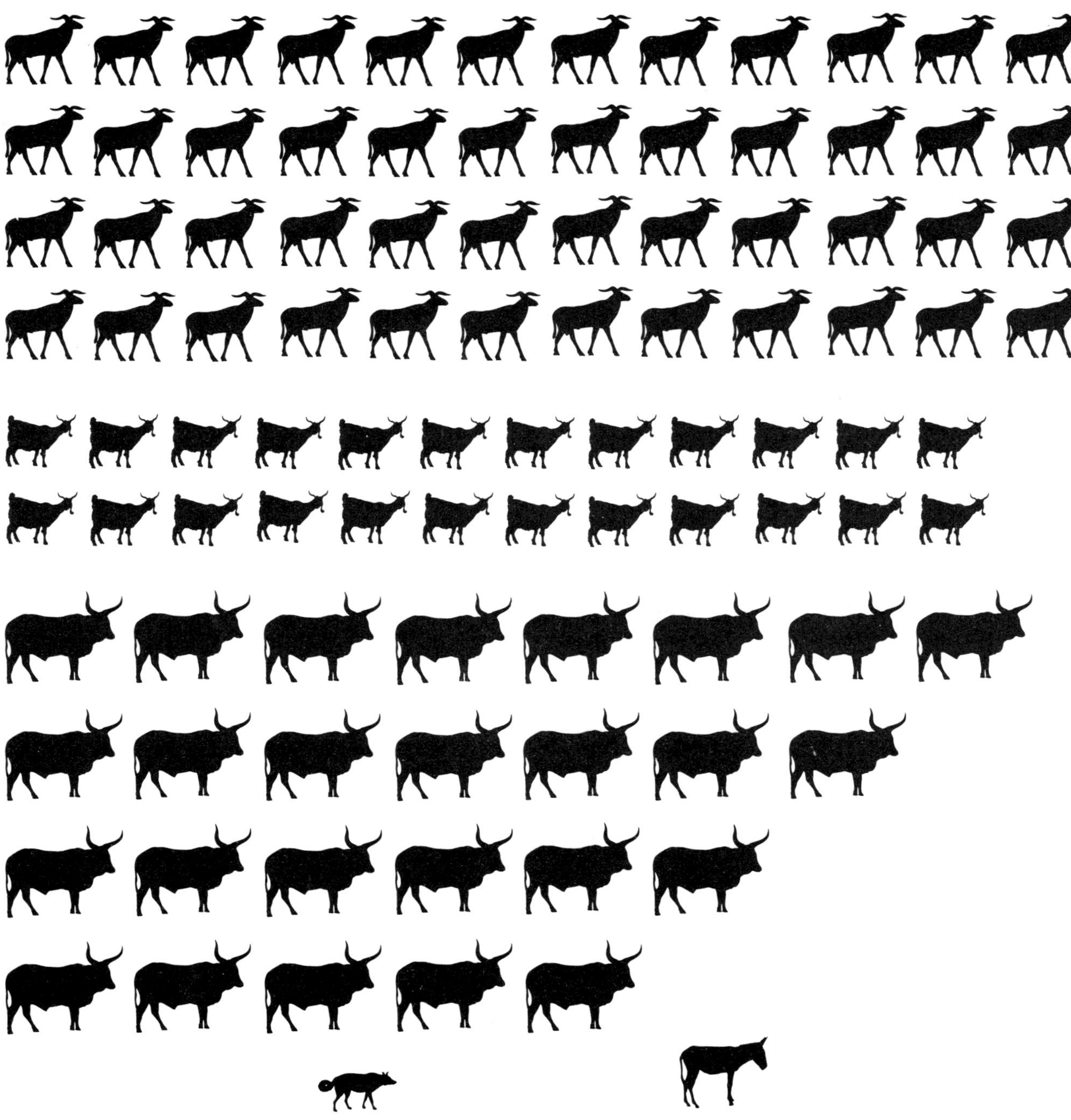

Fig. 1. Compositon of the live-stock in the town of ancient Kerma (each silhouette represents 1%).

The skins of those bovidae were used as mats by the people of Kerma, as is shown by the distortion of the suspension holes, and as shrouds for burials. The hairs which have been preserved show that the oxen had a plain, brown or reddish coat. Piebald coats have not been observed so far.

The detailed study of the Kerma oxen is not yet finished. However, we can see that in the city there is a predominance of adults, although this may simply be the result of differential preservation. In the necropolis, the bucraniums reflect the composition of the live herd, with its bulls, cows, and calves (Chaix 1986b). The presence of castrates is a possibility, but has not yet been proved. The many butchery marks found on the bones show that the bovidae were consumed for their meat. The discovery in the city of the remains of cheese drainers may mean that their milk was used, although it may also have been the milk of caprines.

So far, the presence of zebus has not been substantiated, but some clay statuettes do represent bovidae with the high withers typical of the male (Chaix 1986b; Ferrero 1984).

b. The Caprines (sheep and goats)

These small ruminants represent 45% of the fauna, with the sheep (*Ovis aries L.*) as the predominant species. We know them well, owing to the many complete specimens found in the graves of the necropolis (Chaix and Grant, 1987). The large pieces of meat left in the burial places also come from caprines.

These sheep were tall animals, with a height at the withers of 80 cm. Most of them had horns, but their systematic slaughtering before they reached two years of age precludes a full description of the adults' horns. There seem, however, to have been two types of morphologies, one close to the spiral-horned sheep found in Egyptian sites and described by various authors as 'longpipes', the other similar to those forms called 'platyura', with semi-circular horns pointing forward (Thilenius 1900; Duerst and Gaillard 1902; Pia 1942; Boessneck 1988). Their skeleton is characterized by long, spindly limbs, the most typical elements being the metapodes. Due to the discovery of several complete skeletons, we have been able to show that the average number of caudal vertebrae was fifteen, a primitive trait, which sets them closer to their wild ancestor, the moufflon, than to the sheep found in the area today. Their coats are predominantly all white or beige, but there are also some piebald - black and white - individuals. A detailed study of their hair has shown the appearance at Kerma of sheep with wool-like fleece, in the midst of a population characterized by a total absence of fleece (Ryder 1984, 1987; Ryder and Gabra-Sanders 1987). The age and sex distribution indicates that the purpose of the breeding was the production of meat. The presence of castrates has not yet been observed (Chaix and Grant 1987).

Sheep metapodes were frequently used as awls for leatherwork. Their coats, plucked or complete, were used in the making of various garments, such as skullcaps or loincloths (Choyke 1990). As we shall see later, the sheep also played a major part in funerary rituals.

The goats (*Capra hircus L.*) are present in smaller quantity than the sheep. Several skeletons found in the necropolis show that these goats were small, very much like today's Nubian goat. The height at the withers seems to have been between 60 and 68 cm. They were horned animals with fine, slightly spiralled horns. From the analysis of various coat samples from the graves, we can demonstrate that the goats were black, white or light grey (Ryder, 1984).

Nowadays, the goat population of Sudanese Nubia represents 56.5% of the total livestock, against 32% for the sheep (Adams 1977). It seems that the proportions were the very opposite at the time of the Kushite kingdom, a situation which has not yet been satisfactorily explained.

The goats also seem to have provided the inhabitants of Kerma with raw material for loincloths, various kinds of bags and bone tools, which are often nearly indistinguishable from those made of sheep bones.

c. The Donkey (*Equus asinus L.*)
Only scanty remains of these small equidae have been dug up, both in the various quarters of the city and in an older settlement, described as pre-Kerma, which was discovered in the eastern necropolis. The documents are too few and too fragmented to allow a precise description of these animals.

A detailed study of their dental morphology and the various measurements made seem to indicate that they were domesticated donkeys (Eisenmann 1981). One radius from that species shows marks typical of a dislocation of the shoulder. There is evidence for the consumption of donkey meat during the more recent Napatan period in the same area (Chaix 1988). A rubbish pit structure at Elephantine, dating from the Old Kingdom, also contained donkey bones (Boessneck and von den Driesch 1982).

d. The Dog (*Canis familiaris L.)*
A total of seven dogs has been discovered in the graves and a few isolated bones have been found in the ancient city. The complete specimens found in the graves have enabled us to make a clear description of these animals, in spite of their limited number.

They were of average size, with a height at the withers of 48 to 58 cm. Their bones, particularly those of the limbs, were long and slender. The skull is rather flat, with a narrow muzzle, and the proportions of the neurocranium are very similar to those of the pariah-type dogs which can still be found in that area today. A comparison with a corpus of contemporary specimens from Kerma and Tabo, further south, shows a great likeness beween the fossil specimens and the recent ones. It seems, however, that the dogs of the Kerma era were slightly smaller.

Both males and females were buried with the deceased. In grave 67, a female had been strangled with a leather string which was found around its neck (Chaix 1982). All the specimens discovered had a plain light beige or light brown coat, very similar to the colour of today's 'pariahs'. In two cases, a study of the stomach contents has yielded remains of fish and several bones belonging to young caprines.

B. The Wild Species
We have seen above that wild fauna is very scarce on the Kerma site. Very few bones testify to the presence of entire animals and in most cases they comprise cranial and dental elements which may well have been imported. Among them is an elephant tusk, various canines from hippopotamuses, one canine from a lion and several bony parts of horns from gazelles or wild caprines. The ostrich, whose feathers and egg fragments are numerous in the graves and in the city, is not be found at all in the osteological material.

The inhabitants of Kerma are, however, known to have had a fair knowledge of African fauna, as is shown by numerous decorations, miscellaneous objects and even polychrome frescoes. Instances are zoomorphic ivory plates representing elephants, giraffes, rhinoceros, hippopotamuses, lions, hyenas, foxes, donkeys, ibex and various large unidentified birds. The mica plates present a similar picture. In addition, some representatives of the larger animals are cut out of leather or modelled in clay (Reisner 1923b). The religious role played by some species is confirmed by the hippopotamus friezes which once decorated the walls of the funerary temple K XI.

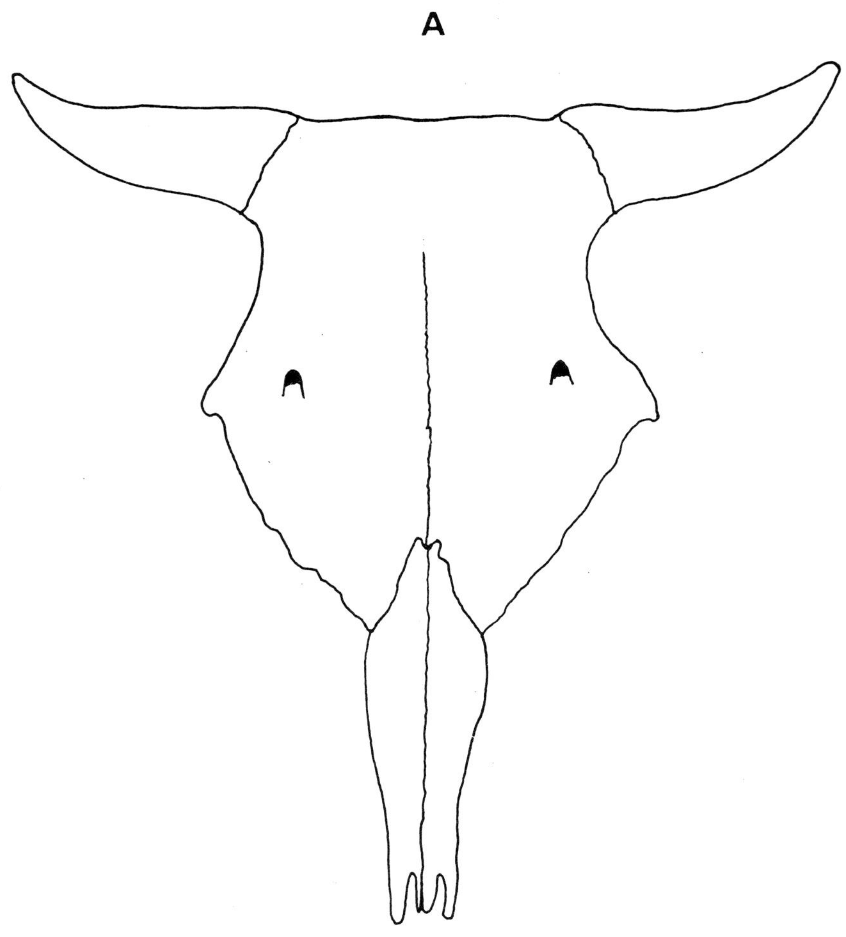

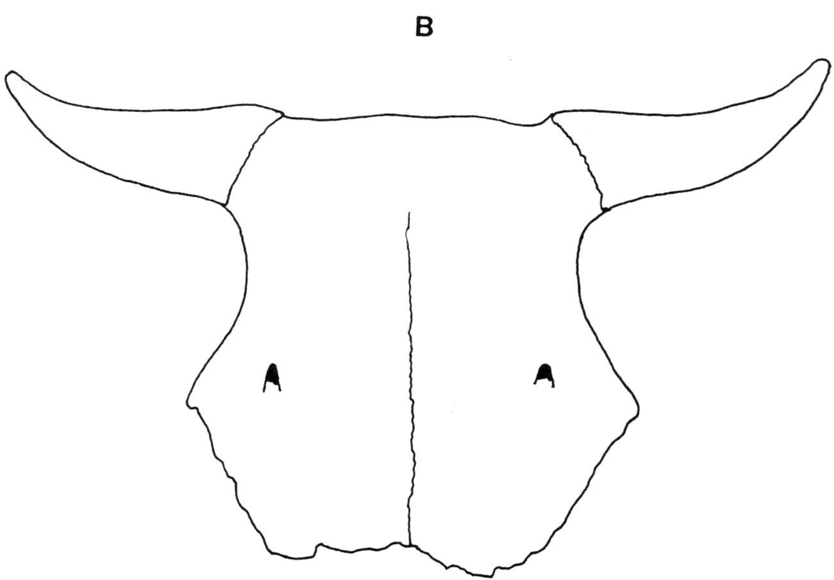

Fig. 2. Schematic shape of the bucraniums
A. Typical ancient Kerma carving
B. Typical middle Kerma carving

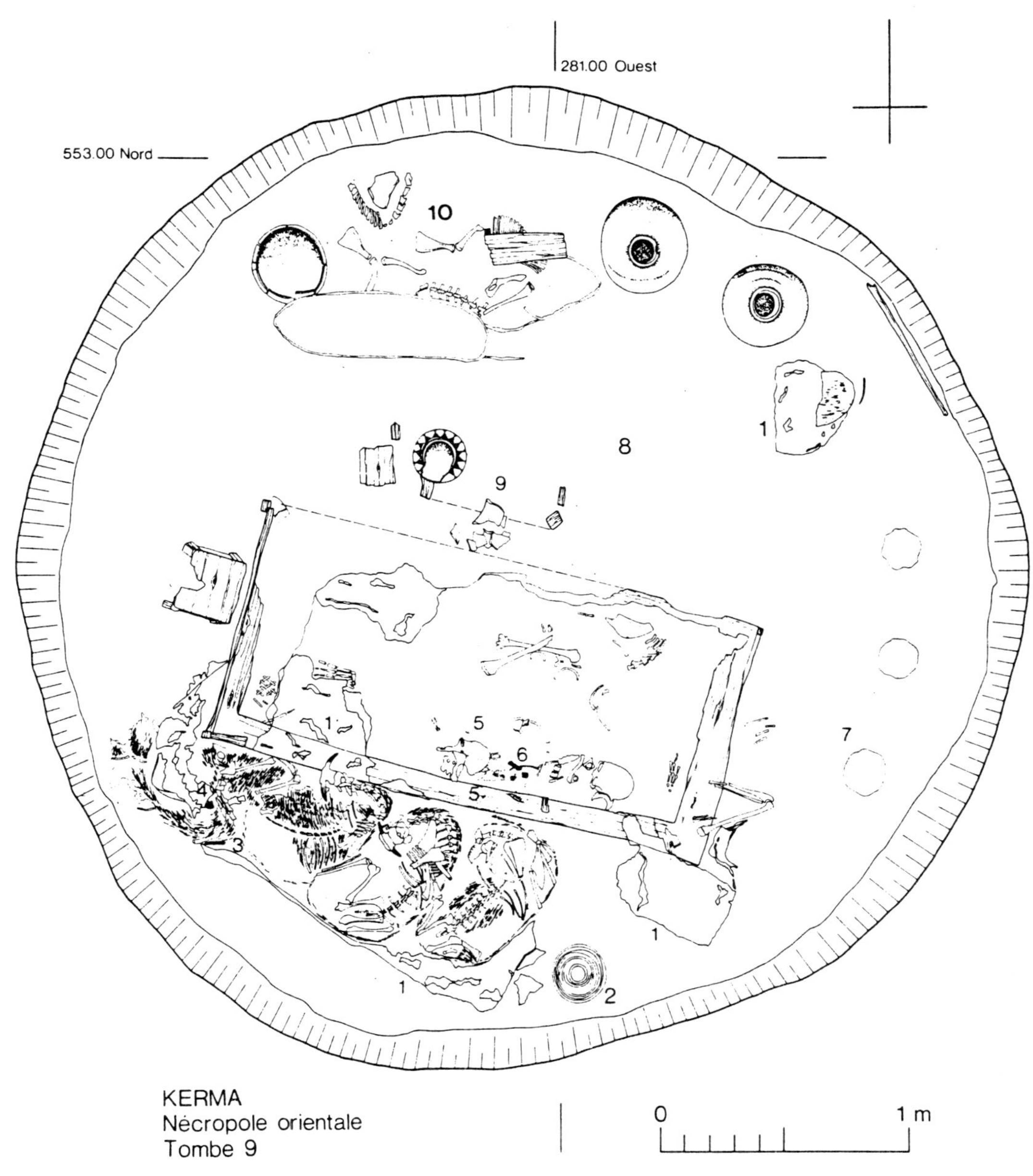

Fig. 3. Typical middle Kerma grave

1. Leather hide covering the deceased
2. Remains of a basket in palm fibres
3. Sacrificed sheep in leather bags
4. Braided leather collar
5. Bronze rivets
6. Fragment of clothes
7. Traces of three little jars
8. Whitewash
9. Table in wood
10. Pieces of meat and vessels

Fig. 4. Reconstruction of a lamb with the disk of ostrich feathers and two ear-pendants (drawing G Deuber).

2. The Animals and the Dead

In addition to the nutritious and technological role played by animals in Kerma, many discoveries point to non-economic aspects of the fauna in that culture. Most of the observations which follow come from the necropolis, which contains several thousand graves and where the animal world is ubiquitous.

As early as the Ancient Kerma phase, between 3000 and 2400 BC, the graves contained horns of bovidae and caprines. The deceased were laid down on shrouds made of tanned ox skins, with a narrow fringe of hair left in place. Suspension holes at one end show through their distortion that they had been used before, probably as mats.

Towards the end of the Ancient Kerma phase, there appear, on the southern edge of the grave cut-out, frontal bones of oxen 'looking' at the deceased. In the more ancient phases, the nasal bones of the bucranium were still in place (Fig. 2).

Later on, during the Middle Kerma phase, animal deposits become more numerous. The horns were replaced by whole caprines (sheep and goats), placed to the south and west of the deceased. This escort consisted mainly of males under two years of age (Chaix and Grant 1987). In some cases, the sheep were placed inside leather bags in a forced position. There are at present no clues as to how these animals were killed. The bucraniums became more numerous, and, in some cases during the later periods, the nasal bones had been removed, a practice which was generalized at the end of the Middle Kerma period. In the part of the grave facing north, large pieces of meat (essentially sheep) were placed, cut in a stereotyped way which is very similar to that practised today in northern Sudan. The skull, however, is systematically missing (Fig. 3). The deposit of whole caprines seems to have been progressively replaced by human sacrifice, which occupies the same area of the grave. Food offerings became more numerous: some burial places contained more than 75 pieces of meat.

The key role played by sheep is reinforced by the repeated discovery of lambs decorated with discs made of ostrich feathers fixed between their horns and with pendants ornamented with stitched pearls and tied to the horned cases, which had previously been perforated (Fig. 4). This layout is reminiscent of the Saharian rock figures of so-called 'spheroid' rams (Camps 1980) and of the ram-headed forms of the god Amon (Wildung 1973).

The final phases of Kerma seem to have been characterized, as was shown by Reisner's excavation, by fewer animal sacrifices and more multiple burials: a hundred, sometimes several hundred, people were immolated and accompanied the high-ranking deceased.

Conclusion

This brief presentation of the animal world of the Kerma culture has shown its great importance both in the economic sphere and in the area of rituals and religious acts (where interpretation is more hazardous).

We have noted, on the one hand, the almost absolute predominance of oxen and caprines, which supplied the breeder and farmer population with the needed proteins, as is also shown by the glossy sickles, the remains of cereals (barley), the rich grinding material and the many bread moulds which have been discovered; on the other hand, a religious world where the animal played an important role, either as a symbol, as in the use of horns or bucraniums, or as a true offering, as in the use of whole sheep or meat pieces in the graves. Numerous other decorative elements confirm the omnipresence of the animal in this culture, at the borderline between Pharaonic Egypt and Africa.

References

Adams, W Y, 1977. *Nubia, Corridor to Africa*. Allen Lane, London.

Boessneck, J, 1988. *Die Tierwelt des Alten Ägypten*. C. H. Beck, München.

Boessneck, J and A von den Driesch, 1982. *Studien an subfossilen Tierknochen aus Ägypten*. MÄS, 40.

Bonnet, C, 1983. 'Kerma: an African Kingdom of the 2nd and 3rd millennia BC', *Archaeology* 36, 6, 125-132.

Bonnet, C, 1986. *Kerma, territoire et métropole. Quatre leçons au Collège de France*. Bibl. Gén. IFAO, 9, Le Caire.

Bonnet, C, 1987. 'Kerma, royaume africain de Haute Nubie', in T. Hägg (ed.), *Nubian Culture: Past and Present,* main papers at the 6th International Conference for Nubian Studies in Uppsala, 11-16 Aug. 1986, Stockholm, 87-111.

Bonnet, C (ed.), 1990. *Kerma, royaume de Nubie. Catalogue de l'exposition.* Genève.

Camps, G, 1980. 'Le bélier à sphéroïde des gravures rupestres de l'Afrique du Nord', *Encyclopédie Berbère,* 26, 1-15.

Chaix, L, 1980. 'Note préliminaire sur la faune de Kerma (Soudan)', *Genava,* n.s., 28, 63-64.

Chaix, L, 1982. 'Seconde note sur la faune de Kerma (Soudan). Campagnes 1981 et 1982', *Genava,* n.s., 30, 39-42.

Chaix, L, 1984. 'Troisième note sur la faune de Kerma (Soudan). Campagnes 1983 et 1984', *Genava,* n.s., 32, 31-34.

Chaix, L, 1986a. 'Quatrième note sur la faune de Kerma (Soudan). Campagnes 1985 et 1986', *Genava,* n.s., 34, 35-40.

Chaix, L, 1986b. 'Les troupeaux et les morts à Kerma (Soudan). (3000 à 1500 avant J.C.)', *Anthropologie physique et archéologie,* Colloque CNRS, Toulouse, Ed. CNRS, Paris, 297-304.

Chaix, L, 1988. 'Cinquième note sur la faune de Kerma (Soudan). Campagnes 1987 et 1988', *Genava,* n.s., 36, 27-29.

Chaix, L, 1990. 'Le monde animal' in Bonnet, C (ed.), *Kerma, royaume de Nubie, Catalogue de l'exposition,* Genève, 109-113.

Chaix, L and A Grant, 1987. 'A study of a prehistoric population of sheep (Ovis aries L.) from Kerma (Sudan) - Archaeozoological and archaeological implications', *Archaeozoologia,* 1, 1, 77-92.

Choyke, A, 1990. 'Travail de l'os et de l'ivoire à Kerma' in Bonnet, C (ed.), *Kerma, royaume de Nubie, Catalogue de l'exposition,* Genève, 140-141.

Degerbøl, M and Fredskild, B, 1970. 'The Urus (*Bos primigenius* BOJANUS) and neolithic domesticated cattle (*Bos taurus domesticus* LINNE) in Denmark', *Det. Kong. Danske Videnskab. Selsk. Biol. Skrift.,* 17, 1.

Duerst, U and Gaillard, C, 1902. 'Studien über des aegyptischen Hausschafes', *Recueil de Travaux,* 24, 44-76.

Eisenmann, V, 1981. 'Etude des dents jugales inférieures des *Equus (Mammalia, Perissodactyla)* actuels et fossiles', *Palaeovertebrata* 10, 3/4, 130-226.

Epstein, H, 1971. *The origin of the domestic animals of Africa.* Africana Publ. Corp., New York, London, Munich.

Ferrero, N, 1984. 'Figurines et modèles en terre mis au jour dans la ville de Kerma', *Genava,* n.s., 32, 21-25.

Gautier, A, 1968. 'Mammalian remains of the northern Sudan and southern Egypt' in F. Wendorf (ed.), *Prehistory of Nubia,* SMU Press, Dallas, 80-99.

Grigson, C, 1976. 'The craniology and relationships of four species of Bos. 3. Basic craniology: Bos taurus L. Sagittal profiles and other non-measureable characters', *Journ. of Arch. Science* 3, 115-136.

Houlihan, P F, 1986. *The Birds of Ancient Egypt.* Aris and Phillips, Warminster.

Lepsius, C R, 1849-1859. *Denkmäler aus Ägypten und Aethiopien,* Plates, 12 vols. Berlin.

Pia, J, 1942. 'Beobachtungen an Schädeln des altägyptischen Hausschafes', *Zeitschr. für Tierzüchtung und Zuchtsbiol.* 53, 171-179.

Reisner, G A, 1923a. *Excavations at Kerma.* Parts I-III. Harvard African Studies, 5. Cambridge, Mass.

Reisner, G A, 1923b. *Excavations at Kerma.* Parts IV-V. Harvard African Sudies, 6. Cambridge, Mass.

Ryder, M, 1984. 'Skin, hair and cloth remains from the ancient Kerma civilization of northern Sudan', *Journ. of Arch. Science* 11, 477-482.

Ryder, M, 1987. 'Sheepskin from Ancient Kerma, Northern Sudan', *Oxford Journ. of Archaeology* 6, 3, 369-380.

Ryder, M and Gabra-Sanders, T, 1987. 'A microscopic study of remains of textiles made from plant fibres', *Oxford Journ. of Archaeology* 6, 1, 91-108.

Thilenius, G, 1900. 'Das aegyptisches Hausschaf', *Recueil de Travaux* 22, 199-212.

Wildung, D, 1973. 'Der widdergestaltige Amun-Ikonographie einer Götterbildes', *Actes du Congrès International des Orientalistes,* Paris.

ASPECTS OF CURRENT ARCHAEOZOOLOGICAL RESEARCH AT THE ANCIENT EGYPTIAN CAPITAL OF MEMPHIS

Barbara Ghaleb

Man's dominant position in the animal kingdom is no better exemplified than through domestication. Archaeological evidence indicates that this process began during the onset of the Holecene period, between 12000 and 10000 years ago, and that relatively few plant and animal species were domesticated. The earliest evidence we have consists primarily of the animal bones and plant remains recovered from settlement sites in Southwest Asia (c. 9000-6000 BC) and it is not until the fourth millennium BC, with the establishment of the first urban civilisations, that the evidence increases significantly in diversity, e.g. artefactual, pictorial and written sources. Among the earliest civilisations ancient Egypt is unique in the quality of the existing evidence for animal exploitation and domestication.

The surviving pictorial and written records of the ancient Egyptians are peculiarly rich in animal representations and they provide tantalising glimpses of the range of animals exploited in dynastic times. These are frequently detailed enough to allow generic-level identification and sometimes identification to species is possible. Painted scenes depicting activities carried out in relation to animals, such as hunting, herding, feeding and butchery, indicate clearly that both domestic and wild animals were exploited extensively in ancient Egypt. Such scenes have earned the Egyptians the title of 'experimenters' in animal domestication. Study of artefactual, textual and pictorial evidence also clearly indicates the myriad roles that animals played in Egyptian life: in social, political and spiritual realms.

Although the artistic representations are an unparalleled source of information on ancient Egyptian animal exploitation, they occur within very specific mortuary contexts and were created with the intention of serving the highest classes of Egyptian society in the after-life. It is, therefore, difficult to assess how representative the tomb-relief scenes may be of animal exploitation by the Egyptian population as a whole.

In contrast to the relative wealth of historical evidence on animal exploitation in ancient Egypt, direct evidence in the form of remains of the animals themselves is not abundant. There are two main reasons for this: a) much of the early work on sacred sites, e.g. tombs and temples, was undertaken when the analysis of animal remains was not considered important; and b) very few well-stratified dynastic settlement sites have been located and excavated. To increase our knowledge of animal exploitation in Egypt, including both indigenous and introduced species, archaeozoological work needs to be carried out on assemblages from well provenanced contexts with specific questions in mind. Results of the analysis of direct evidence of animal exploitation can then be evaluated in the light of the pictorial and written sources of information.

My work in Egypt began in 1988 when I became part of the 'Survey of Memphis' research team, which has been recording the remains of ancient Memphis over the past decade through survey and epigraphic work and excavation. Memphis (23 km south of modern Cairo) is reputed to have been the first capital of ancient Egypt, whose rulers for the first time presided over a unified Upper and Lower Egypt, starting at about 3000 years BC. Study of the historical accounts and surviving architectural remains indicates that, although the site of the royal residences and capital changed, Memphis 'remained the "Balance of the two Lands", the administrative and defensive pivot of the country' (Smith and Jeffreys 1986:89), throughout pharaonic history.

The Survey of Memphis is directed by David Jeffreys, Department of Egyptology, University College London, and is run under the auspices of the London-based Egypt Exploration Society. My participation as archaeozoologist to the project began during the fifth season of excavation at Kom Rabi'a, an area of past settlement located just outside (to the southwest) of the precinct of the Ptah-Temple (Ptah having been the chief God of Memphis), near the modern village of Mit Rahina.

During the 1988 season both New Kingdom and Middle Kingdom contexts were excavated (early Eighteenth Dynasty c. 1300 BC and Second Intermediate Period or Thirteenth Dynasty c. 1800 BC). The seven-week excavation was carried out within a trench approximately 25 metres long (east to west), to a depth of between 1.5-2.5 metres (below the level reached in the previous season). In order to determine the nature and quantity of plant and animal remains from the diversity of archaeological contexts revealed (e.g. living floors, hearths, walls, collapsed walling/roofing), bulk samples of known volume (in litres, with only visible objects and pottery removed) were collected from a hundred of the four hundred contexts excavated (Jeffreys and Giddy 1989).

During the 1989 study season all the bulk samples were first processed in a flotation machine to facilitate the separation and cleaning of the animal and plant remains. Archaeozoological and archaeobotanical work then began on a proportion of the processed bulk samples (see also Murray in this volume). Twenty-four out of a total of seventy-seven Middle Kingdom contexts and six out of a total of twenty-three New Kingdom contexts were initially sorted (see Giddy, Jeffreys and Malek 1990).

Preliminary study of the animal remains from Memphis indicates that the assemblage primarily represents subsistence-related refuse from both primary (e.g. living walls, hearths) and secondary (e.g. ash dumps, brick-rubble foundations, walls) depositional contexts. The condition of the animal remains is generally as one might expect from contexts within settlement areas: highly comminuted and from a range of domestic and wild species. The domesticates are represented by remains of animals exploited for food and/or traction (pig, sheep/goat, cow and donkey), or kept as pets (cats, dogs). Animals that might have been naturally attracted to the site (or possibly eaten), e.g. rodents and lizards, are also present. In addition, a variety of types and sizes of fish and certain birds and molluscs (e.g. oyster and mussel) are part of the archaeozoological assemblage. Although relatively complete bones were found in many of the contexts excavated, they tended to be discreet, solitary finds. The majority of the animal remains recovered were fragmentary with a number of species found together. Study of the superficial damage on the remains (both mammal and fish) has revealed butchery marks, black or white/grey colouration due to exposure to fire, evidence of dog and rodent gnawing and accidental (trampling) marks due to contact between the bone and harder substances (e.g. stone) sometime during the depositional history of the remains.

It is not yet possible to discuss quantitative differences in the kinds of animal species eaten or to offer a comprehensive list of the domesticated and wild species exploited. This is particularly true in relation to the range of fish and bird species exploited. Thus far, very few bird remains have been identified, but the preservation of fish bone is good. The fish remains are also very fragmentary. However, due to the distinctiveness in morphology of a percentage of the surviving fragments, it is clear that a number of species can be identified.

To identify accurately the thousands of fish-bone fragments recovered at Memphis (a large proportion of which range between 1mm-5mm in size), it is necessary to have access to a comprehensive collection of modern fish skeletal material. In the beginning of this century approximately two hundred species of fish were identified from the Nile (Boulenger 1907), but it now seems certain that far fewer species inhabit its waters today (el-Sedfy et al. 1985). My initial decision as to which modern species to collect (for the skeletal remains) was determined by the types identified from study of Old Kingdom tomb (mastaba) reliefs at the necropolis of Memphis, Saqqara (Gaillard 1923). Gaillard's study focused upon the identification of the different types of fish depicted on the walls of two of the Old Kingdom tombs at Saqqara: the mastabas of Ti (Dynasty V, c. 2400 BC) and of Mereruka (Dynasty VI c. 2340 BC). He identified twenty-four fish species (which represent ten families) from scenes of Egyptians fishing with lines, spears, and nets from boats in the marshes or river.

Thus far, species of fish from ten families that inhabit the Nile and one sea-dwelling species (*Sparus* sp.) have been identified from both New and Middle Kingdom contexts at Memphis. Fish species from eight of the families depicted in the tomb scenes at Saqqara identified in Gaillard's study are within the archaeological assemblage, as well as two species which are not depicted. The eight fish families represented both on site and in the tomb reliefs are: Silurdae (*Clarias* sp.), Mochocidae (*Synodontis* sp., at least three species archaeologically), Mormyridae (*Hyperopisus* sp.), Cyprinidae (*Labeo* sp. and *Barbus* sp.), Centropomidae (*Lates niloticus)*, Cichlidae (*Oreochromis (Tilapia)* sp.), Mugilidae (*Mugil* sp.), and Tetraodontidae (*Tetraodon fahaka*). Species of Bagridae (*Bagrus* sp.) are present at Memphis, although rarely identified from tomb scenes (Brewer and Friedman 1989:66), and of Polypteridae (*Polypterus* sp.), for which I have not yet found any historical information. A genus whose species (*Sparus* sp.) inhabits both Mediterranean and Red Sea waters is also present at Memphis. There are, however, two families depicted in the tomb scenes whose species have not yet been identified archaeologically (Citharinidae and Anguillidae), but it must be stressed that the results discussed here are preliminary.

The above discussion clearly illustrates how study of historical data from ancient Egypt can relate directly to archaeozoological enquiry. The depiction of fish in Egyptian tombs and their taxonomic identification provide an excellent starting point for the study of direct evidence of fish exploitation from archaeological contexts. That a greater diversity of fish species was exploited in ancient Egypt than is suggested by the tomb scenes has been demonstrated at Memphis and other dynastic sites (e.g. Wendorf et al. 1980; von den Driesch 1983; van Neer 1984, 1985). However, study of the dynastic written and pictorial material can provide evidence of animal exploitation that may never be recovered archaeologically (e.g. details of technology, butchery, exchange).

A range of animals in addition to fish is artistically represented in the Egyptian tomb reliefs, with different varieties of *Bos* (cow) or *Ovis* (sheep), for example, suggested by differences in horn or body morphology. With the recovery of adequate (in terms of quantity and element types represented) archaeozoological assemblages, it may be possible to test theories of animal domestication in Egypt (e.g. domestication of indigenous species or introduction), which have been based primarily upon study of written and pictorial representations rather than upon study of the animal remains themselves.

As mentioned earlier, the significance of artistic representations of animal species must be considered with reference to the context within which they exist. For example, although the tomb fishing scenes and archaeozoological evidence both suggest that species of 'bulti' (*Oreochromis (Tilapia)*) were as abundant and as popular a food in ancient Egypt as they are today, their artistic portrayal may have conveyed additional meaning. It has been suggested that species of *Oreochromis* became incorporated into the spiritual belief system of the ancient Egyptians on account of observed aspects of their biology. Certain species of *Oreochromis* are mouth brooders: i.e. the female incubates the fertilised eggs in her mouth and provides the hatchlings with

20-30 days of shelter there (Balarin 1984). Hence, in ancient Egypt, these fish may have come to be viewed as symbols of self-propagation and rebirth associated with the creator god Atum (Brewer and Friedman 1989:2; Fazzini 1989). In addition, 'the brilliant breeding colours of Tilapia led to its association with the sun' and it was eventually considered 'as a form of the god Horus, who kills enemies of the sun' (Brewer and Friedman 1989:2). Thus the possibility that the significance of animals depicted extends beyond the economic realm must be addressed where appropriate (e.g. Smith 1969).

Knowledge of the ecology of the animal species identified archaeologically can also provide information relevant to past Egyptian animal exploitation, e.g. the kinds of habitats frequented and the season(s) of capture. Thus species of Silurids (catfish, e.g. *Clarias, Bagrus*), Cyprinids (*Barbus*), Cichlids (*Oreochromis*) and eel are physiologically equipped to inhabit the habitats created on the floodplain during the initial and final periods of Nile inundation (widespread areas of shallow water and gradually reducing pools respectively, see van Neer 1984, 1986). Species of *Clarias* and *Oreochromis* spawn within shallow water over periods of days or weeks and at these times would have been particularly easy to catch. Nile perch, on the other hand, prefer well oxygenated, rapidly flowing waters (Greenwood and Howes 1975) (although *Lates niloticus* have also been caught in shallow swamp-like habitats (Wheeler 1975)), and thus their archaeological presence might suggest fishing within the main river channel (van Neer 1986:104). In addition, the pectoral spines of certain species of catfish have been used as indicators of palaeoclimate and season of capture through study of the growth rings displayed when the spines are cross-sectioned (Wendorf et al. 1980; van Neer 1984).

The site of Memphis is considered to have been that of the first capital of Egypt, which remained one of the most important urban centres throughout dynastic history. Excavation of part of the ancient settlement area has revealed excellent preservation of organic remains and a stratigraphy that thus far spans the Middle and New Kingdoms. Its stratigraphic comprehensiveness makes it an excellent starting point for further comparative study and offers great potential for providing direct evidence of animal exploitation and of other aspects of daily life throughout Egyptian history. Study of direct evidence from a range of archaeological contexts is precisely what is needed to further our understanding of the roles animals played in the world of the ancient Egyptians.

Acknowledgements

The author would like to offer special thanks to Professor Moustafa Zaher, Department of Zoology, Cairo University, Ann Crowley, and the Momtaz and Fahmy families, without whose help and support many aspects of my work in Egypt would not be possible. I would also like to thank Christopher Kirby for assistance in the field and to acknowledge the financial and logistical support of: The Egypt Exploration Society; The Gordon Childe Fund, Institute of Archaeology, University College London; and The Central Research Fund, University of London.

References

Balarin, J D 1984. 'Tilapia', in Mason, I L (ed.), *The Evolution of Domesticated Animals*. Longman Ltd., London and New York, 391-97.

Boulenger, G A V, 1907. 'The Fishes of the Nile', in Anderson, J (ed.), *Zoology of Egypt*, Vols. 1 and 2. London, Egyptian Government, Hugh Rees.

Brewer, D J and Friedman, R F, 1989. *The Natural History of Egypt:* Vol. II, *Fish and Fishing in Ancient Egypt.* Aris and Phillips, England.

von den Driesch, A, 1983. 'Some archaeozoological remarks on fishes in ancient Egypt', in *Animals and Archaeology: (2) Shell Middens, Fishes and*

Birds, C Grigson, and J Clutton-Brock (eds.), BAR International Series 183, 87-110.
Fazzini, R A, Bianchi, R S, Romano, J F and Spanel D B, 1989. *Ancient Egyptian Art in the Brooklyn Museum*. New York and London.
Gaillard, C, 1923. *Recherches sur les poissons représentés dans quelques tombeaux égyptiens de l'Ancien Empire*, Mémoires de l'Institut Français d'Archéologie Orientale du Caire 51, Cairo, 1-136.
Giddy, L, Jeffreys, D G and Malek J, 1990. 'Survey of Memphis 1989', *Journal of Egyptian Archaeology* 76, 1-15.
Greenwood, P H and Howes, F J, 1975. 'Neogene fossil fishes from the Lake Albert-Lake Edwards Rift (Zaire)', *Bull. Br. Mus. (Nat. Hist.), Geol.* 26, 72-127.
Jeffreys, D G and Giddy, L, 1989. 'Memphis, 1988', *Journal of Egyptian Archaeology* 75, 1-12.
van Neer, W, 1984. 'The use of fish remains in African archaeozoology', in *2nd Fish Osteoarchaeology Meeting*, Centre de Recherches Archéologiques Notes et Monographies Techniques no. 16, Desse-Berset, N. (ed.), Editions du CNRS, 155-167.
van Neer, W 1986. 'Some notes on the fish remains from Wadi Kubbaniya (Upper Egypt; late palaeolithic)', in *Fish and Archaeology, Studies in Osteometry, Taphonomy, Seasonality and Fishing Methods*, Brinkhuizen D C and Clason, A T (eds.), BAR International Series 294, 103-113.
el-Sedfy, H M, el-Bolock, A R, Helmy, A R, 1985. 'Some effects of the flood retention on the commercial catch of the Nile fishes', *Egypt. J. Wil. and Nat. Resources* 6, 56-71.
Smith, H S, 1969. 'Animal domestication and animal cult in dynastic Egypt', in *The Domestication and Exploitation of Plants and Animals*, Ucko, P J and Dimbleby F W (eds.). London, Duckworth, 307-314.
Smith, H S and Jeffreys, D G, 1986. 'A Survey of Memphis, Egypt', *Antiquity* 60, 88-95.
Wendorf, F, Schild, R and Close, A E (eds.), 1980. *Loaves and Fishes: The Prehistory of Wadi Kubbaniya*. Southern Methodist University.
Wheeler, A, 1975. *Fishes of the World*. George Rainbird Ltd, London.

POSSIBLE FUTURE DIRECTIONS OF A BIOANTHROPOLOGY OF ANCIENT EGYPT

F W Rösing

Introduction

Trying to see into the future of a science is like tight-rope walking: it is very easy to lose one's balance. The reason is simple: the high variability of present processes opens up the possibility of a virtually unlimited number of different futures. However, it is not hopeless to make the attempt, for two helpful reasons: 'variability' of the present means that there is not only untidiness but a certain amount of order, namely modes and nodes; modes: strings of higher probability than elsewhere; nodes: correlations of seemingly independent strings of events. So the tight-rope walking is reduced to understanding that order. The second helpful element is that there is a limited rate of change in science systems; or at least the more general lines change slowly, particularly in fields with small numbers of functional workers. Naturally this argument must exclude the important paradigm change concept of Thomas Kuhn: revolutions are not predictable. Let us assume for the moment that such a glance at the future is difficult but possible, particularly if we avoid very specific scenarios. Obviously Egypt is one example only; mutatis mutandis, things are the same for the skeletal biology of other regions. Some aspects are even valid for anthropology as a whole.

Methods

There are two methods of forecasting. First, extrapolation from the past. The prerequisite for that is the reduction of a series of newly produced facts to those of its elements that persist despite all evident novelty. Andor Thoma's analysis of Nazlet Khater man is in itself totally new, but it is based on persisting morphological methods and on a comparison with the other known hyperrobust mesolithics. This extrapolation can be done by a single thinker, and it will be the main approach of this essay.

The second method (or better, package of methods) is the delphi procedure, whereby a number of selected experts are asked to draw a picture of the path towards the future. All the single pictures are merged into one and then circulated among the experts. This ping-pong play is repeated as often as is necessary; that is, as long as there are relevant changes to the central picture or still considerable dissent among the experts. This second method is an important extension and slight modification of the first: not just one but many thinkers draw a picture. Further, the pictures are more formalised and shorter. The following essay does not rest on a delphi run, but on a reduced variant of it: most of the core points have been discussed with meta-experts of anthropology.

If this essay was to become a scientific paper, then at this point I would have to elaborate the methods in detail. For instance, the selection process and the criteria for expert selection are important for the direction and relevance of results. As it is, it is enough just to mention the two meta-sciences whose methods are exploited or at least serve as a background. The first is futurology. This little discipline of the fifties and sixties lost its innocence and its intellectual pretentions when it was told that it was unable to cure society. Consequently it was re-named 'technical forecast'. Today this field has almost completely merged into technology assessment - and thus recovered intellectual status but lost much of its initial objective of forecasting.

The other meta-discipline is the science of science, which is the social psychology and system analysis of that process of generating knowledge, augmenting experience and conserving wisdom which we call 'science'. This meta-discipline is part of my own work. Today it has merged into more general higher education research in Europe and institution research in the USA.

On this methodological basis the following passages will cover the past, present and future. The section on the future will be divided into single elements and specific forecasts.

Materials

Just as in real science there is also a paragraph covering materials in this meta-study. We must decide which properties of the science system of anthropology we will use for the forecast. Obviously the first choice is the operative system of the discipline itself, i.e. its preferred methods and objectives, its experience and its hopes.

However, there are alternatives. For example, the focus might be on the overlap with neighbouring disciplines. After all, the preferred partners along with their methods and thinking styles influence the future of a field considerably.

And there is a second alternative: societal expectations, the place of a discipline in public opinion. This is an approach of minor importance because all systems of embedding a science into society (national or institutional systems) have strong constituents of liberty and independence. Therefore, societal expectations do not influence the intrinsic development of a discipline very much.

These two alternatives have had to be mentioned but they will not be used in this essay in order to leave it simple and clear.

Main Trends in the Past

The history of Egyptian bioanthropology in this section will be a history of ideas or paradigms, rather than a history of personalities, schools, strings of events or publication processes. If we proceed in this way, we find two clearly dominant directions of change: typology versus population biology and migration versus in-situ-evolution. Of course, both trends apply to the whole of skeletal biology and history, wherever localised on the globe.

In Egyptian historic bioanthropology the typological approach to morphology was exclusive for almost a century. If we look more closely at the subfields, we see that typology has not very often been applied to high group levels like regions, but to the morphological internal-analyses of local groups. The transition as a whole may be regarded as very nearly complete.

In the case of the migration (allochthony) versus in-situ-evolution (autochthony) antagonism, the transition is even broader and the two concepts are less easily separated. Migrationism appeared to end in the late sixties but sees a slight revival today. At least, there are times (Neolithisation) and case groups (initial settlement), where it is no longer considered aberrant to deviate from the credo of localism. Moreover, the numerous papers by Robert Sokal seem to reopen migrationism as an approach to population history as a whole.

The common element of both processes is the change from a static to a dynamic view of human groups. In the case of typology, the static character is evident; in the case of migration, the populations are regarded as static because usually a local static group is altered only by immigration of an equally static foreign group.

This very brief picture should be enough, because it has already been published in sufficient detail in my recent book on Aswan *(Qubbet el Hawa und Elephantine. Zur Bevölkerungsgeschichte von Ägypten*, Stuttgart/New York, 1990), naturally on the basis of the complete range of previous literature.

A Present System of Disciplinary Objectives

For a temporal perspective the present is unimportant. It is merely one moment in the endless stream of moments called Time. Therefore, we could pass it without a glance. Even so, I should like to make some short remarks, because the still picture of the present also contains considerable process information; the expert of average experience knows when specific parts of the field were introduced. The following passages will make that knowledge explicit, thus leading to an initial assessment of what the future might bring. The specific system is probably personal in some respects; however, when we allow for the usual and necessary imprecision of language, then there should already be sufficient agreement with other experts in the field.

A first group of objectives is taxonomy, the elaboration of genetic similarity between populations. Usual methods are multivariate statistical distance measures, clustering and, today, spatial autocorrelation. Those objectives may be differentiated according to group size: it makes a considerable difference whether races or villages are meant. In between them we might place regional taxonomy (see Friedlaender's Pacific studies). It is at this level that the sub-objective of ethnogenesis of the Soviet anthropologists fits.

A second group of objectives may be subsumed under 'adaptation', thus aiming at the effect of genetic as well as cultural and environmental factors on mankind in general.

A third, large group of objectives might be called population history. Using preferably biological methods, historical insights into specific local groups are obtained. Simple objectives might be migration and genetic drift, isolation and founding effects, assortment and selection, stress and diet or culture and environment - there is no clear order in this list and other entries are possible.

A fourth objective is the reconstruction of genetic families within prehistoric populations (see the successful example of ancient Aswan, quoted above).

A final objective might be the identification of individuals (see, for example, Strouhal's paper in this volume on the members of the Fifth Dynasty royal family) or be concerned with other problems as formulated in mummy research.

Naturally, there is a reason why this particular order of objectives has been chosen. Within taxonomy as well as within the whole system there is a decrease in group size. Further down the list there are fewer individuals who have been studied. This effect is a basis for a forecast element. I formulate the thesis that there is a clear and continuing tendency towards smaller groups. Taxonomy as a whole is slightly dusty, race taxonomy even obsolete, whereas it is the smaller group-sizes of population history etc., which trigger research investment today.

A Present Array of Disciplinary Methods

In order to arrive at those objectives we dispose of many methods. In this case, however, there is no single principle of order which would lead to a picture of the future; therefore, the path towards that picture must be different, the assignment of an assessment of importance to each method:

Present methods	*Modernity*
Anatomy, including species determination and reconstruction of individuals	o
Taphonomy	+
Palaeodemography including sex and age diagnosis	-
Morphology, including	-
epigenetics	+
metrics	-
osteoscopics/topology	o
Chemistry, including	
trace elements	+
serology etc.	-
molecular biology	++
Palaeopathology, comprising	+
diagnosis and	+
epidemiology	++
Statistics	+

++ assessed to be highly modern/important
+ modern
o indifferent or not yet assessable
- outdated/not important for qualitative innovation

There is no need to delve into the details of this list, examples suffice. Metrics and serology may be regarded as only classical; on the other hand, there are at least two methods which act as a magnet for project investment: molecular biology and epidemiology, in that order. There seems to be one common element in at least two of these : a technical character as opposed to a human character. The personal, visual assessment of similarity is not considered as leading to the future, but the use of machinery is. Possibly the term machinery only reflects a chance coincidence, but more probably it is a real common element.

At this point another limitation to this study must be named. The present discipline may also be described by institutional constituents, for instance, present anthropology is increasingly divided into regional blocks: Europe, India, East Asia, the Soviet Union and North America. This is most clearly seen in the meagre amount of 'foreign' literature, which is read in the other blocks. A sad record in this tendency towards isolation has been achieved by North America. This point is also passed over, not for reasons of simplicity, but because it has not yet been discussed enough in the science of science, nor even sufficiently measured.

Trends for the Future

In keeping with the strategy of restricting the aims of this study to the general level for reasons of safety, there are two past and present trends which may be extended into the future. They have been prepared by the preceding text on present disciplinary objectives and methods.

The first trend is from large to small groups, from races via village affinity assessment to the dynamics of a local population and down to the individual level. Probably the number of skeletons analysed is just a technical variable but it is highly correlated to the objectives and to other properties of bioanthropological studies. One such property might be (although this is not certain) the ideographic directions of adaptation and environment as opposed to the more nomothetical population genetics. In any case, the trend seems to be large enough to be extended into the future. The last chapter will treat this extension with an example.

The second trend is derived from the method section whereby the future may be characterised by more technical methods. In this case, there is a small doubt about extending this trend into the future; it may be that the necessary quantification of anthropology has reached saturation level. However, this seems improbable. The immediate present, at least, is still characterised by an increase in the use of quantification and technology and even a further wish for it.

Besides those trends, there is a stable core in the anthropology of ancient Egypt which is the variability of man in its historic dimension. In the course of this essay that is just an aside; for the field as a whole, it is the most important part.

Now to the final and specific section on the future. Coincidences and associations in it are deliberate.

Two Books

We are now in the year 2015 - only 22 years away from 1993 - and two books have just been published, *The Happy and Painful Life of Nakht MMM 733* and *Environment and Culture as Determinants of the Life and Death of Nakht MMM 733*. They appear to be similar because one element of the title is common to both: the person of Nakht who is one of the Manchester University Museum mummies. Although the same person is the subject in both, the two books are rather different, taking, in fact, opposing positions.

The first book concentrates on the human being himself. There are two main chapters, the first one on genetics. The analysis of nucleotide sequence has been subdivided into the structural and regulatory part. The structure reveals the cause of death: his specific HLA haplotype, including loci E and F, determined a chronic diarrhoea. Regrettably it was proven that there was no family connection to the Pharaohs, as Dr Rosalie David, a former Keeper of Egyptology in the museum, once presumed. The regulation part concentrates on those genes which were active before death, especially on activated cell proteins and enzymes, and finally, as a mark of respect to older colleagues, on the morphology and histology of joints, muscle attachments, etc. Thus it is shown that Nakht never worked very hard, that he obviously liked to gesticulate while talking, and that above all he liked to feast and drink. So the characteristics of this book easily allow us to classify it as belonging to the school of the adaptationists.

The other book obviously comes from the environmentalists. Now mathematical function models take over. The chapter on the natural environment shows that climate and bacteria density in their interaction caused the untimely death of Nakht. The cultural environment chapter proves that a person like this does not have to work very hard but has a rather easy-going life. This analysis is, of course, also based on the new societal motivation flow models.

The first reviews of the books show that the one on genetics is quite successful. After all it contains numerous new technical procedures. The one on environment clearly sells less well, but it is the one that is discussed, not just used. So the former division of the discipline into a technical and a human direction is found to recur, although at a different level, one which the former period would have deemed entirely technical.

The scientific public is at first a bit annoyed by the parallelisms and the contradictions between the books. But then pride takes over, pride for the versatility of this modern year 2015. The discipline clearly feels that it stands not on the shoulders of giants but of dwarfs, just as we feel in our deeply historic year 1993.

Summary
This forecast rests first on the extrapolation of present trends of objectives towards smaller groups. The largest group level, race, is obsolete; the smallest group levels, villages and families, are most prestigious. The second trend applies to methods and is directed towards the increased use of more quantification and technology. These trends find expression in the concepts of two books in the year 2015.

Acknowledgment
I am most grateful to Jennifer Hartog MA, who revised the manuscript text.

THE PLATES

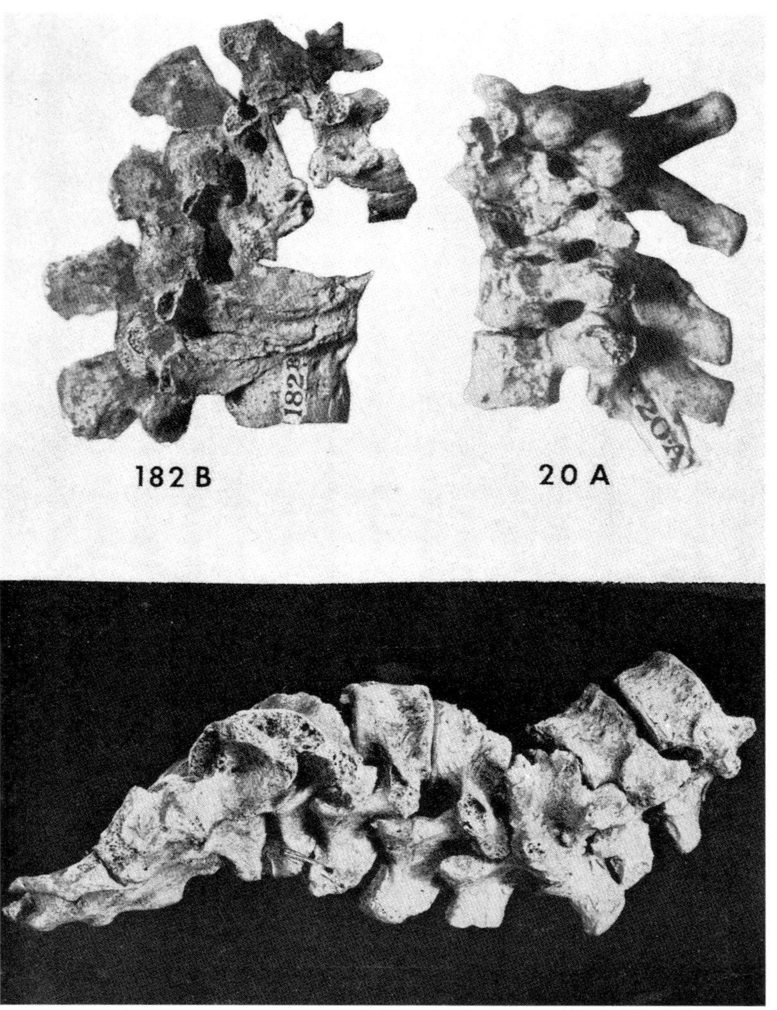

1. Specimens 182B, 20A and 182C. Reproduction of Figure 8 from Morse, Brothwell and Ucko (1964:532). Published by permission of the authors and the *American Review of Respiratory Diseases*.

2. Specimens 182E. Reproduction of Figure 9 from Morse, Brothwell and Ucko (1964:533). Published by permission of the authors and the *American Review of Respiratory Diseases*.

PLATE 2

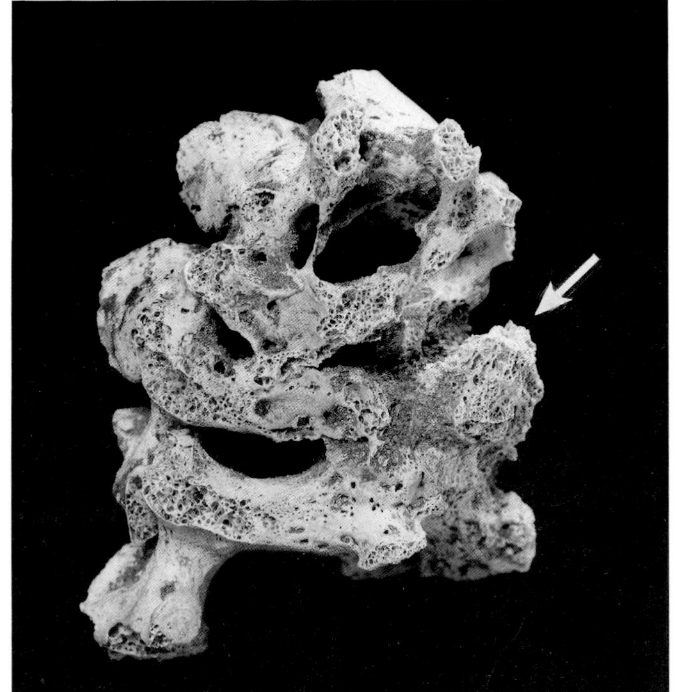

1. Specimen 182E. View from right side. Arrow indicates location of displaced and rotated centrum for T8.

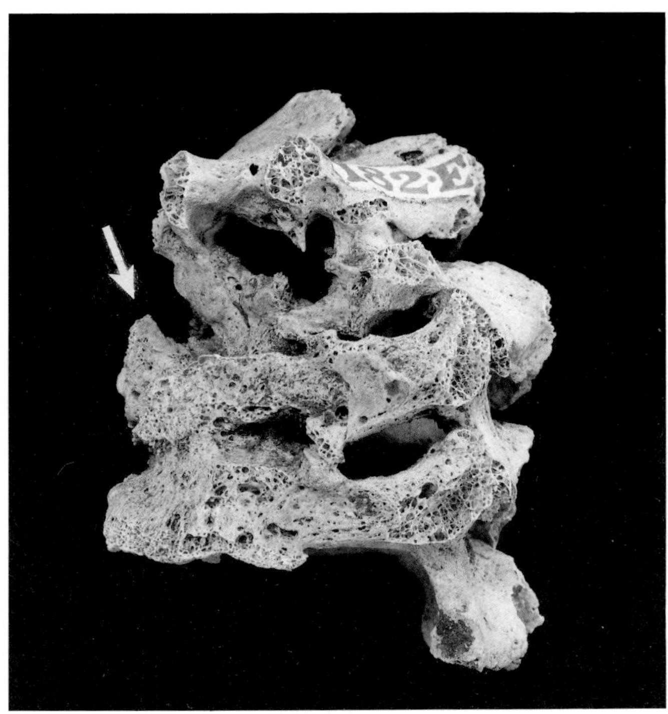

2. Specimen 182E. View from left side. Arrow indicates location of displaced and rotated centrum for T8.

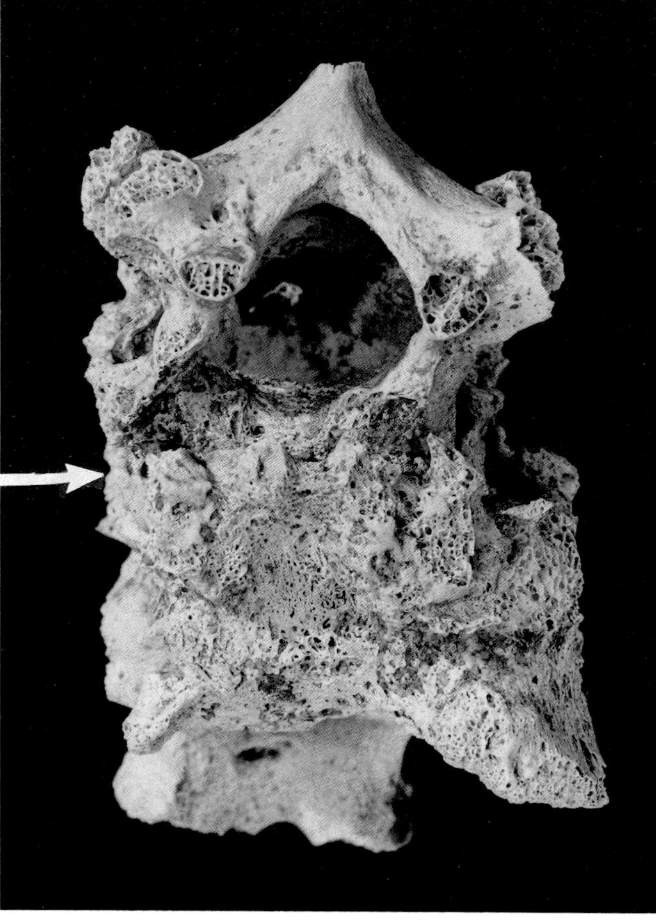

3. Specimen 182E. View from ventral aspect. Arrow indicates location of displaced and rotated centrum for T8.

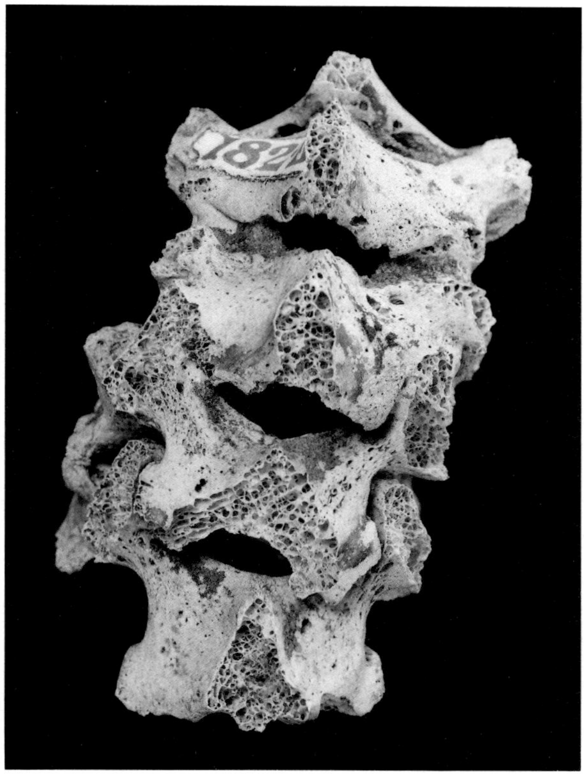

1. Specimen 182E. Dorsal aspect, illustrating scoliosis.

2. Specimen 182E. Inferior aspect of L1.

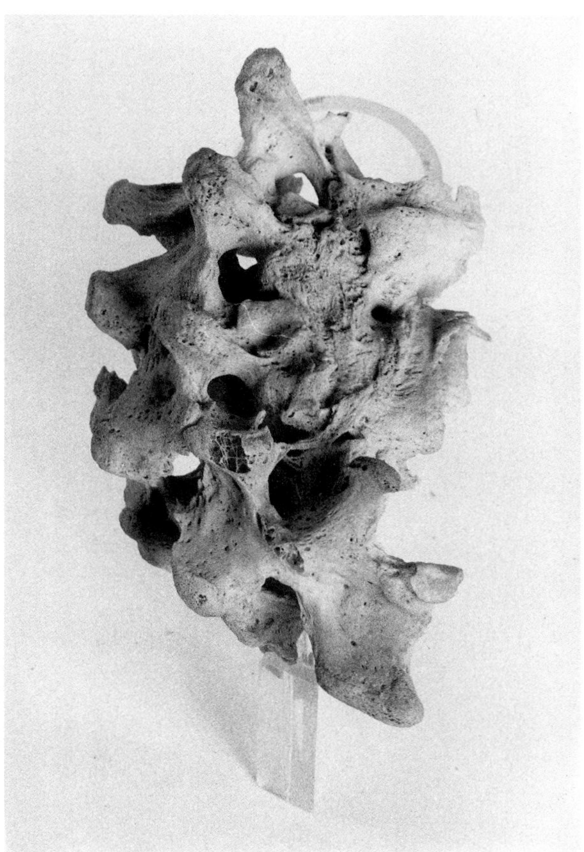

3. Iurudef Specimen 51, illustrating tuberculosis-like spinal disease in T8-L1.

4. Iurudef Specimen 51, total spinal column.

PLATE 4

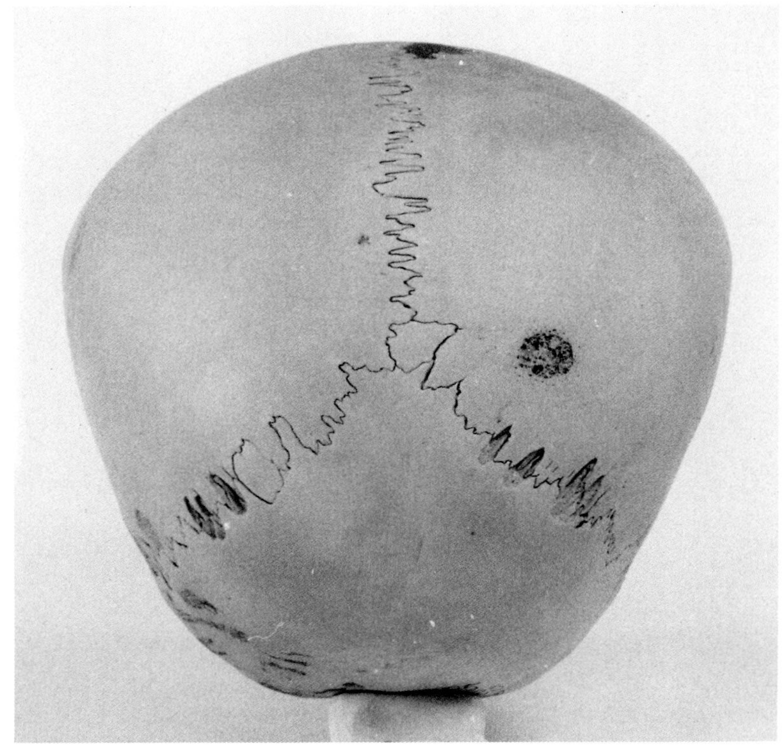

1. Iurudef Specimen 25. View from posterior aspect of skull illustrating lesion on right parietal near lambda.

2. Iurudef Specimen 25. Close-up of lesion seen in Plate 4,1.

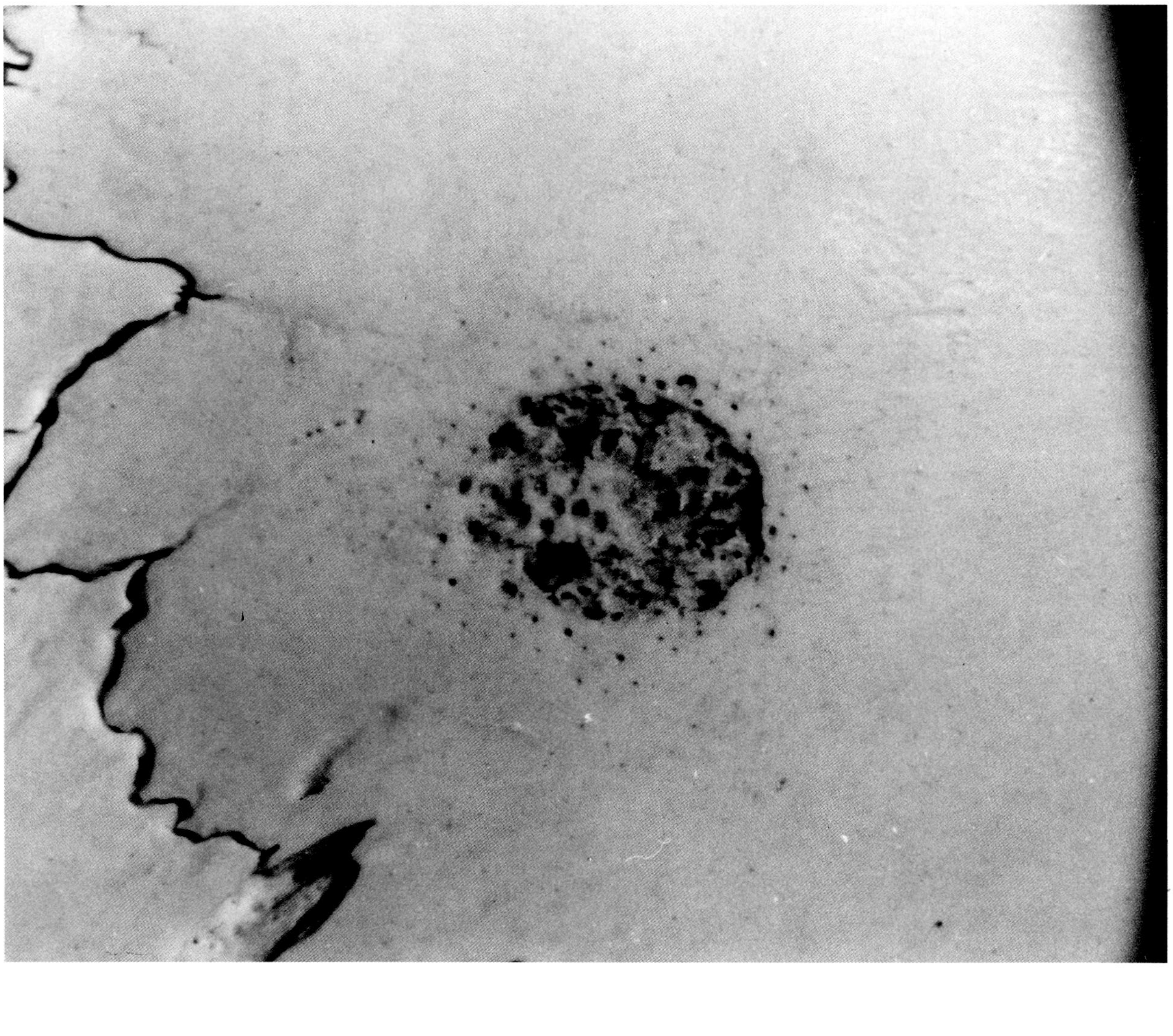

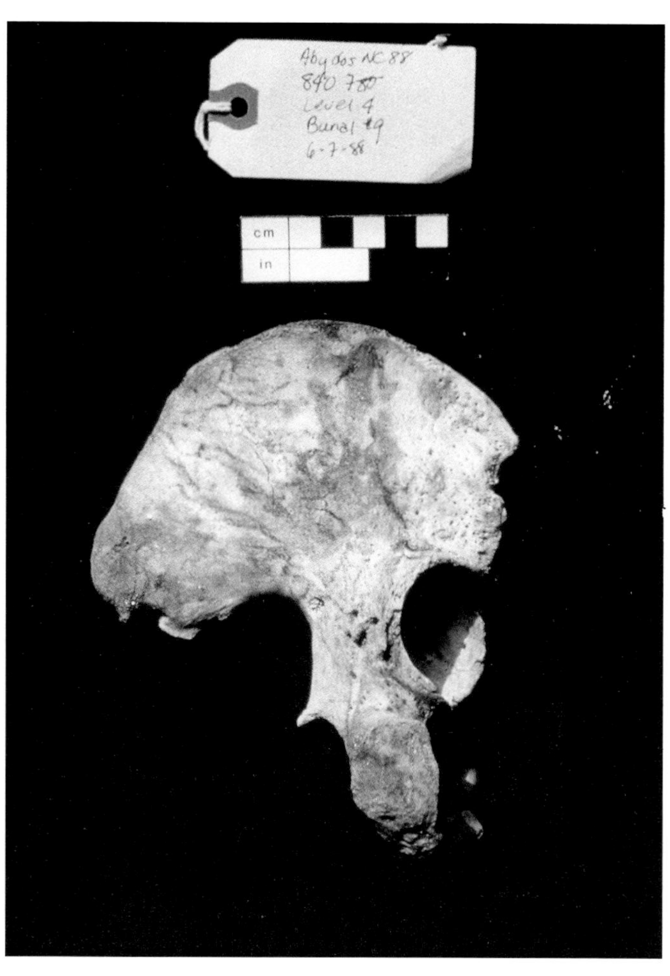

1. Abydos, Northern Cemetery, E840N780 Burial 9. Right hip bone, lateral view, presenting a focus of infection and remodelling around the acetabulum and anterior iliac spine.

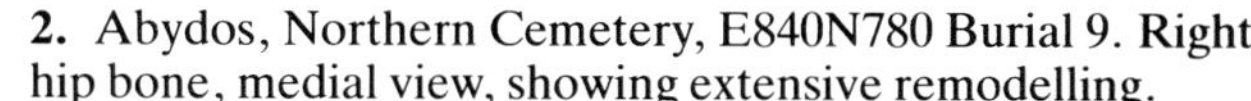

2. Abydos, Northern Cemetery, E840N780 Burial 9. Right hip bone, medial view, showing extensive remodelling.

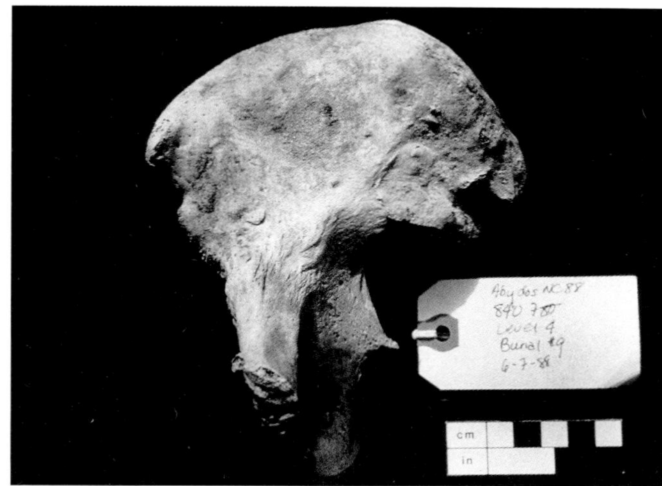

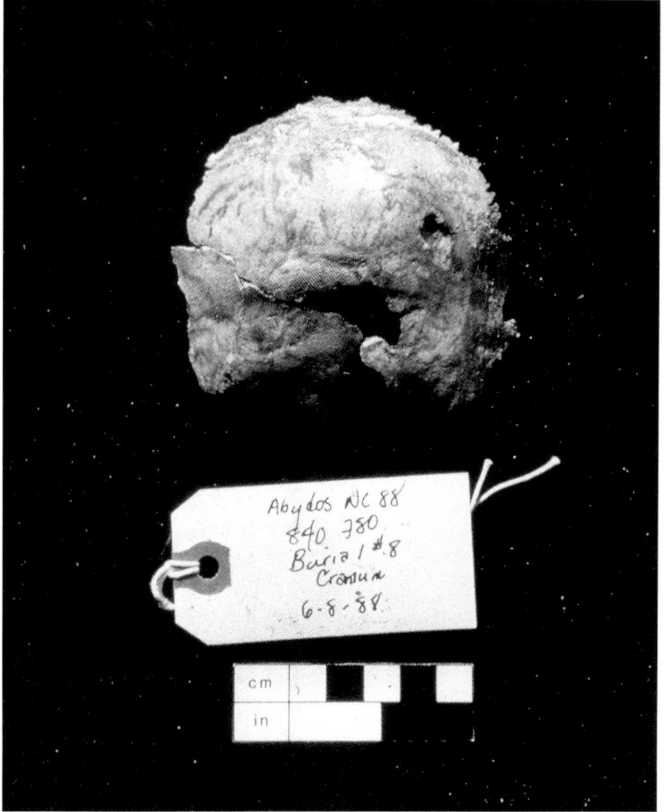

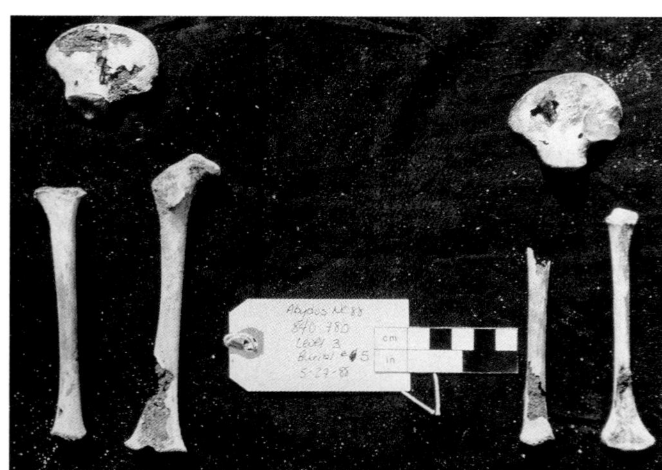

3. *(left)* Abydos, Northern Cemetery, E840N780 Burial 8. Occipital bone of a child (3-4 years old) displaying a large area of bone destruction nearly filled in with new bone growth.

4. Abydos, Northern Cemetery, E840N780 Burial 5. Long bones and ilia of an infant, showing areas of destruction and periosteal reaction. The right ilium (upper right) is remodelled adjacent to the area of destruction.

Predynastic adolescent with active schistosomiasis infection. From Gebelein. British Museum, EA 32753.

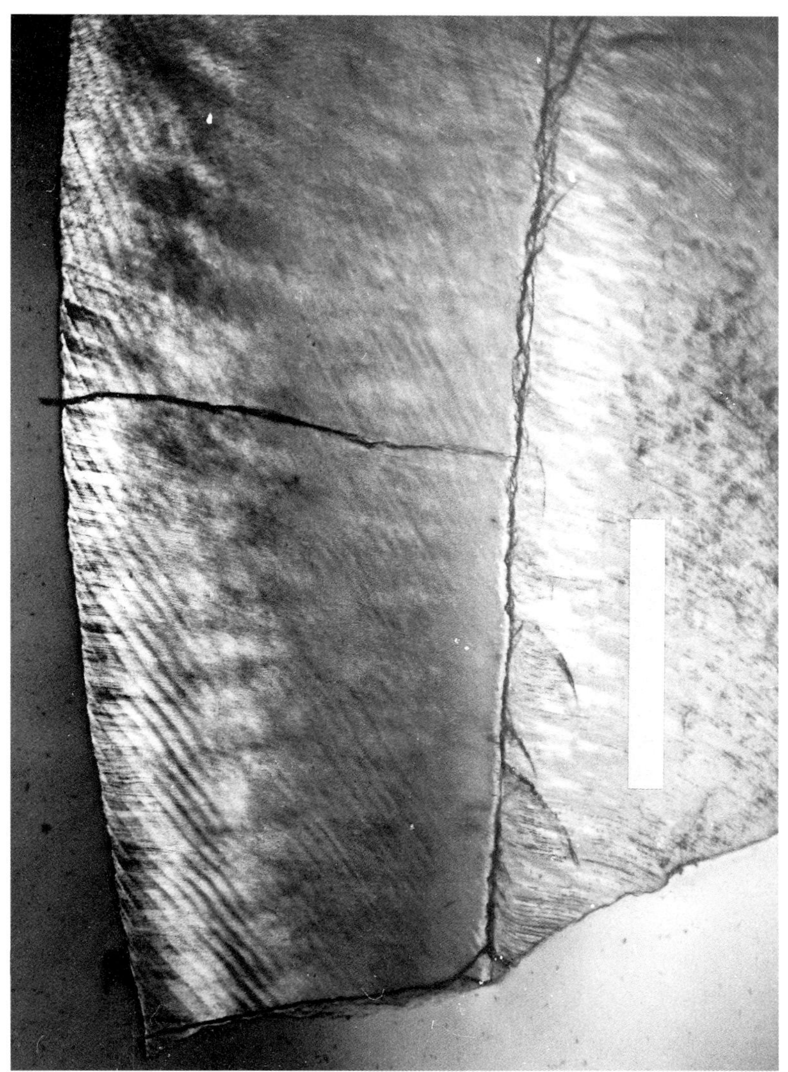

1. Brown striae of Retzius in a human canine tooth fragment from Kerma, Nubia. Ground, polished section viewed in polarizing microscope, length of scale bar 500μm. The fragment is from the lower part of the crown side and the occlusal region of the crown would have been some way above the upper margin of the photograph. The crown surface is outlined on the left of the photograph and the clear, dark line running from top to bottom on the right is the enamel-dentine junction. Brown striae of Retzius are visible running up diagonally from the junction line to the crown surface.

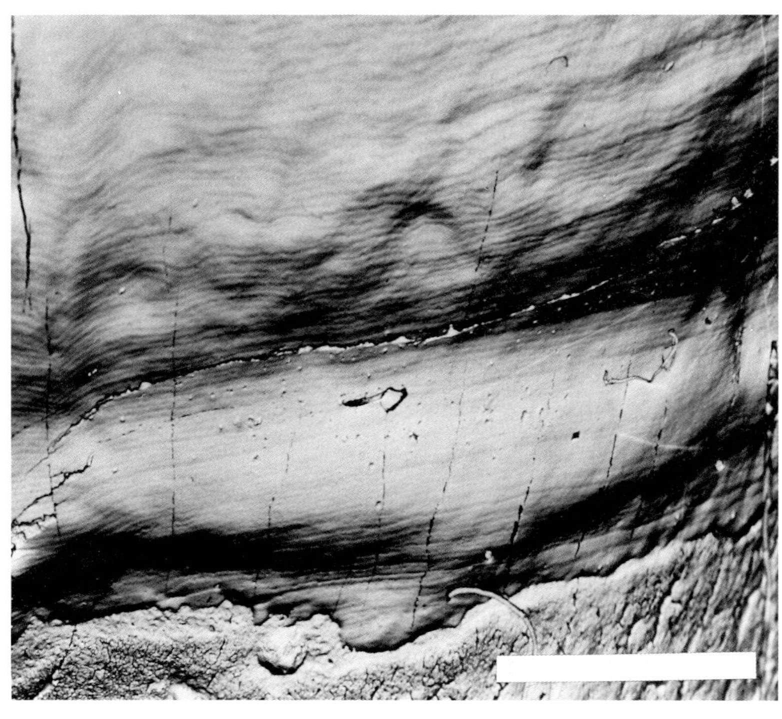

2. Perikymata on the buccal crown surface of a human permanent second molar from the predynastic cemeteries at Badari, Egypt. A prominent groove-like enamel hypoplasia defect is visible. Scanning electron microscope operated in secondary electron mode at 15kV. Scale bar represents 1mm. Material from the collection of the Duckworth Laboratory, Faculty of Archaeology and Anthropology, University of Cambridge.

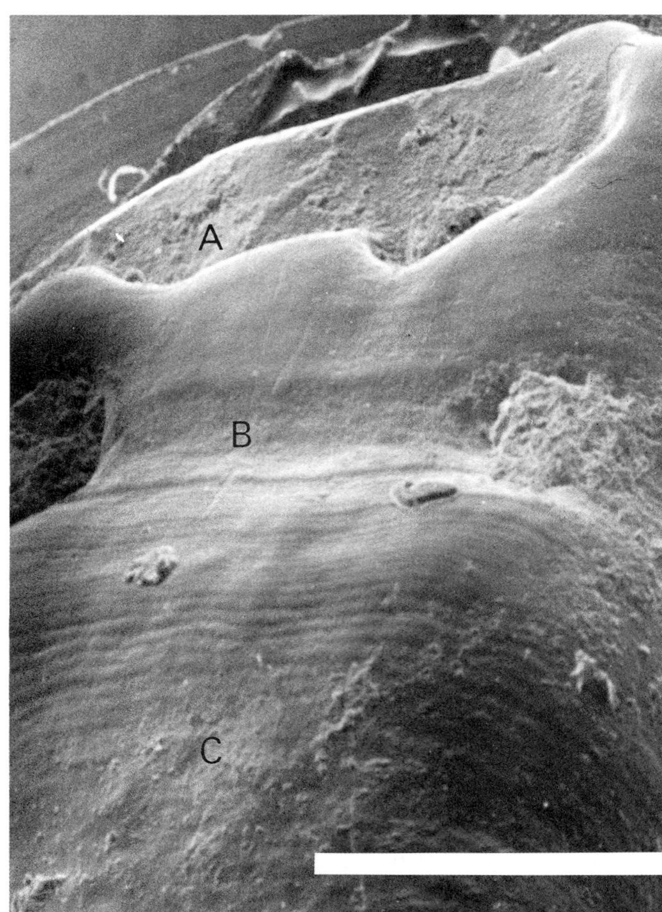

1 and 2. Enamel hypoplasia defects in a child from the cemetery at Liverpool Street/Broad Street in London. Part of a study of material held by the Greater London Environmental Archaeology Section, Museum of London. Plate 8,1, shows a permanent lower first molar crown with the cusps at the top of the photograph and the neck of the tooth at the bottom. Plate 8,2, shows a permanent lower second incisor in the same orientation. Scanning electron micrographs of epoxy resin replicas. SEM operated at 15kV. Scale bar 2mm in both photographs. The three defects described in the text are labelled A to C.

1. The anterior half of King Djedkare Isesi's calva.

2. The incomplete squama occipitalis *(left)* and right parietale *(right)*.

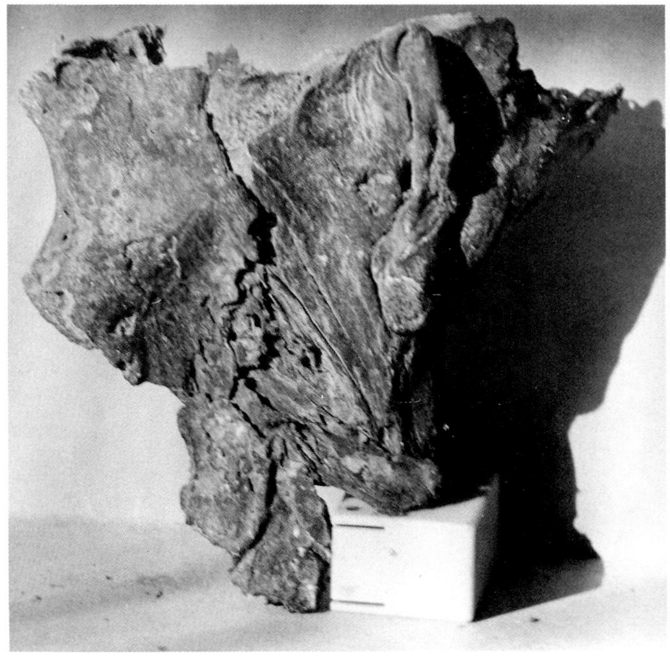

3. Left posterior inferior part of the face with preserved soft tissue structures including the auricle.

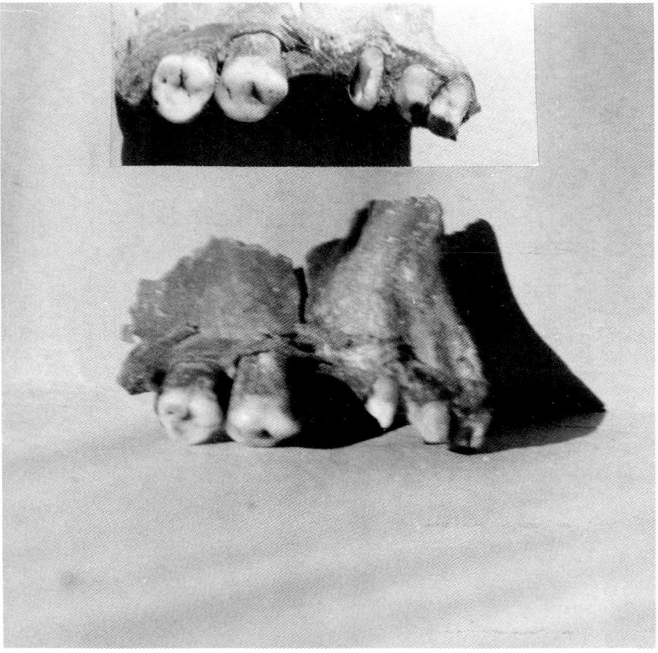

4. Part of the right maxillar alveolar process with teeth in situ.

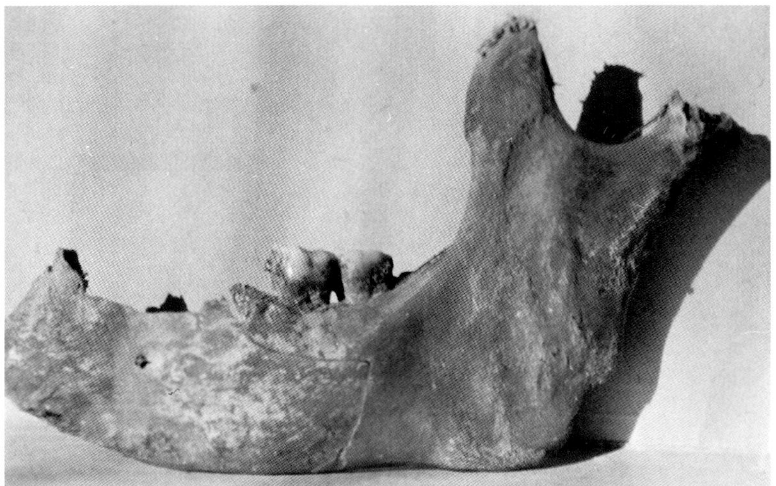

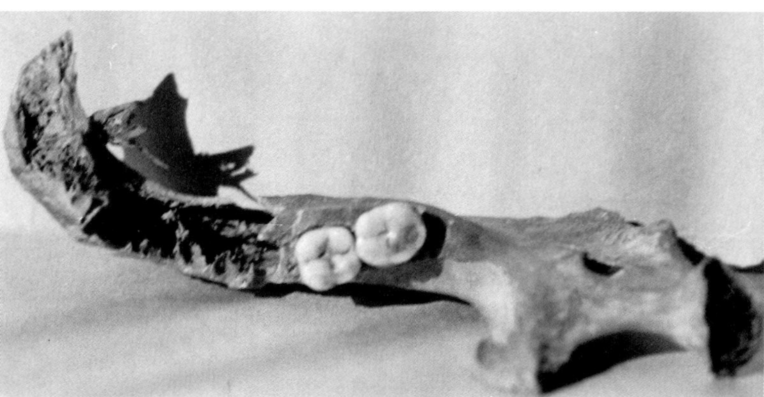

1. The left half of the mandible with two molars in situ.

2. The sternum covered by soft tissue, textile and stucco with attached left clavicle and rib fragments, external view.

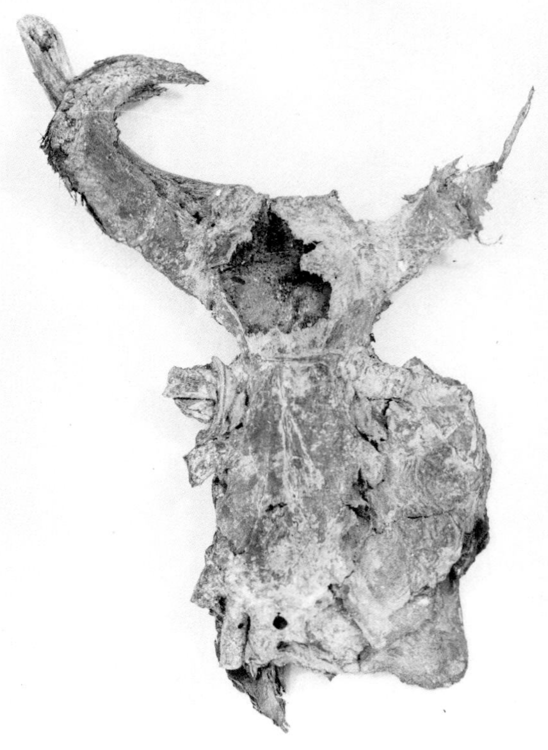

3. The sternum covered by soft tissue with attached left clavicle and rib fragments, internal view.

1. The cervical and upper thoracic spine with preserved soft tissue and isolated fragment of distal T_6 connected with T_7 in anterior view.

2. Fragment of the distal end of the left humerus and proximal ends of both antebrachial bones.

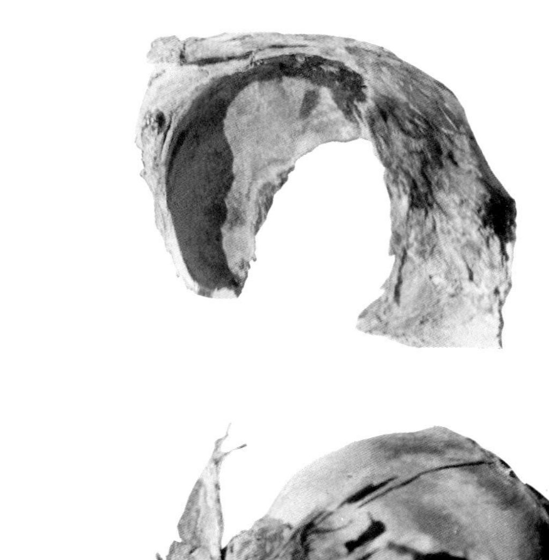

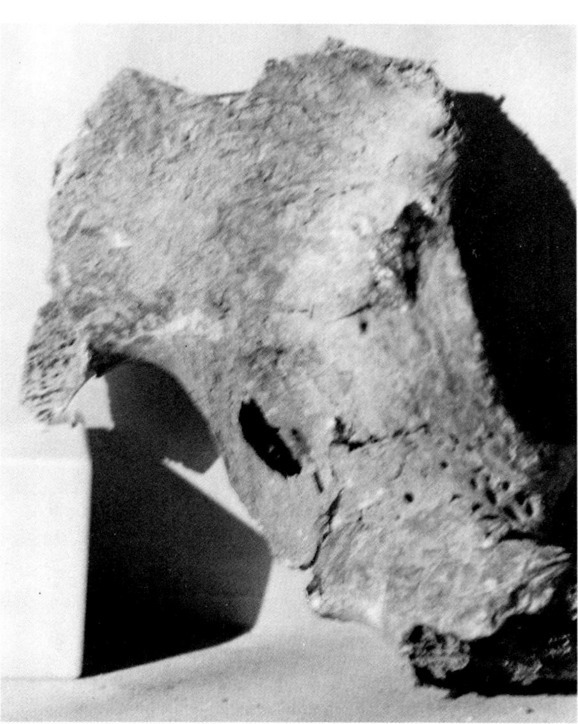

3. Fragment of the right os coxae.

4. Fragment of the proximal end of the left femur with the originally adhering rim of the left acetabulum.

1. The lateral half of the left foot from naviculare to tips of toes.

2. Pieces of soft tissue, possibly from one of the shoulder regions, external *(top)* and internal views.

3. Fragment of a textile pad covered by lime with an adhering tubular bead *(left)*, and an isolated lumbar vertebral disc *(right)*.

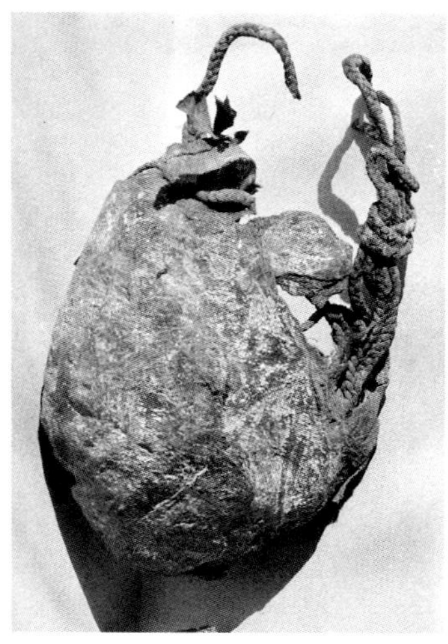

4. A spheric oblong object (bag?), external *(left)* and internal *(right)* views.

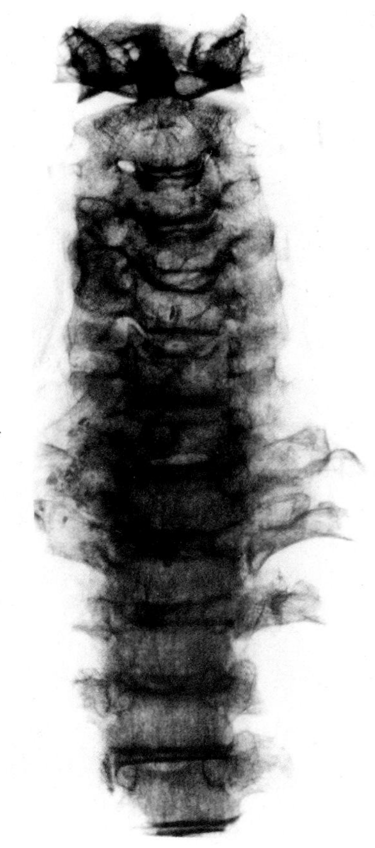

1. Antero-posterior radiogram of the cervical and upper thoracic spine.

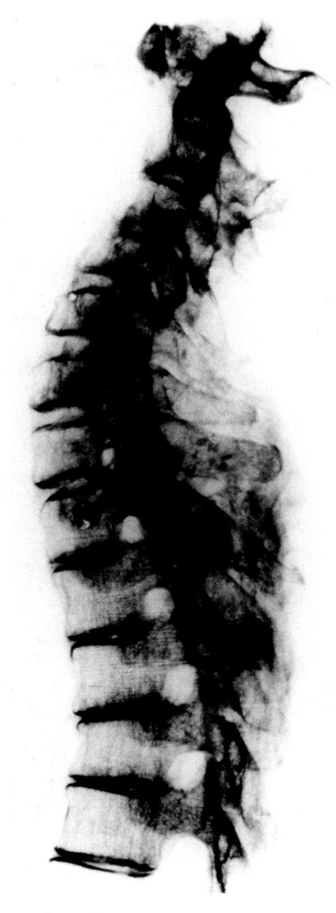

2. Lateral radiogram of the cervical and upper thoracic spine.

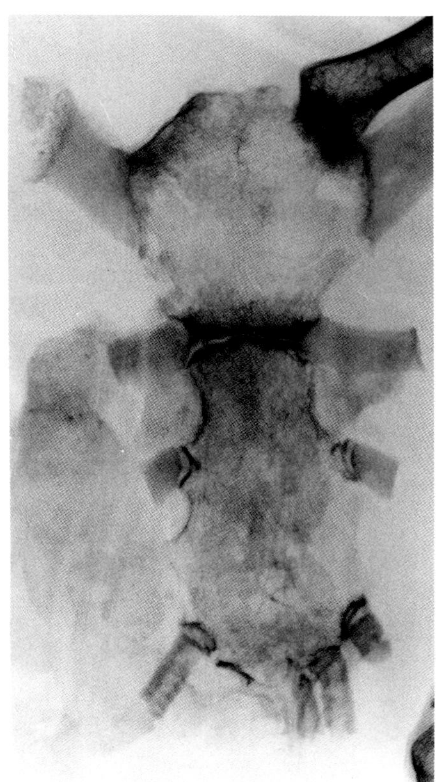

3. Radiogram of the sternum with attached left clavicle and ribs.

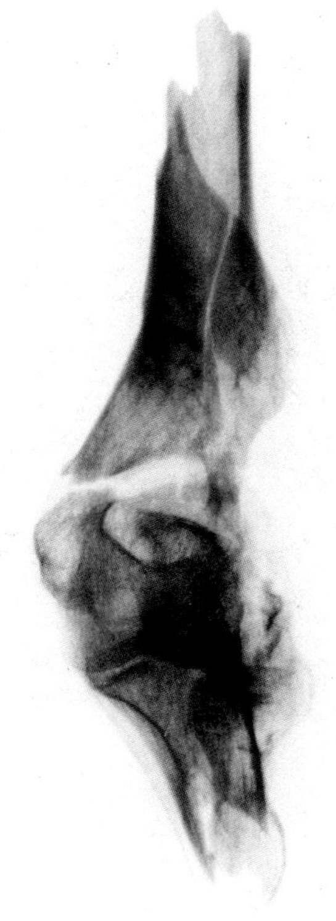

4. Radiogram of the postmortally-broken left cubit region.

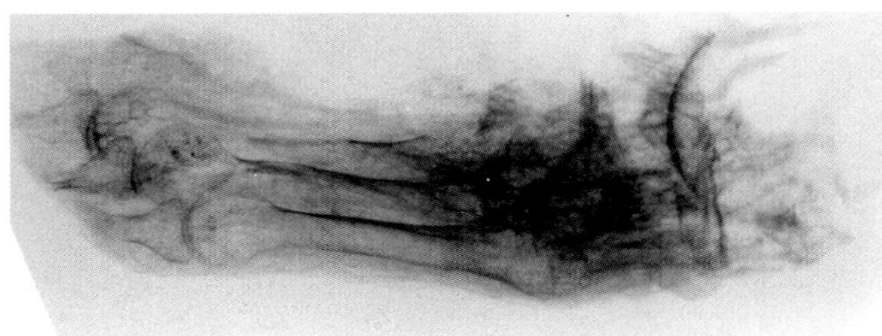

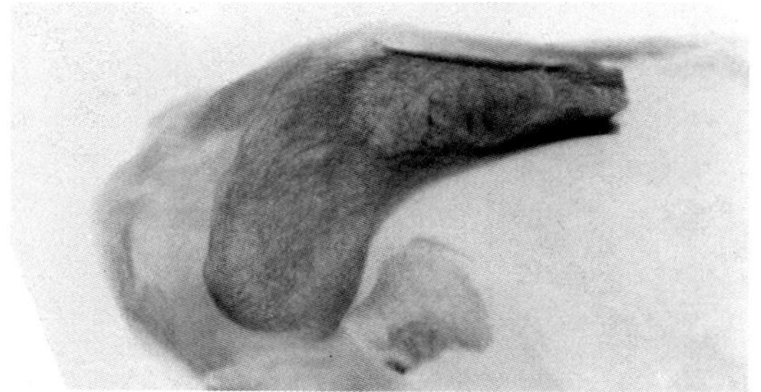

1. Radiogram of the lateral half of the left foot *(above)* and of the shoulder fragment with the lateral part of a clavicle *(below)*.

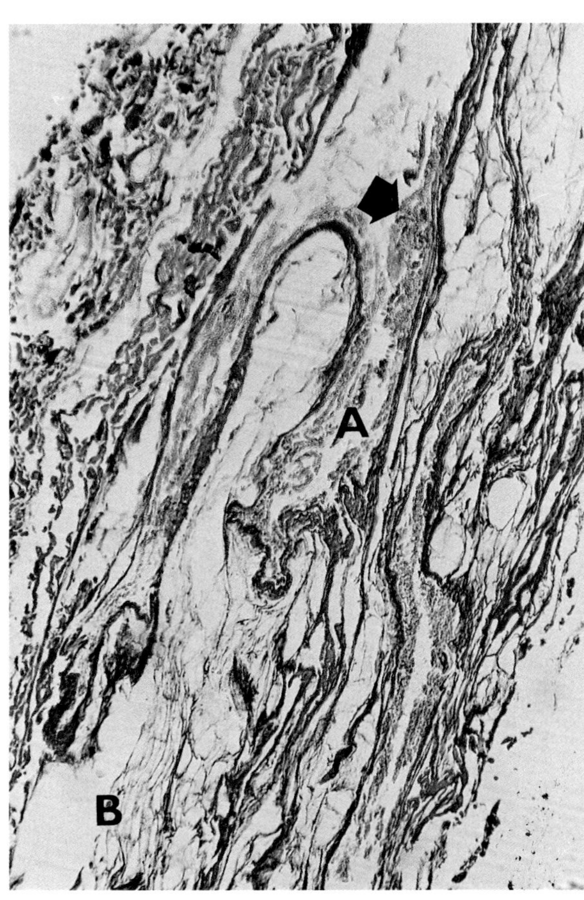

2. Microphotograph of a thin section of skin stained by Verhoeff's haematoxylin. A = reticular layer of dermis and a blood vessel. B = subcutaneous adipose tissue (40x).

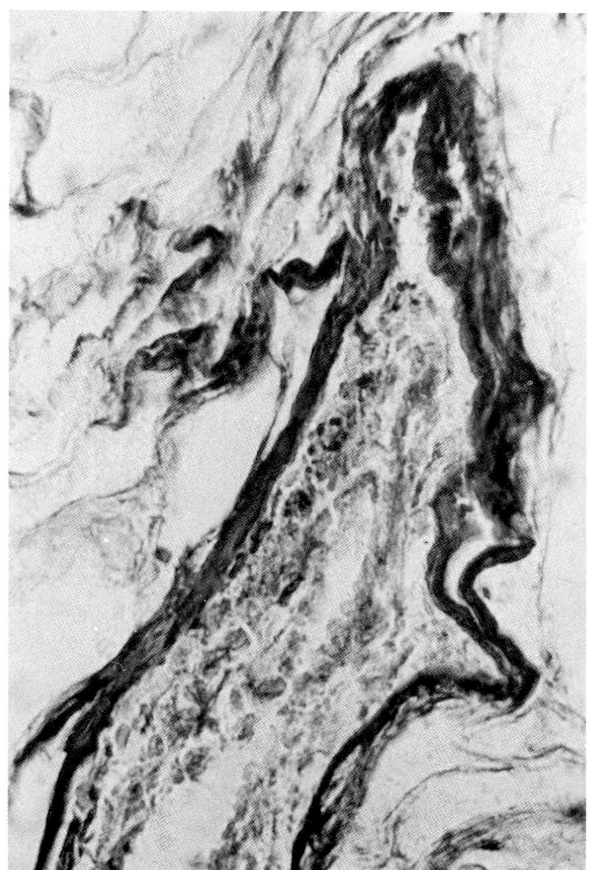

3. Microphotograph of a thin section of skin stained by Verhoeff's haematoxylin, showing a vessel with degenerative changes of its wall and obliteration of its lumen caused by atherosclerosis (190x).

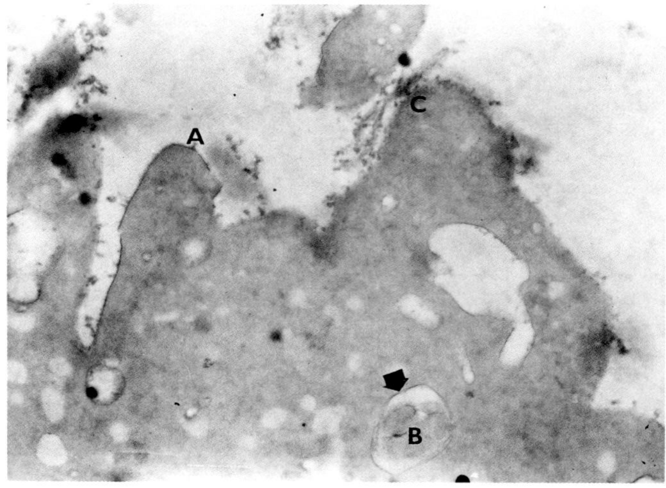

4. Electron micrograph of skin. The cells of the stratum germinativum with cells' membrane, A; keratohyalin granules, B; and desmosome, C (10,000x).

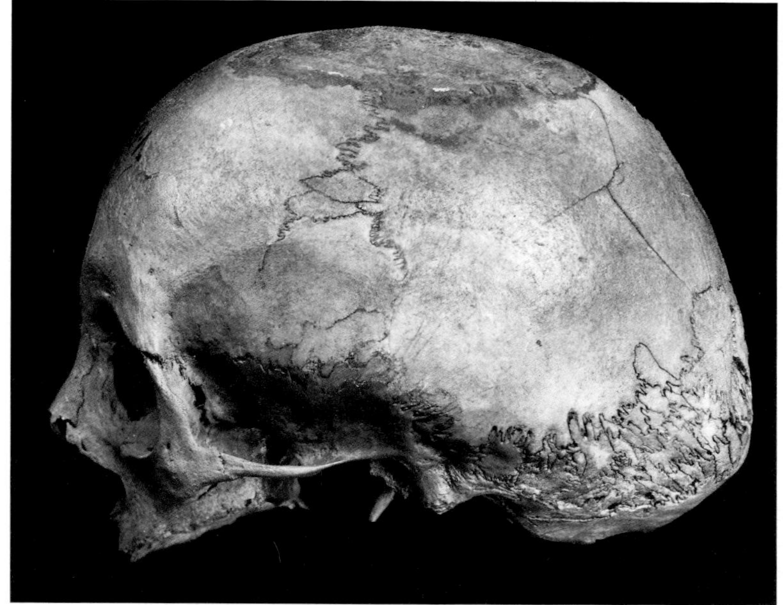

1. A 'Tam O'Shanta' skull typical of bone changes in osteogenesis imperfecta. 1968.8.8.17. One of the skulls presented by University College London.

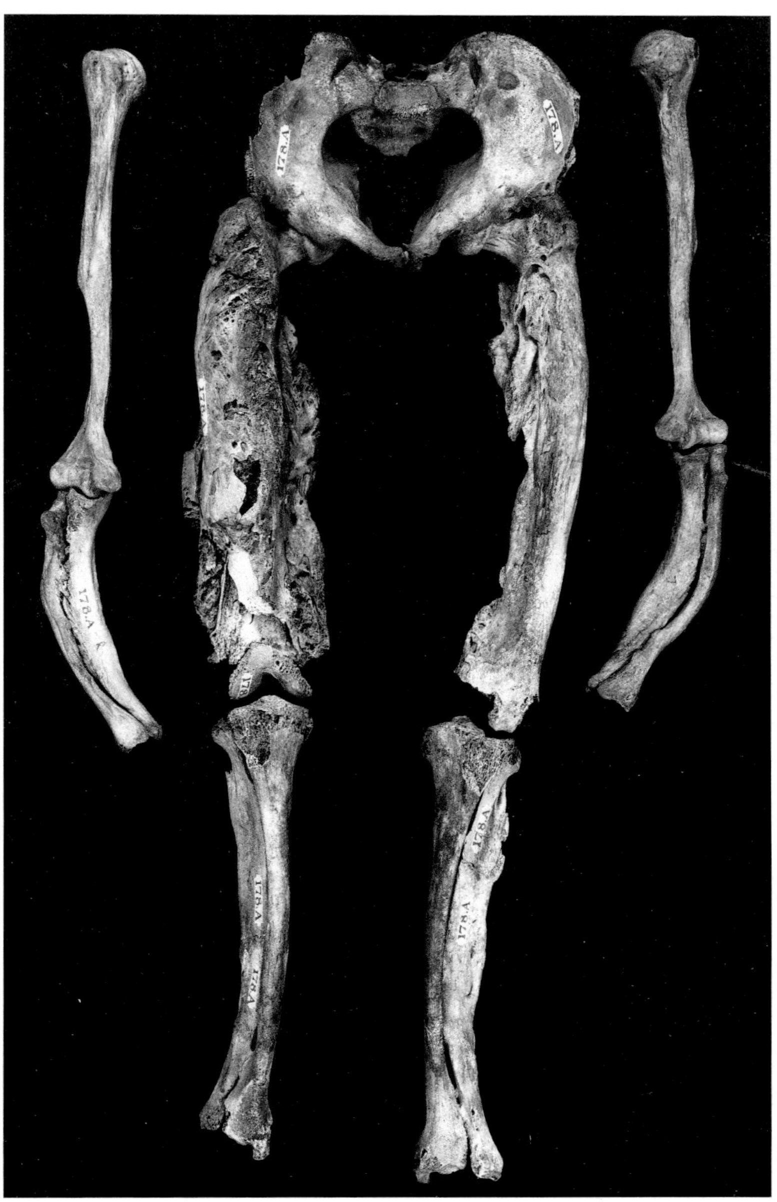

2. Part of a post cranial skeleton with marked changes, which have been variously attributed to rickets or osteogenesis imperfecta. RCS 178A. No provenance.

Artistic evidence. The ancient Egyptians ground cereals on saddle querns, as shown in this wooden tomb model *(left)* from Asyut. In this case the quern stone is placed on the floor. Other depictions show them mounted on raised quern emplacements. Middle Kingdom, about 1900 BC. Maximum height 22 cm; length 41 cm. British Museum, EA 45197.

1. Ethnographic evidence. A woman from Bechuanaland grinds cereal on a saddle quern. By courtesy of the Haddon Collection, Museum of Archaeology and Anthropology, University of Cambridge.

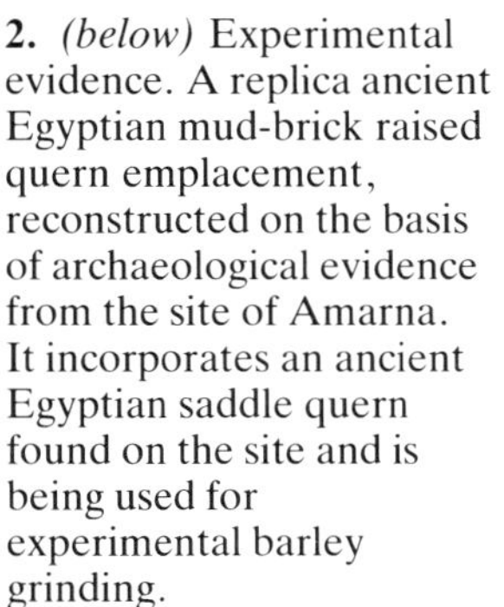

2. *(below)* Experimental evidence. A replica ancient Egyptian mud-brick raised quern emplacement, reconstructed on the basis of archaeological evidence from the site of Amarna. It incorporates an ancient Egyptian saddle quern found on the site and is being used for experimental barley grinding.

Garland of Marjoram from Hawara. Liverpool Museum acc. no. 56.20.755. Photo: National Museums and Galleries on Merseyside.